Power Engineering: Novel Concepts and Applications

Power Engineering: Novel Concepts and Applications

Edited by
Tim Kurian

Willford Press
www.willfordpress.com

Published by Willford Press,
118-35 Queens Blvd., Suite 400,
Forest Hills, NY 11375, USA

ISBN: 978-1-68285-363-4

Cataloging-in-Publication Data

Power engineering : novel concepts and applications / edited by Tim Kurian.
p. cm.
Includes bibliographical references and index.
ISBN 978-1-68285-363-4
1. Power (Mechanics). 2. Power transmission. 3. Electric power systems.
4. Electrical engineering. I. Kurian, Tim.

TJ163.9 .P69 2017
621.042--dc23

For information on all Willford Press publications
visit our website at www.willfordpress.com

Willford Press

Printed in the United States of America.

Contents

Preface

This book was inspired by the evolution of our times; to answer the curiosity of inquisitive minds. Many developments have occurred across the globe in the recent past which has transformed the progress in the field.

Power engineering is the study, design and building of technology that is able to transmit, receive and generate electric power. This book on power engineering deals with the most advanced applications of power systems which are used in large-scale power generation and transmission. Some of the diverse topics covered in this book address the varied branches that fall under this category. It elucidates new techniques and their applications in a multidisciplinary approach. While understanding the long-term perspectives of the topics, the book makes an effort in highlighting their impact as a modern tool for the growth of the discipline. It will serve as a valuable source of reference for graduate and post graduate students. The chapters covered in this book offer the readers new insights in the field of power engineering.

This book was developed from a mere concept to drafts to chapters and finally compiled together as a complete text to benefit the readers across all nations. To ensure the quality of the content we instilled two significant steps in our procedure. The first was to appoint an editorial team that would verify the data and statistics provided in the book and also select the most appropriate and valuable contributions from the plentiful contributions we received from authors worldwide. The next step was to appoint an expert of the topic as the Editor-in-Chief, who would head the project and finally make the necessary amendments and modifications to make the text reader-friendly. I was then commissioned to examine all the material to present the topics in the most comprehensible and productive format.

I would like to take this opportunity to thank all the contributing authors who were supportive enough to contribute their time and knowledge to this project. I also wish to convey my regards to my family who have been extremely supportive during the entire project.

Editor

1

A Hybrid Algorithm for Optimal Location and Sizing of Capacitors in the presence of Different Load Models in Distribution Network

R. Baghipour*, S.M. Hosseini

Department of Electrical Engineering, Babol Noshirvani University of Technology, Babol, Iran

ABSTRACT

In practical situations, distribution network loads are the mixtures of residential, industrial, and commercial types. This paper presents a hybrid optimization algorithm for the optimal placement of shunt capacitor banks in radial distribution networks in the presence of different voltage-dependent load models. The algorithm is based on the combination of Genetic Algorithm (GA) and Binary Particle Swarm Optimization (BPSO) algorithm. For this purpose, an objective function including the cost of energy loss, reliability, and investment cost of the capacitor banks is considered. Also, the impacts of voltage-dependent load models, considering annual load duration curve, is taken into account. In addition, different types of customers such as industrial, residential, and commercial loads are considered for load modeling. Simulation results for 33-bus and 69-bus IEEE radial distribution networks using the proposed method are presented and compared with the other methods. The results showed that this method provided an economical solution for considerable loss reduction and reliability and voltage improvement.

KEYWORDS: GA, BPSO, GA/BPSO, Reliability improvement, Loss reduction, Load modeling.

1. INTRODUCTION

The utilization of shunt capacitors in distribution networks is essential for many reasons, including power flow control, loss minimization, power factor correction, voltage profile management, and system stability improvement. However, to achieve these objectives, keeping in mind the overall economy, capacitor planning must determine the optimal number, location, and size of capacitors to be installed in the distribution system. Many approaches have been proposed to solving the capacitor planning problem. Duran [1] considered capacitor sizes as discrete variables and employed dynamic programming to the problem solving. Grainger and Lee [2] developed a nonlinear programming-based method, in which capacitor location and capacity were expressed as continuous variables. Baran and Wu [3] presented a method with mixed integer programming. Also, some authors [4, 5] proposed genetic algorithm approach for determining the optimal placement of capacitors based on the mechanism of natural selection.

Srinivasas et al. [6] Suggested an approach consisting of two parts; in the first part, loss sensitivity factors were used to select the candidate locations for the capacitor placement and, in the second part, plant growth simulation algorithm was used to estimate the optimal size of capacitors at the optimal buses determined in part one. Tabatabaei and Vahidi [7] represented a methodology based on fuzzy decision making which used a new evolutionary method. In this method, the installation node was selected by the fuzzy reasoning supported by the fuzzy set theory in a step by step procedure. In addition, an evolutionary algorithm known as bacteria foraging algorithm was used for solving the multivariable objective optimization problem and the optimal node for capacitor placement was

*Corresponding author:
R. Baghipour (E-mail: reza.baghipur@yahoo.com)

determined. Federico et al. [8] proposed an extended dynamic programming approach that lifted Duran's restricting assumptions and restored the perspective of achieving global optimality for capacitor allocation on radial distribution feeders. Hamouda and Sayaha [9] represented a fast heuristic method to solve the capacitor sizing and locating problems. In the proposed method, the candidate locations of the capacitor were determined by means of node stability indices and capacitor optimal sizes were determined by solving a non-linear constrained problem. Abul'wafa [10] suggested a loss sensitivity technique for selecting the candidate locations for the capacitor placement. The size of the optimal capacitor at the compensated nodes was determined simultaneously by optimizing the loss saving equation with respect to the capacitor currents. The authors in [11] proposed a dynamic model, in which load growth rate, load factor, and cost of power and energy losses were incorporated to maintain the voltage profile, considering the multiperiod capacitor allocation problem of radial distribution system. Huang and Liu [12] reported a plant growth-based optimization approach for capacitor placement in power systems.

Reliability improvement is one of the important goals of power utilities in system planning; but, it has not been sufficiently investigated in the studies related to the problem of capacitor placement thus far. In the above-mentioned works, loss reduction and/or voltage improvement has been considered for the objective problem. Etemadi and Fotuhi [13] proposed a new objective function for capacitor placement problem in order to improve reliability; however, similar to most of the above-mentioned works, a constant power load model was considered in the distribution system, which is far away from the real situation in distribution systems.

In this paper, a combined optimization algorithm is proposed for determining the optimal number, location, and size of capacitor banks in distribution networks including different load models for loss reduction and reliability improvement. In this method, optimum location of capacitors is determined by GA algorithm and their optimum size is determined by BPSO. At first, the initial population for the capacitor size and location is randomly produced. Then, the cost function is calculated using distribution system load flow for each random size and location and the size of capacitor is optimized using BPSO algorithm. In the next step, the new location of capacitor is optimized using GA by cost function minimization. In other words, in the proposed algorithm in this paper, the size of capacitor was optimized by BPSO which reduced the search area for the GA. Then, the location of capacitor is optimized by GA. The results demonstrated that the proposed combined GA/BPSO method is better than GA and BPSO methods when they are used individually.

The remaining part of the paper is organized as follows: In Sec. 2, the allocation and sizing of capacitors are formulated. The optimization algorithm and its implementation are explained in Sec. 3. Results and conclusions are given in Sec. 4 and 5, respectively.

2. FORMULATION OF THE CAPACITOR PLACEMENT

2.1. Reliability analysis of distribution system

Analysis of the customer failure statistics of most utilities shows that distribution system makes the greatest individual contribution to the unavailability of supply to customers [13]. Most distribution systems have been operated as radial networks; consequently, the principles of series systems can be directly applied to them. Three basic reliability indices of the system, as average failure rate, λ_s, average outage time, r_s, and annual outage time U_s, are given by:

$$\lambda_s = \sum_i \lambda_i \tag{1}$$

$$U_s = \sum_i \lambda_i r_i \tag{2}$$

$$r_s = \frac{U_s}{\lambda_s} \tag{3}$$

Where λ_i, r_i, and $\lambda_i r_i$ are average failure rate, average outage time, and annual outage time of the ith component, respectively. Energy not Supplied (ENS) as one of the most important reliability indices of a distribution system is evaluated and is included as a part of the objective function. This index reflects total ENS by the system due to the faults in the system components. ENS can be calculated for each load point i using the following equation:

$$ENS_i = L_{a(i)} U_i \quad (4)$$

where, $L_{a(i)}$ is the active load connected to load point i.

2.2. Impact of capacitor placement on reliability enhancement

Customer interruptions are caused by a wide range of phenomena including equipment failure, animals, trees, severe weather, and human error. Feeders in distribution systems deliver power from distribution substations to distribution transformers. A considerable portion of customer interruptions is caused by equipment failure in distribution systems consisting of underground cables and overhead lines [13]. Resistive losses increase the temperature of feeders, which is proportional to the square of the current magnitude flowing through the feeder. For underground cables, there is a maximum operating temperature which, if exceeded, would cause the insulation problem and an increase in component failure rates [13].

Life expectancy of the insulation material exponentially decreases as the operating temperature is raised [14]. On the other hand, a major reliability concern pertaining to the underground cables is water treeing. Severity of treeing is strongly correlated with thermal age since moisture absorption occurs more rapidly at high temperatures [15].

Temperature also has impacts on the reliability of overhead lines. High currents will cause lines to sag, reduce ground clearance, and increase the probability of phase conductors swinging into contact. Higher currents can cause conductors to anneal, reduce tensile strength, and increase the probability of a break occurrence [16].

Capacitor placement can supply a part of the reactive power demands. Therefore, due to the reduction in the current magnitude, the resistive losses would decrease. As a result, the destructive effects of temperature on the reliability of overhead lines and underground cables are moderated. These impacts on reliability are taken into consideration as the failure rate reduction of distribution feeder components. Before capacitor placement, any feeder i has an uncompensated failure rate of λ_i^{uncomp}. If the reactive component of a feeder is fully compensated, its failure rate is reduced to λ_i^{comp}. If the reactive component of current is not completely compensated for, a failure rate is defined with linear relationship to the percentage of compensation. Thus, the compensation coefficient of the *ith* branch is defined as:

$$\alpha_i = \frac{I_r^{new}}{I_r^{old}} \quad (5)$$

where, I_r^{new} and I_r^{old} are the reactive component of the ith branch current after and before compensation, respectively. The new failure rate of the *i*th branch is computed as follows:

$$\lambda_{i-new} = \alpha_i (\lambda_i^{uncomp} - \lambda_i^{comp}) + \lambda_i^{comp} \quad (6)$$

2.3. Objective function and constraints

The objective of capacitor placement in the distribution system is the minimization of the annual cost of the system. In this paper, capacitor placement problem was formulated as the minimization of the total defined system cost including the capacitor investment cost, cost of energy losses, and cost of reliability improvement. The objective function which is minimized subject to the constraints (Sec. 2.3.2) is defined as follows:

$$TCOST = K_e \sum_{j=1}^{L} T_j P_{T,j} + K \sum_{k=1}^{L} ENS_k + \sum_{i=1}^{ncap} K_c Q_{ci} \quad (7)$$

where, *TCOST* is the total annual cost of the system ($/year), $P_{T,j}$ is the power loss for each load level *j*, T_j is the time duration of the jth load level, ENS_k is total ENS because of the occurrence of faults in overhead lines and underground cables for each load level (kWh), Q_{ci} is the size of the capacitor at node i, ncap is the number of candidate locations for capacitor placement, *L* is the number of load level, K_c is the cost of capacitor per kVAr, *K* is the price of ENS ($/kWh), and K_e is the factor to convert energy loss into dollar ($/kWh).

2.3.1. Power losses

Generally, distribution systems are fed at one point and have a radial structure. The load flow equations of radial distribution network are computed by forward/backward method because of its low memory requirements, computational efficiency, and robust convergence characteristic. The active and reactive power losses of the line section connecting buses *i* and *i+1* may be computed as:

$$P_{Loss}(i, i+1) = R_{i,i+1} I_{i,i+1}^2 \quad (8)$$

$$Q_{Loss}(i, i+1) = X_{i,i+1} I_{i,i+1}^2 \quad (9)$$

Where $I_{i,i+1}$ is the magnitude of the current of the line section connecting buses *i* and *i+1*, $R_{i,i+1}$, and $X_{i,i+1}$ are resistance and reactance of the line section connecting buses *i* and *i+1*, respectively. Also, $P_{Loss}(i, i+1)$ and $Q_{Loss}(i, i+1)$ are active and reactive power losses of the line section connecting buses *i* and *i+1*, respectively. The total active and reactive power losses of the feeders in the system are determined as follows:

$$P_{T,Loss} = \sum_{i=1}^{NB} P_{Loss}(i, i+1) \quad (10)$$

$$Q_{T,Loss} = \sum_{i=1}^{NB} Q_{Loss}(i, i+1) \quad (11)$$

where, *NB* is the number of line section in the distribution system and $P_{T,Loss}$ and $Q_{T,Loss}$ are total active and reactive power losses in the distribution system, respectively.

2.3.2. Operational constraints

The magnitude of bus voltage for all buses and current magnitude for all branches in distribution system should be maintained within the acceptable range. These constraints are expressed as follows:

$$V_{min} < |V_i| < V_{max} \quad (12)$$

$$|I_i| < I_{i,max} \quad (13)$$

Where $|V_i|$ is the voltage magnitude of bus *i*, V_{min} and V_{max} are minimum and maximum bus voltage limits, respectively, $|I_i|$ is current magnitude, and $I_{i,max}$ is maximum current limit of branch i.

2.4. Load model

Practical voltage-dependent load models, i.e., residential, industrial, and commercial ones, were considered in this paper. The load models can be mathematically expressed as [17]:

$$P_i = P_{oi}\left(\frac{V_i}{V_{oi}}\right)^{\alpha} \quad (14)$$

$$Q_i = Q_{oi}\left(\frac{V_i}{V_{oi}}\right)^{\beta} \quad (15)$$

where, V_i is the voltage at bus *i*, V_{oi} is the nominal operating voltage at bus *i*, P_i and Q_i are active and reactive power for load point *i* with bus voltage V_i, respectively, P_{oi} and Q_{oi} are the active and reactive power for load point *i* with bus voltage V_{oi}, respectively, and α and β are real and reactive power exponents, respectively. In the constant power model conventionally used in power flow studies, α = β = 0 is assumed. The values of the active and reactive exponents used in this paper for industrial, residential, and commercial loads are given in Table 1 [17].

In practical situations, loads are the mixtures of different load types, depending on the nature of the area being supplied. Thus, in this study, three different types of load consisting of residential, industrial, and commercial loads are considered, in which every bus of the system had one type of load. On the other hand, distribution system load varies in different seasons of the year. Thus, in this paper, load condition is considered in three stages as low level for summer night period, medium level for summer day and winter night periods, and peak-load-level for winter day period. Load levels in different periods of the year for determining size and location are presented in Fig. 1 [17].

Table 1. Typical load types and exponent values.

Season	Load type	Day		Night	
		α	β	α	β
Summer	Residential	0.72	2.96	0.92	4.04
	Commercial	1.25	3.50	0.99	3.95
	Industrial	0.18	6.00	0.18	6.00
Winter	Residential	1.04	4.19	1.30	4.38
	Commercial	1.50	3.15	1.51	3.40
	Industrial	0.18	6.00	0.18	6.00

Load duration data is listed in Table 2.

Table 2. Load duration data.

Level	Network situation	Duration time (h)
1	Low load	2190
2	Medium load	4380
3	Peak load	2190

It should be noted that power loss and ENS index for each load level are calculated considering its duration time. In other words, the sum of power loss and ENS index for three load levels is considered in the objective function.

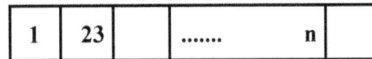

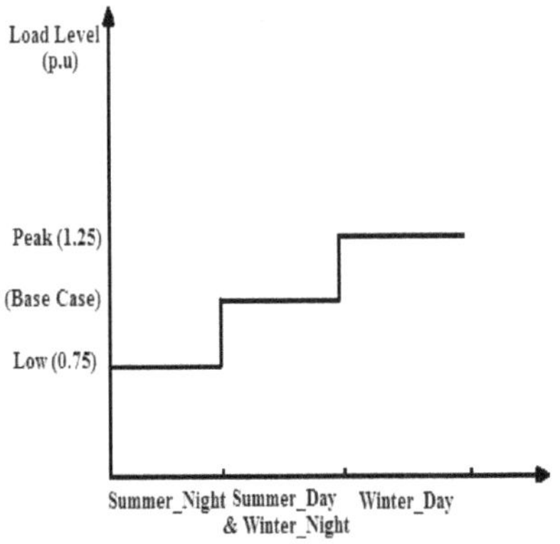

Fig. 1. Load levels in different periods of the year.

3. PROPOSED HYBRID OPTIMIZATION ALGORITHM FOR CAPACITOR ALLOCATION

3.1. Genetic algorithm

GA is an effective search technique for solving the optimization problem, which provides a solution for an optimization problem using the population of individuals representing a possible solution [18]. Each possible solution is termed a "chromosome". New points of the search space are generated through GA operations, known as reproduction, crossover and mutation. These operations consistently produce fitter offspring through successive generations, which rapidly lead the search to global optima.

3.2. Particle swarm optimization algorithm

3.2.1 Classical approach abstract

The PSO method is an optimization technique which is motivated by the social behaviors of organisms such as fish schooling and bird flocking [19]. It provides a population-based search procedure, in which individuals called "particles" change their positions (states) over time. In a PSO system, particles fly around in a multidimensional search space. During the flight, each particle adjusts its position according to its own experience and the experience of neighboring particles, making use of the best position encountered by itself and its neighbors. The swarm direction of a particle is defined by the set of particles neighboring the particle and its history experience.

3.2.2. Binary particle swarm optimization

In order to solve optimization problems in discrete search spaces, Kennedy and Eberhart (1997) developed a binary version of PSO [20]. In this version, the particle is characterized by a binary solution representation and the velocity must be transformed into the change of probability for each binary dimension to take the value of one.

3.3. Proposed hybrid algorithm

This searching technique is developed for optimal capacitor locating and sizing. The problem consists of two parts; the first part is to determine the optimal location of capacitor and the second one is to determine the optimal sizing of capacitor. Location of capacitor which is one of buses in the distribution system is an integer parameter. Therefore, an integer-based optimization algorithm such as GA is needed. In this paper, the structure of each chromosome coding for the location of capacitor used in GA is shown in Fig. 2.

1	23		.... n	

Fig. 2. Structure of each chromosome coding for location of capacitor used in GA.

It can be seen in this figure that the chromosome was formed of "n" bit strings, each of which corresponded to a system bus for capacitor installation. It should be noted that each bit showed the situation of the capacitor bank installation on the specific bus so that the values of 0 and 1 demonstrated the absence and presence of capacitor at the specific bus, respectively. The location of capacitor determined by GA is used in BPSO algorithm to optimize the sizing of capacitor. BPSO has the fast convergence ability that is a suitable method for a large iterative and time-consuming problem. Fig. 3 shows the structure of the particles used for BPSO.

C_1	C_2....	. C_{NB}	

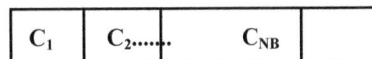

Fig. 3. Structure of a particleused for BPSO.

It is observed in Fig. 3 that this particle is composed of *NB* cells with the value of C_i, which is the rating of capacitor (*NB* is the number of candidate buses determined by GA.). In fact, each candidate bus for capacitor installation determined by GA is assigned by a cell in BPSO and the rating of the capacitor at the relative candidate bus is the value of the corresponding cell.

For example, C_i is the rating of the capacitor installed at bus i.

The parameters of GA/BPSO method used for solving the optimization problems are presented in Table 3. Flowchart of the GA/BPSO method for optimal location and sizing of capacitor is shown in Fig. 4 in the following steps:

1. Initializing: Set the time counter t=0 and randomly generate "n" chromosomes which represent "n" initial candidates for the location of capacitor.
2. Calculating fitness function using BPSO: Evaluate each chromosome in terms of determining the optimal sizing of capacitor.
   - Initialize particle population for the sizing of capacitor.
   - Calculate the objective function.
   - Determine the minimum value of the objective function as the overall, global best of the group and record the best candidate of particle for the sizing of capacitor.
   - Update velocity (v) and position parameters of BPSO.
   - Check the stop criterion.
3. Time updating: Update the time counter $t=t+1$.
4. New population: Create a new population for the location of capacitor using the following operations by GA: Selection - Crossover – Mutation.
5. Calculating fitness function using BPSO and time updating.
6. Checking the stop criterion.

4. TEST RESULTS

Two standard distribution systems consisting of 33-bus and 69-bus test systems are used to validate the effectiveness of the proposed optimization algorithm. For this purpose, the GA, BPSO, and proposed GA/BPSO algorithms are used for determining the optimal sitting and sizing of capacitor in the test systems, the results of which are compared and discussed in this section. For calculating the reliability indices and determining optimal capacitor placement, it is assumed that the section with the highest resistance had the biggest failure rate of 0.5 f/year and the section with the smallest resistance had the least failure rate of 0.1 f/year [13]. Based on this assumption, failure rates of other sections are linearly calculated proportional to these two values according to their resistance.

Furthermore, it is assumed that, if the reactive component of a section's current is fully compensated for, its failure rate would reduce to 85% of its uncompensated failure rate [13]. Also, for partial compensation, the failure rate is calculated using (6). Moreover, in both test systems, it was assumed that there is only one breaker at the beginning of the main feeder and also there is one sectionalizer at the beginning of each section.

Besides, for each line, the repair time and total isolation and switching time are considered 8 and 0.5 h, respectively. Also, other components such as transformers, busbars, breakers and disconnects are assumed to be fully reliable in this paper. The parameters of the test systems are listed in Table 4.

Table 3. GA/BPSO, GA and BPSO parameters.

Method	Pop. size		Selection method	Crossover	Mutation		Algorithm stopcriterion
GA/BPSO	Chromosome	30	Normalized geometric selection	Simple Xover	Binary mutation		Maximum number of generation(200)
	Particle	20					
GA	40		Normalized geometric selection	Simple Xover	Binary mutation		Maximum number of generation(300)
BPSO	30		C_1	C_2	r_1	r_2	Maximum number of generation(250)
			2	2	1	1	

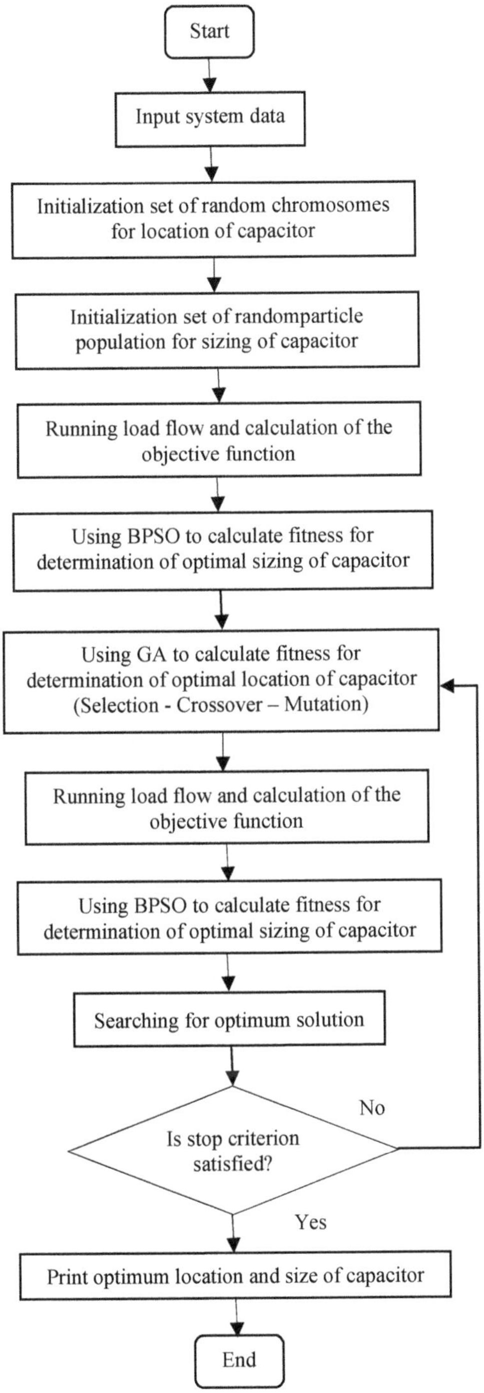

Fig. 4. Flowchart of the GA/BPSO method for optimal sitting and sizing of capacitor.

Table 4. The value of constant coefficients [21], [22].

Parameter	Value
K_e	0.06
K	0.1
K_c	3

4.1. 33 Bus radial distribution system

Figure 5 shows the single line diagram of the 12.66 kV, 33-bus, 4-lateral radial distribution system. Total load of the system in the base case is (3715+ j2300) kVA. The load flow data of the system were taken from [23] and load type data of each bus are presented in Table A.1. in the Appendix.

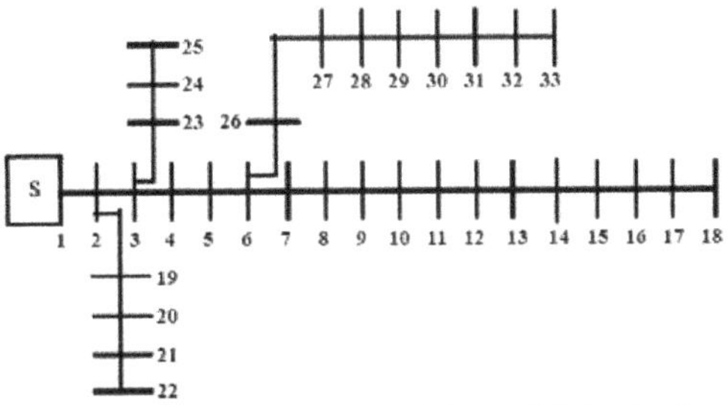

Fig. 5. Single line diagram of 33-bus radial distribution system.

Results of the optimal placement of capacitor banks using GA, BPSO, and GA/BPSO algorithms in 33-bus test system are presented in Table 5.

Table 5. The results of optimal placement of capacitor banks in 33-bus test system.

Method	Bus no.	Size(kVAr)
GA/BPSO	5	972.8
	8	101.4
	21	101
GA	2	140.6
	3	876.2
	5	129.6
BPSO	2	305.6
	5	952.5
	30	127.9

In this case, optimal number of capacitor is obtained to be three. Table 6 demonstrates the effects of optimal capacitor placement by GA, BPSO, and GA/BPSO algorithms on active power loss, reactive power loss, ENS, and total annual cost in 33-bus test system. It can be seen from this table that determination of optimum size and location of capacitors by GA, BPSO, and GA/BPSO algorithms had a considerable effect on active and reactive power loss, ENS, and total annual cost. For example, it is observed in Table 6 that the installation of capacitors decreased total cost from 45884.5 \$/year in the base case to 42404.3 \$/year,

Table 6. The results of optimal placement of capacitor banks using GA, BPSO and GA/BPSO algorithms in 33-bus test system.

Case		Total Cost ($/year)	ENS (kWh/year)	$P_{T,Loss}$ (kW)	$Q_{T,Loss}$ (kVAr)
Before installation		45884.5	57286	76.4	50.7
After installation	GA/BPSO	42404.3	53454	63.8	42.2
	GA	43953.5	54568	66.7	44.3
	BPSO	43119.4	53755	63.9	42.3

Table 7. The percentage of loss reduction, reliability improvement and total cost reduction using three optimization algorithms in 33-bus test system.

	Total Cost (%)	ENS (%)	$P_{T,Loss}$ (%)	$Q_{T,Loss}$ (%)
GA/BPSO	7.5847	6.6892	16.4921	16.7653
GA	4.2084	4.7446	12.6963	12.6232
BPSO	6.0262	6.1638	16.3612	16.5680

43953.5 $/year, and 43119.4 $/year by GA/BPSO, GA and BPSO algorithms, respectively. Moreover, it is observed in Table 7 that optimum capacitor installation by combined GA/BPSO algorithm caused more loss reduction, reliability improvement, and total cost reduction compared to GA and BPSO algorithms. For example, it can be seen in Table 7 that capacitor installation caused 7.5847%, 4.2084%, and 6.0262% reduction in total cost by GA/BPSO, GA and BPSO algorithms, respectively, compared to the base case.

In addition to minimizing power losses and reliability improvement in distribution networks, proper capacitor planning can improve the overall network voltage profiles. For instance, Fig. 6 shows the voltage profile improvement before and after capacitor installation by GA/BPSO, GA and BPSO algorithms for summer day and winter night load levels in the 33-bus system.

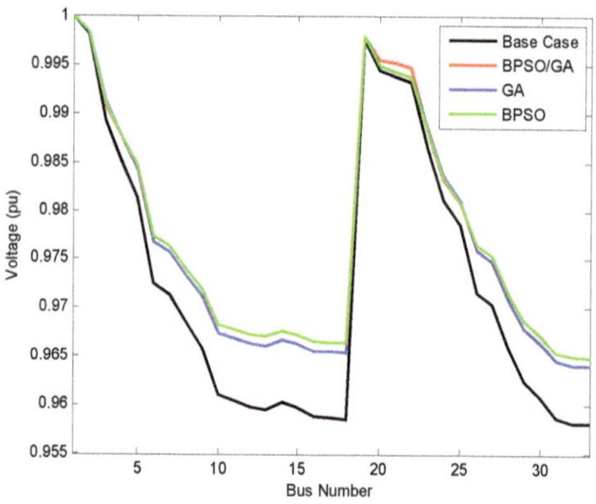

Fig. 6. Voltage profile for Summer-Day and Winter-Night load levels in 33-bus system.

69-bus radial distribution system with the total load of 3.80 MW and 2.69 MVar in the base case is considered as another test system, which has 7 laterals, 69 buses, and 68 branches. The load flow data of the system are taken from [24] and load type data of each bus are presented in Table A.2. in the Appendix. The results for the optimal number of four capacitor banks of GA/BPSO, GA, and BPSO algorithms are demonstrated in Table 8.

Table 8. The results of optimal placement of capacitor banks in 69-bus test system.

Method	Bus no.	Size(kVAr)
GA/BPSO	17	110.5
	31	95.2
	47	368.4
	61	76.2
GA	2	300
	14	399.2
	29	357.1
	52	237.6
BPSO	28	140.7
	36	132.5
	38	336
	52	277.4

The effects of optimal capacitor placement by GA, BPSO and GA/BPSO algorithms on active and reactive power losses, ENS, and total annual cost in 69-bus test system are presented in Table 9. Also, improvement percent of these parameters after capacitor installation is shown in Table 10.

Similar to 33-bus test system, it is illustrated in Tables 9 and 10 that optimal capacitor placement by GA/BPSO, GA and BPSO algorithms led to loss

Table 9. The results of optimal placement of capacitor banks using GA, BPSO and GA/BPSO algorithms in 69-bus test system.

Case		Total Cost ($/year)	ENS (kWh/year)	$P_{T,Loss}$ (kW)	$Q_{T,Loss}$ (kVAr)
Before installation		63630.7	92837	103.4	50.1
After installation	GA/BPSO	20507.9	85180	19.1	11.1
	GA	22952	85583	20	11.9
	BPSO	21068.7	85644	18.73	11

Table 10. The percentage of loss reduction, reliability improvement and total cost reduction using three optimization algorithms in 69-bus test system.

	Total Cost (%)	ENS (%)	$P_{T,Loss}$ (%)	$Q_{T,Loss}$(%)
GA/BPSO	67.770	8.247	81.528	77.844
GA	63.929	7.813	80.657	76.247
BPSO	66.889	7.748	81.885	78.044

reduction and reliability improvement of 69-bus distribution system. However, it is observed in these tables that optimum capacitor installation by combined GA/BPSO algorithm caused more total cost reduction than GA and BPSO algorithms.

Results of Table 10 show that capacitor installation caused 67.770%, 63.929% and 66.889% reduction in total cost by GA/BPSO, GA and BPSO algorithms, respectively, compared to the base case.

Also, optimal placement of capacitors led to the voltage improvement of 69-bus distribution system. Figure 8 shows the voltage profile improvement before and after capacitor installation by GA/BPSO, GA, and BPSO algorithms for summer day and winter night load levels.

Totally, while comparing the effect of capacitor installation in the two test systems, it can be concluded that optimal sitting and sizing of capacitor in the systems have considerable effects on loss reduction, reliability enhancement, and voltage improvement. In particular, in combined GA/BPSO algorithm, the search space is reduced and a tight distribution was obtained for the search results. The combined method is converted into a solution in the minimum number of iterations and the BPSO was the least. However, the running time for the BPSO is faster than the other two and was the least for GA.

4.3. Comparing variances of objective functions calculated by GA/BPSO, GA, and BPSO algorithms

Figure 9 shows the variance of the objective function (per unit) determined by three optimization algorithms in 33-bus and 69-bus test systems. The variances are calculated for the 30 initial populations. It can be seen in this figure that the variances of the objective function for GA and BPSO are 0.0831 and 0.032 and also 0.0826 and 0.0317 in 33-bus and 69-bus systems, respectively, while variances of the objective function for

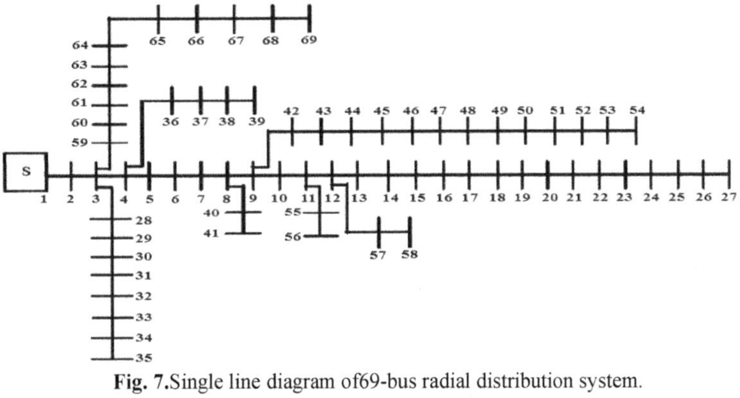

Fig. 7. Single line diagram of69-bus radial distribution system.

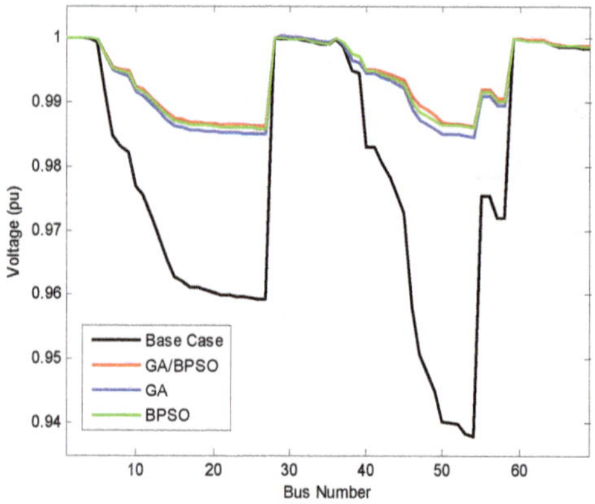

Fig. 8. Voltage profile for Summer-Day and Winter-Night load levels in 69-bus system.

GA/BPSO are 0.0017 and 0.0016 in 33-bus and 69-bus systems, respectively. Very small variance in the results of GA/BPSO algorithm. demonstrated that this method led to more uniformity and reliable results than GA and BPSO algorithms.

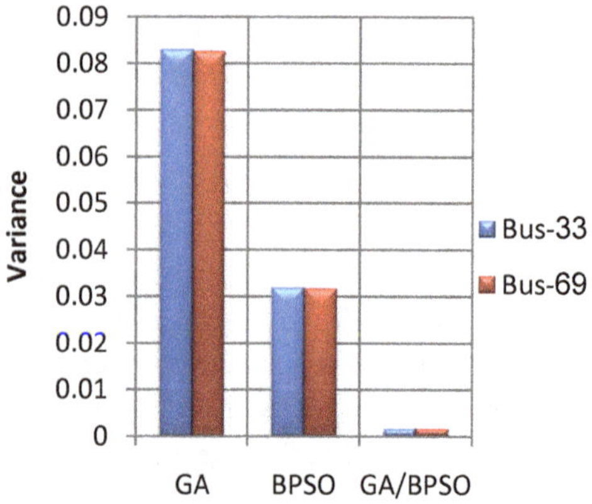

Fig. 9. Variances of objective function calculated by GA/BPSO, GA and BPSO algorithms.

5. CONCLUSIONS

In the paper, a combined optimization method is presented for determining the optimum location and capacity of shunt capacitors in distribution systems. The proposed method is based on the combination of GA and BPSO algorithms for optimal sitting and sizing of capacitors. Also, the objective function consisted of the cost of energy loss, reliability and investment cost of the capacitor banks for the optimization problem. In addition, the impacts of types of customers for load modeling, voltage-dependent load models and annual load duration curve are considered for the optimization problem. The proposed GA/BPSO method is implemented on 33-bus and 69-bus test systems for the purpose of loss minimization and reliability improvement considering cost factor. Also, the obtained results are compared with those of GA and BPSO algorithms. Simulation results demonstrate that GA, BPSO and the proposed method (GA/BPSO) may result in considerable loss reduction, reliability and voltage profile improvement and annual cost reduction. However, the results indicate that optimum capacitor installation by combined GA/BPSO algorithm is caused more total cost reduction than GA and BPSO algorithms. In addition, the proposed method has less variance than GA and BPSO algorithms. Therefore, the results obtained from this method are more reliable than those of GA and BPSO algorithms in terms of finding optimum solutions.

APPENDIX

Table A.1. Load type dataof 33-bus system.

Bus no.	Customer type	Bus no.	Customer type
1	Commercial	18	Commercial
2	Commercial	19	Commercial
3	Commercial	20	Commercial
4	Residential	21	Commercial
5	Residential	22	Commercial
6	Industrial	23	Industrial
7	Industrial	24	Industrial
8	Residential	25	Residential
9	Residential	26	Residential
10	Residential	27	Residential
11	Residential	28	Commercial
12	Commercial	29	Industrial
13	Commercial	30	Commercial
14	Residential	31	Industrial
15	Residential	32	Commercial
16	Residential		
17	Residential		

Table A.2. Load type dataof 69-bus system.

Bus no.	Customer type	Bus no.	Customer type
1	-	35	-
2	-	36	Industrial
3	-	37	Commercial
4	-	38	Commercial
5	Residential	39	Commercial
6	Residential	40	Industrial
7	Residential	41	Industrial
8	Residential	42	Industrial
9	-	43	Industrial
10	Residential	44	-
11	Residential	45	-
12	Residential	46	-
13	Residential	47	Residential
14	-	48	-
15	Residential	49	Industrial
16	Residential	50	Industrial
17	Residential	51	-
18	-	52	Commercial
19	Industrial	53	Commercial
20	Commercial	54	Industrial
21	Commercial	55	Industrial
22	-	56	Residential
23	Commercial	57	Residential
24	-	58	Commercial
25	Residential	59	Commercial
26	Residential	60	-
27	Residential	61	Industrial
28	Residential	62	Industrial
29	-	63	Industrial
30	-	64	-
31	-	65	Industrial
32	Industrial	66	-
33	Residential	67	Residential
34	Industrial	68	Residential

REFERENCES

[1] H. Duran, "Optimum number, location and size of shunt capacitors in radial distribution feeders: a dynamic programming approach," *IEEE Transaction on Power Apparatus and Systems*, vol. 87, no. 9, pp. 1769-1774, 1968.

[2] J.J. Grainger and S.H. Lee, "Optimum size and location of shunt capacitors for reduction of losses on distribution feeders," *IEEE Transaction on Power Apparatus and Systems,* vol. 100, no. 3, pp. 1105-1118, 1981.

[3] M.E. Baran and F.F. Wu, "Optimal sizing of capacitors placed on a radial distribution system," *IEEE Transaction on Power Delivery*, vol. 4, no. 1, pp. 735-743, 1989.

[4] S. Sundhararajan and A. Pahwa, "Optimal selection of capacitors for radial distribution systems using genetic algorithm," *IEEE Transaction on Power Systems,* vol. 9, no. 3, pp. 1499-1507, 1994.

[5] M.A.S. Masoum, M. Ladjevardi, A. Jafarian, and E.F. Fuchs, "Optimal placement, replacement and sizing of capacitor banks in distorted distribution networks by genetic algorithms," *IEEE Transaction on Power Delivery,* vol. 19, no. 4, pp. 1794-1801, 2004.

[6] R. SrinivasasRao, S.V L. Narasimham and M. Ramalingaraju, "Optimal capacitor placement in a radial distribution system using plant growth simulation algorithm," *International Journal of Electrical Power and Energy Systems*, vol. 33, no. 5, pp. 1133-1139, 2011.

[7] S. M. Tabatabaei and B. Vahidi, "Bacterial foraging solution based fuzzy logic decision for optimal capacitor allocation in radial distribution system," *Electric Power Systems Research,* vol. 81, no. 4, pp. 1045-1050, 2011.

[8] J.F.V. González, C. Lyra, and F.L. Usberti, "A pseudo-polynomial algorithm for optimal capacitor placement on electric power distribution networks," *European Journal of Operational Research,* vol. 222, no.1, pp. 149-156, 2012.

[9] A. Hamouda and S. Samir, "Optimal capacitors sizing in distribution feeders using heuristic search based node stability-indices," *International Journal of Electrical Power and Energy Systems*, vol. 46, pp. 56-64, 2013.

[10] A.R. Abul'Wafa, "Optimal capacitor allocation in radial distribution systems for loss reduction: A two stage method," *Electric Power Systems Research,* vol. 95, pp. 168-174, 2013.

[11] D. Kaur and J. Sharma, "Multiperiod shunt capacitor allocation in radial distribution systems," *International Journal of Electrical Power and Energy Systems*, vol. 52, pp. 247-253, 2013.

[12] S.J. Huang and X.Z. Liu, "A plant growth-based optimization approach applied to capacitor placement in power systems,"*IEEE Transactions on Power Systems,* vol. 27, no. 4, pp. 2138-2145, 2012.

[13] A.H. Etemadi and M. Fotuhi-Firuzabad, "Distribution system reliability enhancement using optimal capacitor placement," *IET Generation Transmission & Distribution,* vol. 2, no. 5, pp. 621-631, 2008.

[14] P.L. Lewin, J.E. Theed, A.E. Davies and S.T. Larsen, "Method for rating power cables buried in surface troughs," *IET Proceedings on Generation,*

Transmission and Distribution, vol. 146, no. 4, pp. 360-364, 1999.

[15] S.V. Nikolajevic, "The behavior of water in XLPE and EPR cables and its influence on the electrical characteristics of insulation," *IEEE Transactions on Power Delivery*, vol. 14, no. 1, pp. 39-45, 1999.

[16] R.E. Brown, *Electric power distribution reliability*, Marcel Dekker Inc., New York, Basel, 2009.

[17] K. Qian, C. Zhou, M. Allan and Y. Yuan, "Load modelling in distributed generation planning," *International Conference on Sustainable Power Generation and Supply*, pp. 1-6, 2009.

[18] G. Boone and H.D. Chiang, "Optimal capacitor placement in distribution systems by genetic algorithm," *International Journal of Electrical Power and Energy Systems*, vol. 15, no. 3, pp. 155-161, 1993.

[19] R. Eberhart and J. Kennedy, "A new optimizer using particle swarm theory," *Proceedings of the Sixth International Symposium on Micro Machine and Human Science*, pp. 39-43, 1995.

[20] R. Baghipour and S.M. Hosseini, "Placement of DG and capacitor for loss reduction, reliability and voltage improvement in distribution networks using BPSO," *International Journal of Intelligent Systems and Applications*, vol. 4, no. 12, pp. 57-64, 2012.

[21] D. Das, "Optimal placement of capacitors in radial distribution system using a Fuzzy-GA method," *International Journal of Electrical Power and Energy Systems*, vol. 30, no. 6, pp. 361-367, 2008.

[22] M. Gilvanejad, H.A. Abyaneh and K. Mazlumi, "Fuse cutout allocation in radial distribution system considering the effect of hidden failures," *International Journal of Electrical Power and Energy Systems*, vol. 42, no. 1, pp. 575-582, 2012.

[23] M.A. Kashem, V. Ganapathy, G.B. Jasmon and M.I. Buhari, "A novel method for loss minimization in distribution networks," *Proceedings of the International Conference on Electric Utility: Deregulation and Restructuring and Power Technologies*, pp. 251-255, 2000.

[24] M.E. Baran and F.F. Wu, "Optimal capacitor placement on radial distribution systems," *IEEE Transactions on Power Delivery*, vol. 4, no. 1, pp. 725-734, 1989.

2

Reduction the Number of Power Electronic Devices of a Cascaded Multilevel Inverter Based on New General Topology

S. Laali, E. Babaei*, and M.B.B. Sharifian

Faculty of Electrical and Computer Engineering,University of Tabriz,Tabriz, Iran

ABSTRACT

In this paper, a new cascaded multilevel inverter by capability of increasing the number of output voltage levels with reduced number of power switches is proposed. The proposed topology consists of series connection of a number of proposed basic multilevel units. In order to generate all voltage levels at the output, five different algorithms are proposed to determine the magnitude of DC voltage sources. Reduction of the used power switches and the variety of DC voltage sources magnitudes are two main advantages of the proposed topology. These results are obtained by comparison of the proposed inverter with the H-bridge cascaded multilevel inverter and one of recently presented topologies. The remarkable ability of the proposed topology with its algorithms in generating all voltage levels (even and odd) is verified through PSCAD/EMTDC simulation and experimental results of a 17-level inverter.

KEYWORDS: Multilevel inverters, Conventional cascaded multilevel inverter, Bidirectional switches.

1. INTRODUCTION

Recently, the multilevel inverters have received more and more attention in researches because of their capabilities in high power and medium voltage applications. High efficiency, high power quality, lower order harmonics, better electromagnetic interference, lower dv/dt stress on switches, and lower switching losses are some of the advantages of the multilevel inverters [1-7]. There are three main topologies for multilevel inverters: diode clamp multilevel inverter, flying capacitor multilevel inverter and cascaded multilevel inverter [2-4]. The cascaded multilevel inverters have received special attention due to the modularity and simplicity of the control. In addition, the ability of the cascaded inverters in generating higher number of levels with minimum number of semiconductor devices in comparison with other main topologies increases the trend to these types of inverters. In this inverter, the desired AC output waveform is synthesized from several steps of DC source as inputs [6-8]. The cascaded multilevel inverters are mainly classified into two groups: symmetric; with the equal magnitude of dc voltage source and asymmetric with different values of DC voltage sources, which leads to increasing the number of output voltage levels [8-10].

Up to now, different cascaded topologies have been presented in literatures. In [10-11], two different topologies for cascaded multilevel inverters have been presented that are known as symmetric ones. It has been also presented H-bridge cascaded multilevel inverter in [12]. Here, two different algorithms to determine the values of DC voltage sources have been represented that are known as symmetric and asymmetric ones. These symmetric topologies have the minimum variance of the value of the DC voltage sources that is the most important advantage of them. While, the higher number of switches and insulated gate bipolar transistors (IGBTs) are required for generation specific output voltage levels because of the low amplitude of the used DC voltage sources.

*Corresponding author:
E. Babaei (E-mail: e-babaei@tabrizu.ac.ir)

Moreover, each switch requires a driver circuit. As a result, increasing in the installation space and total cost of the symmetric inverters are the most important disadvantages of them. Therefore, different asymmetric cascaded multilevel inverters have been suggested in [14-15]. In addition, different asymmetric H-bridge cascaded multilevel inverters have been also reported in [8-9] and [13]. The remarkable advantage of these topologies is the high number of generated output voltage levels with minimum number of used power electronic devices, but the high variety of the value of the DC voltage sources is their main disadvantage.

Although, all of the presented topologies in the literatures have their own advantages and disadvantages, but in this paper, a cascaded multilevel inverter based on the new basic unit is proposed. This inverter increases the number of output voltage level by using minimum number of power switches, driver circuit and IGBTs and less variety of the value of the DC voltage sources. Then, five different algorithms for generating all voltage levels are presented. In addition, the proposed topology with its algorithms is compared with the H-bridge cascaded multilevel inverter and the presented topology in [15] to investigate advantages of the proposed topology. Finally, the obtained results of PSCAD/EMTDC simulation and experimental prototype on a 17-level inverter reconfirm the correct performance of the proposed topology in generating all voltage levels.

2. PROPOSED TOPOLOGY

The basic multilevel unit is shown in Fig. 1. As Fig. 1 shows, the proposed basic unit consists of two DC voltage sources, one bidirectional switch (S_2) and two unidirectional ones (S_1 and S_3). It is important to note that each unidirectional switch consists of an IGBT with an anti-parallel diode and a driver circuit, however, the bidirectional ones include of two IGBTs with two anti-parallel diodes and a driver circuit if the switch with common emitter configuration is used. Therefore, the number of driver circuit for the bidirectional switches is as same as unidirectional ones in the proposed basic unit. The proposed basic unit is able to generate three voltage levels of V_1, 0 and V_1+V_2 at the output. According to Fig. 1, the switches (S_1, S_2) or (S_1, S_3) or (S_2, S_3) or (S_1, S_2,S_3) can't be turned on simultaneously, because a short circuit across the DC voltage sources would be produced. Table 1 shows the output voltage levels of the proposed basic unit based on different switching patterns. In this Table, 1 and 0 indicate the on and off states of the switches, respectively. It is obvious from Table 1 that the proposed unit is only able to generate positive levels at the output.

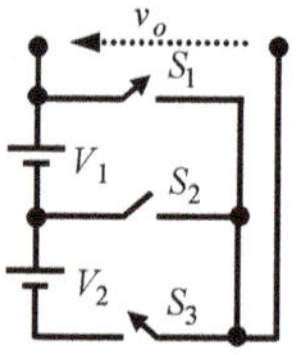

Fig.1. The proposed basic unit.

Table 1. The output voltageof the proposed basic unit based on different switching pattern.

State	S_1	S_2	S_3	v_o
1	0	1	0	V_1
2	0	0	1	V_1+V_2
3	1	0	0	0

A new cascaded multilevel inverter could be made by series connection of the n number of the basic unit shown in Fig. 1. This new proposed cascaded multilevel inverter is shown in Fig. 2. The output voltage of the proposed inverter is equal to adding the output voltage of each unit and can be written as follows:

$$v_o(t)=v_{o,1}(t)+v_{o,2}(t)+\cdots+v_{o,n}(t) \quad (1)$$

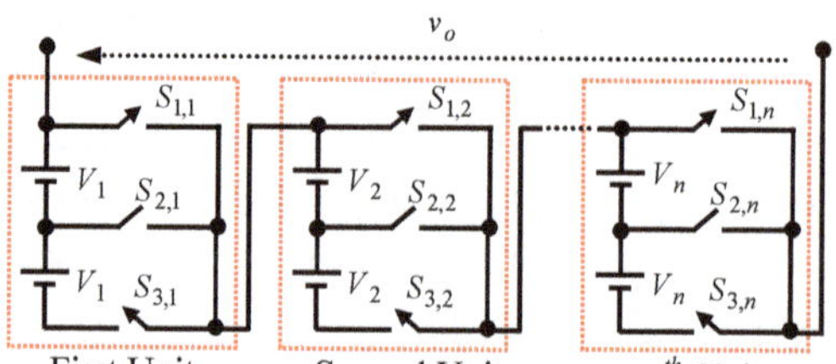

Fig. 2. Series connection of n number of basic unit.

As mentioned before, this inverter is only able to generate positive levels at the output so in order to generate all voltage levels (positive and negative), it is required to use of the H-bridge at the output. The completed cascaded multilevel inverter is shown in Fig. 3. As this figure indicates, T_1 to T_4 are the unidirectional switches. The output voltage level will be positive and negative, if the switches T_1, T_2 and T_3, T_4 are turned on, respectively. In addition, by turning on the switches T_1 and T_3 or T_2 and T_4 the output voltage will be zero.

It is pointed out that the value of blocked voltage of power switches in each basic unit of the proposed topology is low that leads to use of power switches with low nominal voltage range. This is one of the main advantages of the proposed inverter. However, the used power switches in H-bridge, examine high value of blocked voltage because of cascading several basic units. This is the main disadvantage of the proposed inverter.

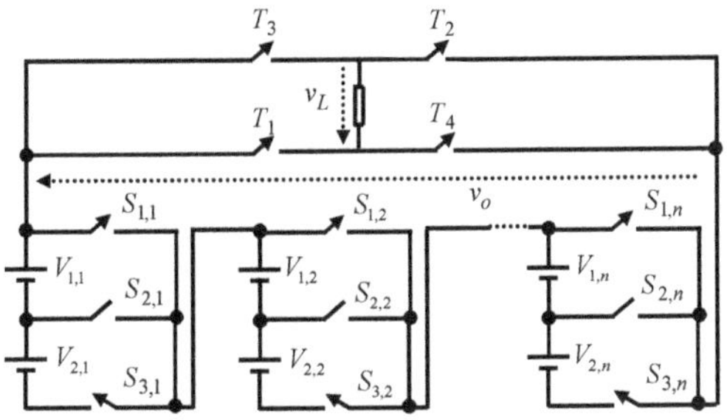

Fig. 3. The proposed cascaded multilevel inverter.

In the proposed cascaded multilevel inverter, the number of switches (N_{switch}), IGBTs (N_{IGBT}), driver circuits (N_{driver}) and dc voltage sources (N_{source}) are calculated as follows:

$$N_{switch} = 3n + 4 \quad (2)$$

$$N_{IGBT} = 4n + 4 \quad (3)$$

$$N_{driver} = 3n + 4 \quad (4)$$

$$N_{source} = 2n \quad (5)$$

At the following sub-sections in order to generate all voltage levels at the output, five different algorithms are proposed to determine the magnitude of the DC voltage sources.

2.1. First proposed algorithm (P_1)

In this sub-section, the amplitude of the DC voltage sources of the basic unit shown in Fig. 3 is considered as follows:

$$V_{1,j} = V_{2,j} = V_{dc} \qquad for \quad j = 1, 2, \cdots, n \quad (6)$$

This inverter is known as symmetric cascaded multilevel inverter. In this algorithm, the number of output voltage levels (N_{level}) and the maximum amplitude of the producible output voltage $(V_{o,\max})$ are respectively equal to:

$$N_{level} = 4n + 1 \quad (7)$$

$$V_{o,\max} = \sum_{j=1}^{n} V_j = (2n)\, V_{dc} \quad (8)$$

Although the number of used power electronic devices such as the number of the power switches, IGBTs and driver circuit have significant influence on the installation space and total cost of the inverter, but the variety of the magnitude of DC voltage sources is another important feature in determining this index. By reducing the variety of the value of the DC voltage sources, the total cost of the inverter decreases. Therefore, the variety of the values of the DC voltage sources $(N_{variety})$ in the proposed algorithm is calculated as follows:

$$N_{variety} = 1 \quad (9)$$

Considering (8), the low variety of the amount of the DC voltage sources is an advantage for the proposed algorithm.

2.2. Second proposed algorithm (P_2)

In the second proposed algorithm, the magnitudes of the DC voltage sources are determined as follows:

$$V_{1,1} = V_{2,1} = V_{dc} \quad (10)$$

$$V_{1,j} = V_{2,j} = 2V_{dc} \qquad for \quad j = 2, 3, \cdots, n \quad (11)$$

Considering this proposed algorithm, the number of output voltage levels, the maximum magnitude of the output voltage and the variety of the values of DC voltage sources are calculated as follows:

$$N_{level} = 8n - 3 \quad (12)$$

$$V_{o,\max} = (4n - 2)\, V_{dc} \quad (13)$$

$$N_{variety} = 2 \quad (14)$$

2.3. Third proposed algorithm (P_3)

In this sub-section, the values of the DC voltage sources are selected by below equations:

$$V_{1,1} = V_{2,1} = V_{dc} \quad (15)$$

$$V_{1,j} = V_{2,j} = 3V_{dc} \qquad for \quad j = 2, 3, \cdots, n \quad (16)$$

In this condition, the number of output voltage levels, the maximum magnitude of the output voltage and the variety of the values of DC voltage sources are written as follows:

$$N_{level} = 12n - 7 \quad (17)$$

$$N_{variety} = 2 \tag{18}$$

$$V_{o,\max} = (6n-4)\,V_{dc} \tag{19}$$

2.4. Fourth proposed algorithm (P_4)

In the fourth proposed algorithm, the magnitudes of the DC voltage sources of the proposed cascaded topology are selected as follows:

$$V_{1,1} = V_{2,1} = V_{dc} \tag{20}$$

$$V_{1,j} = V_{2,j} = 2^{j-1} V_{dc} \quad for \;\; j = 2, 3, \cdots, n \tag{21}$$

In the proposed algorithm, the number of output voltage levels, the maximum magnitude of the output voltage and the variety of the values of DC voltage sources are equal to:

$$N_{level} = 2^{n+2} - 3 \tag{22}$$

$$V_{o,\max} = (2^{n+1} - 2)\,V_{dc} \tag{23}$$

$$N_{variety} = n \tag{24}$$

2.5. Fifth proposed algorithm (P_5)

In this sub-section, the values of the DC voltage sources in the proposed cascaded multilevel inverter are written as follows:

$$V_{1,1} = V_{2,1} = V_{dc} \tag{25}$$

$$V_{1,j} = V_{2,j} = 3^{j-1} V_{dc} \quad for \;\; j = 2, 3, \cdots, n \tag{26}$$

Considering the fifth proposed algorithm, the number of output voltage levels, the maximum magnitude of the output voltage and the variety of the values of DC voltage sources are calculated as follows:

$$N_{level} = 2 \times 3^n - 1 \tag{27}$$

$$V_{o,\max} = (3^n - 1)\,V_{dc} \tag{28}$$

$$N_{variety} = n \tag{29}$$

3. COMPARING THE PROPOSED TOPOLOGY WITH THE CONVENTIONAL TOPOLOGIES

The most important aim of introducing the new cascaded multilevel inverter and its proposed algorithms is increasing the number of output voltage levels by using less number of power electronic devices such as switches, IGBTs, driver circuits and so on. In this section, the proposed cascaded multilevel inverter with its presented algorithms is compared with H-bridge multilevel inverter from the number of used power electronic devices point of view. This investigation is done to determine the advantages and disadvantages of the proposed topology.

The proposed topology and its algorithms considered as $P_1 - P_5$ in this investigation. In [12], the H-bridge cascaded multilevel inverter and two different algorithms have been presented. One of the represented algorithms in [12] cause the symmetric cascaded inverter and another presented algorithm leads to asymmetric one. In this comparison, these two different algorithms are considered as R_1 and R_2, respectively. Then, in order to increase the number of output voltage levels in the H-bridge cascaded multilevel inverter, another algorithm based on ternary method has been presented in [13]. This algorithm is considered by R_3 in this investigation. The main advantage of the presented algorithm in [13] is increasing the output voltage levels with minimum number of used H-bridges. However, the high variety of the value of DC voltage source for the presented algorithm in [13] leads to the introduction of two other algorithms in the literatures. These algorithms have been presented in [8-9] and are considered by $R_4 - R_5$ in this comparison. Moreover, the presented topology in [15] is considered by R_6. Figure 4 indicates the H-bridge cascaded multilevel inverter.

Figure 5 shows the comparison of the proposed topology and its algorithms with the H-bridge cascaded multilevel inverter with its different algorithms and the presented topology in [15] from the number of switches point of view. As shown in this figure, the number of required switch for the fifth proposed algorithm is lower than the number of switches in other presented algorithm in the literature. In addition, this algorithm has even better performance than other presented algorithms for the proposed topology.

As mentioned before, the number of switches in the proposed cascaded multilevel inverter is as same as the number of driver circuits in the proposed topology. As a result, this topology needs a less number of driver circuits than other presented algorithms in literature.

Because of using bidirectional switches in the proposed topology, which consist of two IGBT with two anti-parallel diodes, it is necessary to compare the number of required IGBTs in this topology with the H-bridge cascaded multilevel inverter and the presented topology in [15]. This comparison is shown in Fig. 6. As Fig. 6 shows, the proposed cascaded topology uses a lower number of IGBTs than other topologies except the algorithm that is indicated by R_3. From the viewpoint of required IGBTs number, the fifth proposed algorithm has also best performance between other proposed algorithms. Moreover, the number of diodes in the proposed inverter is lower than H-bridge cascaded inverter and the presented algorithm in [15]. The minimum number of required IGBTs, diodes and driver circuit in the proposed topology cause the reduction in the required installation space and total cost of the system.

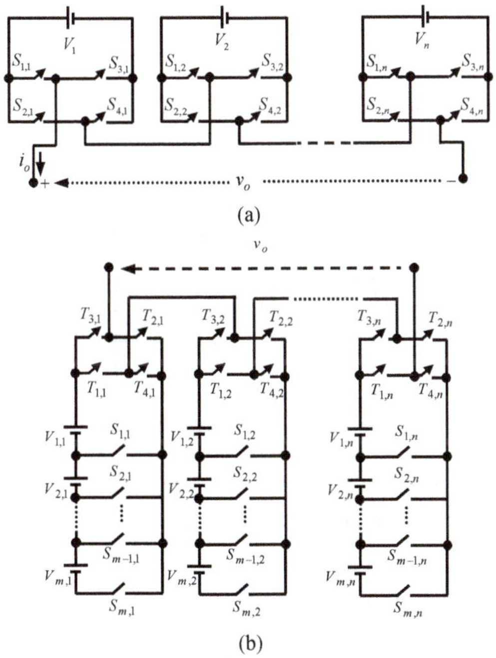

Fig. 4. The conventional cascaded multilevel inverter;(a) H-bridge cascaded inverter R_1 for $V_1 = V_2 = V_3 = \cdots = V_n = V_{dc}$, R_2 for $V_1 = V_{dc}$,$V_2 = 2V_{dc} \cdots, V_n = 2^{n-1}V_{dc}$, R_3 for $V_1 = V_{dc}$, $V_2 = 3V_{dc}, \cdots, V_n = 3^{n-1}V_{dc}$, R_4 for $V_1 = V_{dc}$, $V_2 = V_3 = \cdots = V_n = 2V_{dc}$, R_5 for $V_1 = V_{dc}$, $V_2 = V_3 = \cdots = V_n = 3V_{dc}$; (b) presented topology in [15].

Figure 7 compares the number of DC voltage sources in the proposed topology with the H-bridge cascaded multilevel inverter and the presented topology in [15]. As it is obvious, the number of used DC voltage sources in the proposed topology especially, by using the fifth algorithm, is less than other proposed algorithms and presented algorithms for the H-bridge cascaded inverter. However, this index in R_2 and R_3 is less than other proposed algorithms. This result is obtained from the differences between the topology of basic unit of the proposed cascaded inverter with the H-bridge cascaded inverter and the presented topology in [15]. In other word, there are two DC voltage sources in the proposed basic unit while one DC voltage source is required in each unit of the H-bridge cascaded inverter. However, this feature is considered as the main disadvantage of the proposed multilevel inverter, but because of increasing interest in using renewable energy sources such as solar cells, wind power and so on, this disadvantage could be easily eliminated.

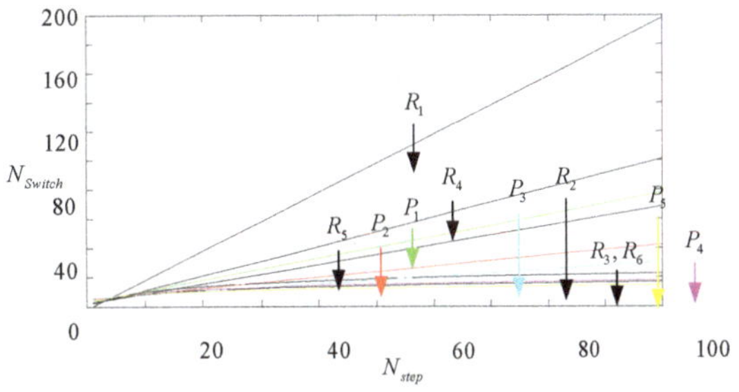

Fig. 5. Variation of N_{switch} versus N_{level} .

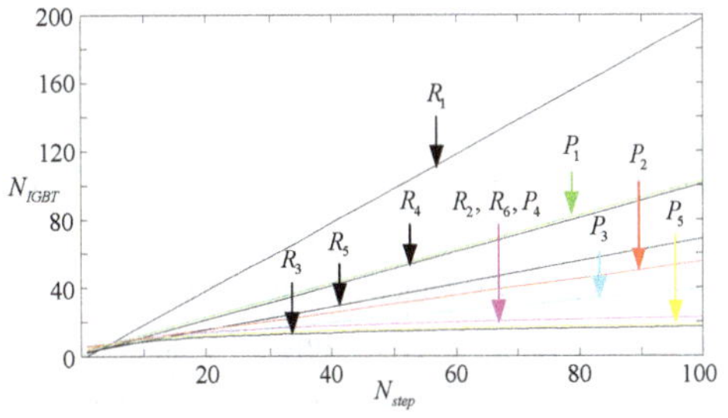

Fig. 6. Variation of N_{IGBT} versus N_{level} .

The variety of the value of the used DC voltage sources in the proposed topology with the H-bridge cascaded multilevel inverter and the presented topology in [15] is shown in Fig. 8. As Fig. 8 shows, this index in the proposed cascaded inverter is lowerthan other presented topologies in the literature. In addition, it is important to note that the fifth algorithm has also the best performance from this point of view in comparison with other proposed algorithms.

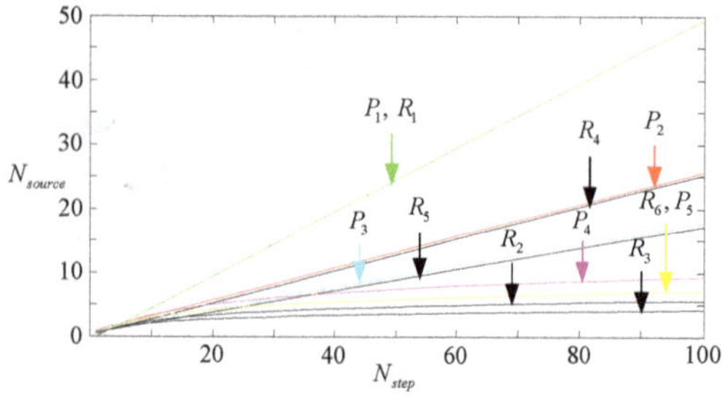

Fig. 7. Variation of N_{source} versus N_{level} .

Therefore, it is possible to consider that the proposed cascaded multilevel inverter with its new presented algorithms has better performance in comparison with the H-bridge cascaded multilevel inverter from the number of switches, driver circuits, IGBTs and diodes and the variety of the value of the DC voltage sources points of view. However, the number of required DC voltage source could be considered as the most important disadvantage of the proposed topology. This disadvantage is eliminated by using renewable energy sources.

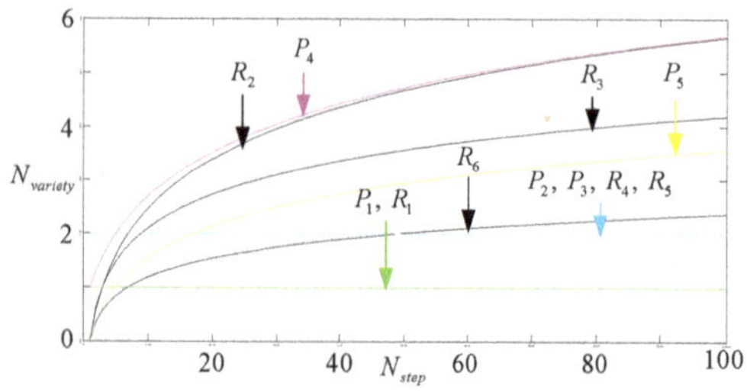

Fig. 8. Variationof $N_{variety}$ versus N_{level} .

4. SIMULATION AND EXPERIMENTAL RESULTS

The suitable performance of the proposed cascaded multilevel inverter in generating all voltage levels at the output is verified through simulation and experimental results. The simulation results are obtained by using EMTDC/PSCAD software program and the experimental results are obtained by using a laboratory prototype. These results are obtained from a 17-level inverter based on the proposed basic unit that is shown in Fig. 3. The proposed 17-level inverter is shown in Fig. 9. As it is obvious from Fig. 9, this inverter consists of two basic units, which are connected in series and the magnitude of its DC voltage sources are determined by using fifth proposed algorithms. Therefore, based on (25) and (26), the values of the used DC voltage sources are considered $V_{1,1} = V_{2,1} = 25V$ and $V_{1,2} = V_{2,2} = 75V$ in the first and the second unit, respectively. According to (27) and (28) this inverter is able to generate 17 levels (eight positive levels, eight negative levels and one zero level) with the maximum amplitude of 200 V at the output. It is important to note that the IGBTs used on the prototype are HGTP10N40CID (with an internal anti-parallel diode). The 89C52 microcontroller by ATMEL company has been used to generate all switching pattern. The load connected to the inverter is considered a resistive-inductive load with the values of $R = 100\,\Omega$ and $L = 55\,mH$. Moreover, the fundamental frequency control method is used in this inverter. This selection is based on low switching losses in comparison with the other control methods. This feature is because of low switching frequency in this control method.

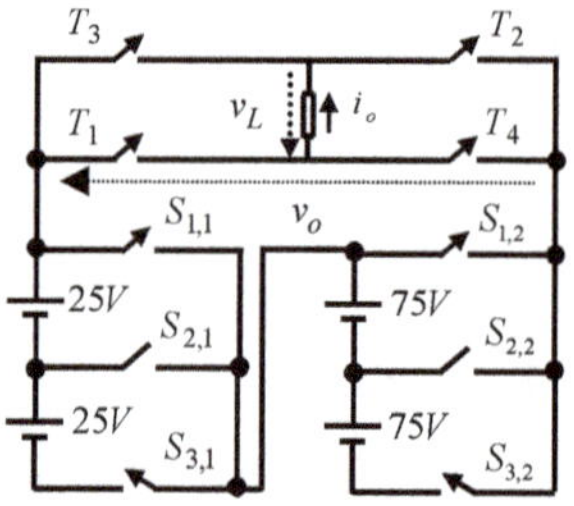

Fig. 9. The proposed cascaded 17-level inverter.

Figures 10(a) and 10(b) show the simulation and experimental output voltage waveforms (v_o) of the proposed topology, respectively. As it is obvious from Fig. 10(a), this inverter is only able to generate positive level at the output. Therefore, eight steps with the maximum amplitude of 200 V is generated at the output. In addition, Fig. 10(b) reconfirm the obtained results from simulation result. In order to have all voltage levels (positive and negative) at the output, the H-bridge is added to the proposed inverter. Figure 11 indicates the simulation and experimental current and voltage waveforms of the load, respectively. As shown in Fig. 11(a), this inverter generates a step 17-level waveform the same as sinusoidal one with the maximum amplitude of 200 V and 1.96 A on the load, respectively. Moreover, comparing the voltage waveform with current waveform shows that the current waveform is closer to ideal sinusoidal one in addition to the existence of a phase shift between voltage and current waveforms. These features are due to resistive-inductive load feature, which behaves as a low-pass filter. The experimental results of current and voltage waveforms that is shown in Fig. 11(b) reconfirms the obtained results from simulation ones.

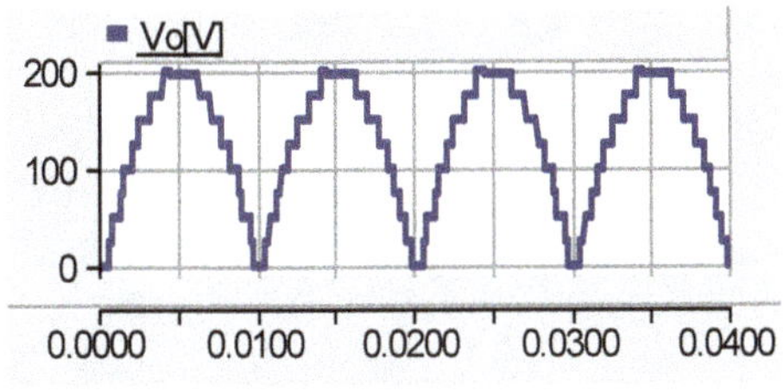

(a)

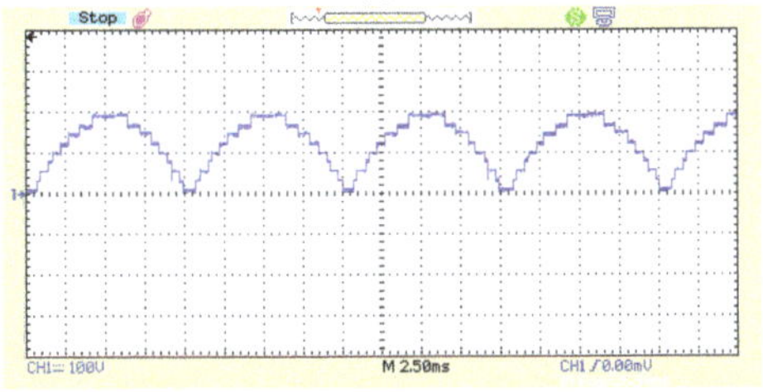

(b)

Fig. 10. The output voltage of the proposed inverter; (a) simulation result; (b) experimental result.

As mentioned before, this inverter consists of one bidirectional switch and two unidirectional ones from voltage point of view. In other word, although both of them conduct current in two dirrections, but the unidirectional switches block voltage in one direction while bidirectional one can block voltage in two directions. In order to investigate these facts, the voltage on switches of the first basic unit of the proposed cascaded inverter are shown in Fig. 12. Figure 12(a) and Fig. 12(e) show the voltage on switches $S_{1,1}$ and $S_{3,1}$ based on simulation results, respectively. As shown in these figures, the magnitudes of the blocked voltage on switches are either positive or zero, so there is not any negative amount on them. In addition, the amount of blocked voltage is equal to adding the magnitude of the used DC voltage sources in first basic unit. As a result, the existence of two unidirectional switches is reconfirmed in the proposed cascaded multilevel inverter. Figure 12(b) and Fig. 12(f) are obtained from experimental prototype. These figures also reconfirm the existence of two unidirectional power switches in the proposed topology. Fig. 12(c) indicates the blocked voltage on the switch $S_{2,1}$ is based on simulation results. As shown in this figure, there are positive and negative amount of voltages on it. This fact verifies that the switch $S_{2,1}$ is a bidirectional one. Fig. 12(d) is obtained from experimental prototype. This figure also reconfirms the existence of a unidirectional power switch in the proposed topology.

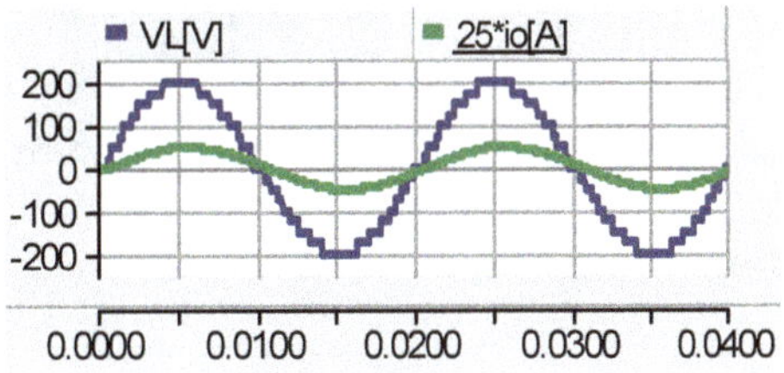

(a)

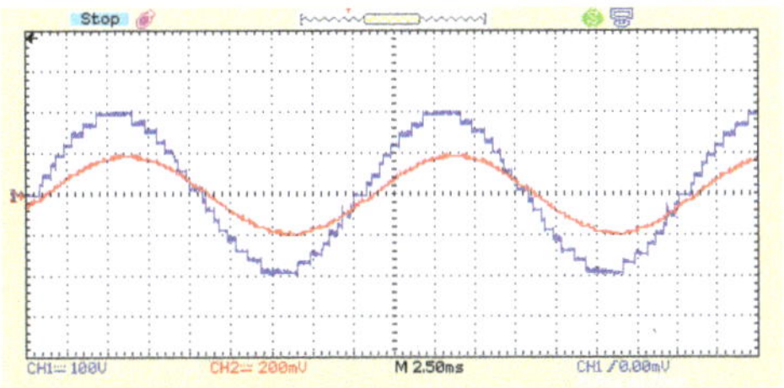

(b)

Fig. 11. Current and voltage waveforms of the load in the 17-level proposed inverter; (a) simulation result; (b) experimental result.

{R} In addition, the simulation results of the proposed topology in steady state operation by considering the load as high inductive one with the value of $L = 55\,mH$ are obtained. Figure 13 shows the simulation output voltage waveforms (v_o) of the proposed topology. As it is obvious from Fig. 13, this inverter is only able to generate positive level at the output. Therefore, eight steps with the maximum amplitude of 200 V is generated at the output. Fig. 14 indicates the simulation of current and voltage waveforms of the load. As shown in Fig. 14, this inverter generates a step 17-level waveform same as

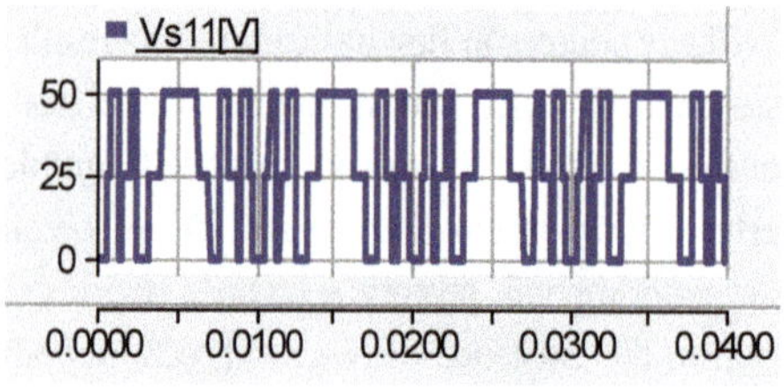

(a)

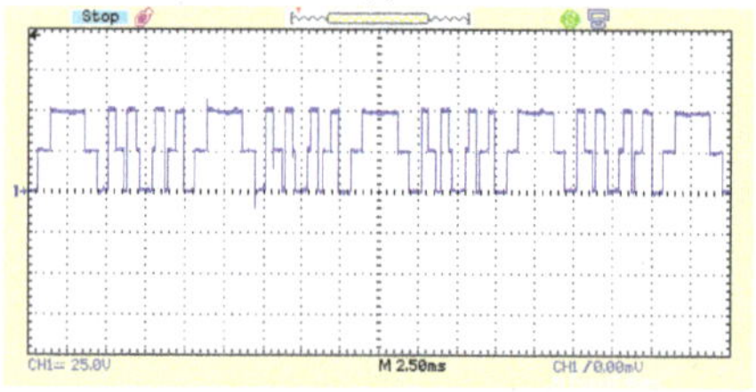

(b)

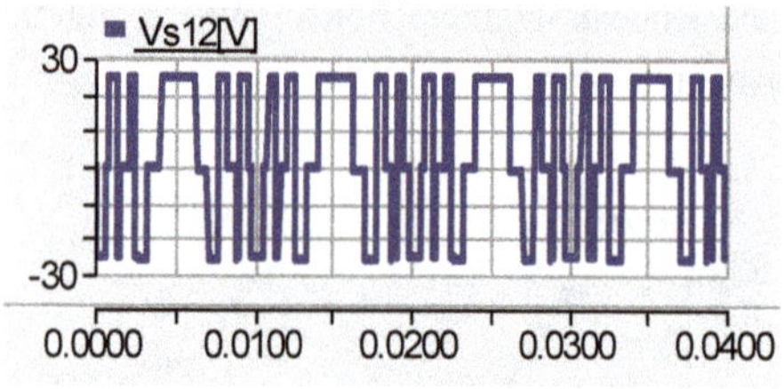

(c)

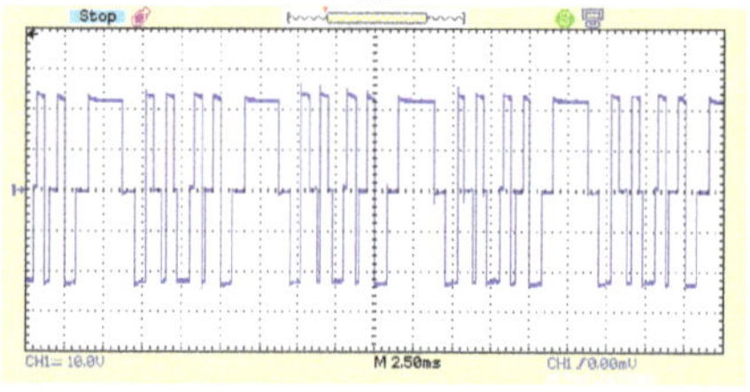

(d)

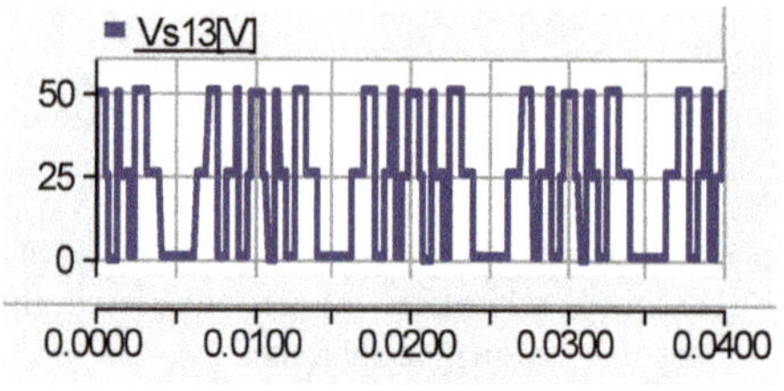

(e)

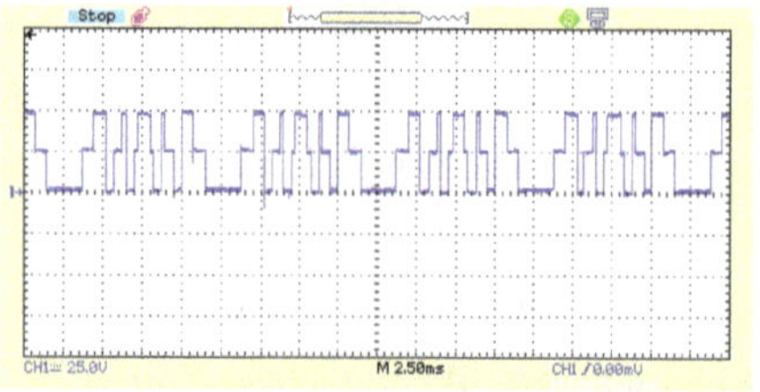

(f)

Fig. 12. The simulation results of the blocked voltage on the power switches in the first basic unit; (a) simulation result for $S_{1,1}$; (b) experimental result for $S_{1,1}$; (c) simulation result for $S_{2,1}$; (d) experimental result for $S_{2,1}$; (e) simulation result for $S_{3,1}$; (f) experimental result for $S_{3,1}$

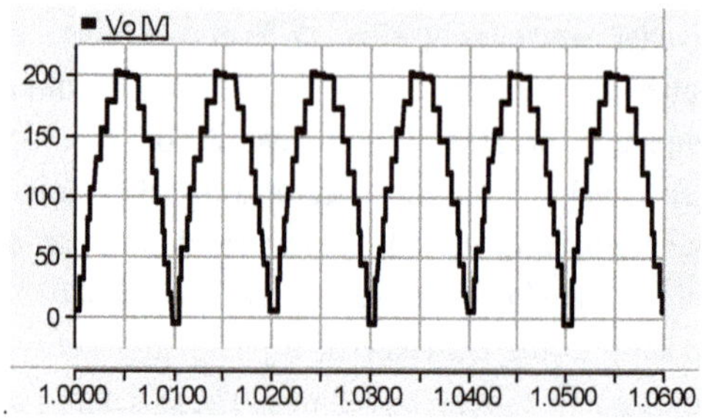

Fig. 13. The simulation output voltage of the proposed inverter.

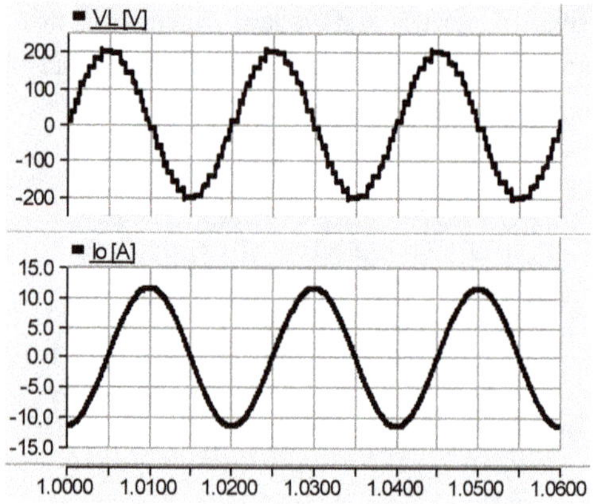

Fig. 14. Simulation result of current and voltage waveforms of the load in the 17-level proposed inverter.

sinusoidal one with the maximum amplitude of 200 V and 11.5 A on the load, respectively. Moreover, comparing the voltage and current waveforms shows that there is 90° a phase difference between voltage and current waveforms. As shown in these figures, the proposed inverter can operate correctly in high inductive loads.

Figure 15(a) and Fig. 15(c) show the voltage on switches $S_{1,1}$ and $S_{3,1}$ based on simulation results, respectively. As shown in these figures, the magnitudes of the blocked voltage on switches are equal to 0, 25 V and 50 V, so there isn't any negative amount on them. In addition, the amount of blocked voltage is equal to adding the magnitude of the used DC voltage sources in first basic unit. As a result, the existence of two unidirectional switches is reconfirmed in the proposed cascaded multilevel inverter. Figure 15(c) indicates the blocked voltage on the switch $S_{2,1}$ based on simulation results. As shown in this figure, there is the voltage from $-30V$ to $30V$ on it. Therefore, there are positive and negative amount of voltages on it. This fact verifies that the switch $S_{2,1}$ is a bidirectional one.

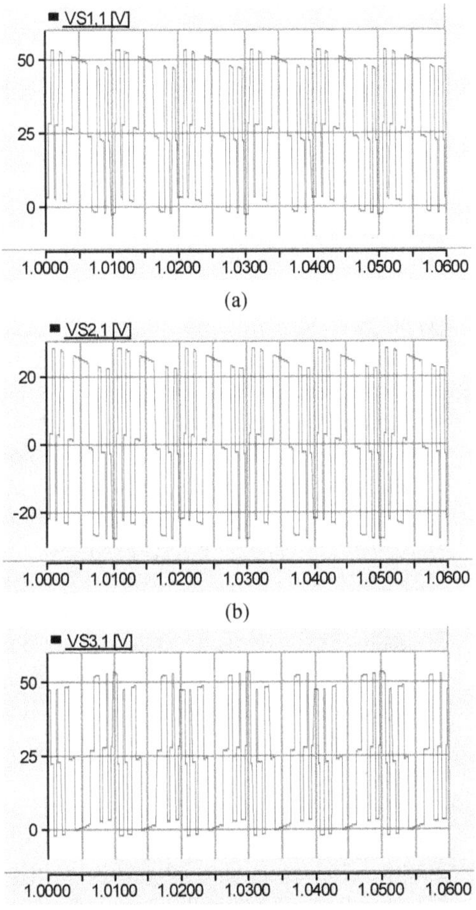

(c)

Fig. 15. The simulation results of the blocked voltage on the power switches in the first basic unit; (a) $S_{1,1}$; (b) $S_{2,1}$; (c) $S_{3,1}$.

5. CONCLUSION

In this paper, a cascaded multilevel inverter based on a new basic unit is proposed. The proposed unit is only able to generate positive levels at the output. Therefore, in order to generate all voltage levels (positive and negative) the H-bridge is added to the proposed topology. Then, fifth different algorithms to determine the magnitude of DC voltage sources are presented. Moreover, a comparison between the proposed topology with the H-bridge cascaded multilevel inverter and the presented inverter in [15] are done from the number of power electronic devices point of view. The most significant advantages of the proposed topology are that the number of switches, driver circuits and IGBTs and the variety of the value of the DC voltage sources are lower than the H-bridge cascaded inverter. These features cause to reduce the installation space and total cost of the inverter. For instance, to generate 161 levels at the output, the proposed cascaded multilevel inverter based on the fifth proposed algorithm needs an inverter with four numbers of basic topologies, $N_{switch} = 16$, $N_{IGBT} = 20$, $N_{driver} = 16$ and $N_{variety} = 4$. This selection is based on the best performance in comparisons with other proposed algorithms. However, in order to generate the same steps at the output in the H-bridge cascaded inverter, based on the presented algorithm in [13] it is required five numbers of the H-bridges, $N_{switch} = N_{IGBT} = N_{Driver} = 20$ and $N_{variety} = 5$ whereas based on the presented algorithm in [12] it is needed eight numbers of the H-bridges, $N_{switch} = N_{IGBT} = N_{Driver} = 32$ and $N_{variety} = 8$. These algorithms are shown by R_3 and R_2 in comparisons and have the best performance between all of the presented algorithms for H-bridge cascaded inverters. As it is obvious, the proposed inverter needs less number of power electronic devices. The main disadvantage of the proposed cascaded inverter is the high number of required DC voltage sources. For instance, to generate 161 levels at the output the proposed inverter needs $N_{source} = 8$ while the presented algorithm for the H-bridge inverter as R_3 requires $N_{source} = 5$. Moreover, it is suggested to use renewable energy sources as necessity dc voltage sources in the proposed topology to overcome this problem. Finally, the simulation and experimental results reconfirm the capability of the proposed cascaded inverter in generating all voltage levels through a 17-level inverter.

REFERENCES

[1] S. Laali, K. Abbaszades and H. Lesani, "New Hybrid control methods based on multi-carrier PWM techniques and charge balance control methods for cascaded multilevel converters," *Proceedings of the 24th Canadian Conference on Electrical and Computer Engineering,* Ontario, Canada, pp. 243-246, 2011.

[2] J. Napoles, A.J. Watson, and J.J. Padilla, "Selective harmonic mitigation technique for cascaded H-bridge converter with nonequal dc link voltages," *IEEE Transactions on Industrial Electronics,* vol. 60, no. 5, pp. 1963-1971, May 2013.

[3] K. Ding, K.W.E. Cheng, and Y.P. Zou, "Analysis of an asymmetric modulation methods for cascaded multilevel inverters," *IET Power Electronics*, vol. 5, no. 1, pp. 74-85, 2012.

[4] N. Farokhnia, S.H. Fathi, N. Yousefpoor, and M.K. Bakhshizadeh, "Minimisation of total harmonic distortion in a cascaded multilevel inverter by regulating of voltages dc sources," *IET Power Electronics*, vol. 5, no. 1, pp. 106-114, 2012.

[5] S. Mekhilef, M.N. Abdul Kadir, and Z. Salam, "Digital control of three phase three-stage hybrid multilevel inverter,"*IEEE Transactions on Industrial Informatics,* vol. 9, no. 2, pp. 719-727, 2013.

[6] K. Ramani and A. Krishan, "New hybrid multilevel inverter fed induction motor drive - A diagnostic study," *International Review of Electrical Engineering,* vol. 5, no. 6, part. A, pp. 2562-2569, 2010.

[7] S. Laali, K. Abbaszadeh, and H. Lesani, "Control of asymmetric cascaded multilevel inverters based on charge balance control methods," *International Review of Electrical Engineering*, vol. 6, no. 2, pp. 522-528, 2011.

[8] E. Babaeiandand S.H. Hosseini, "Charge balance control methods for asymmetrical cascaded multilevel converters," *Proceedings of the International Conference on Electrical Machines and Systems*, pp. 74-79, Korea, 2007.

[9] S. Laali, K. Abbaszadeh, and H. Lesani, "A new algorithm to determine the magnitudes of dc voltage sources in asymmetrical cascaded multilevel converters capable of using charge balance control methods," *Proceedings of the International Conference on Electrical Machines and Systems*, pp. 56-61, Incheon, Korea, 2010.

[10] W.K. Choi and F.S. Kang, "H-bridge based multilevel inverter using PWM switching function," *Proceedings of the31st Internat-ional Conference on Telecommunications Energy*, pp. 1-5, 2009.

[11] G. Waltrich, and I. Barbi, "Three-phase cascaded multilevel inverter using power cells with two inverter legs in series", *IEEE Transactions on Industrial Applications*, vol. 57, no. 8, pp. 2605-2612, 2010.

[12] M. Manjrekar, and T.A. Lipo, "A hybrid multilevel inverter topology for drive application," *Proceedings of the 30st International Conference onApplied Power Electronics Conference and Exposition,* pp. 523-529, 1998.

[13] A. Rufer, M. Veenstra, K. Gopakumar, "Asymmetric multilevel converter for high resolution voltage phasor generation," *Proceedings of theEuropean Conference on Power Electronics and Applications, Switzerland,* 1999.

[14] E. Babaei, M.F. Kangarlu, M. Sabahi, and M.R. Alizadeh Pahlavani, "Cascaded multilevel inverter using sub-multilevel cells," *Electric Power Systems Research*, vol. 96, pp. 101-110, March 2013.

[15] J. Ebrahimi, E. Babaei, and G.B. Gharehpetian, "A new multilevel converter topology with reduced number of power electronic components," *IEEE Transactions on Industrial Electronics*, vol. 59, no. 2, pp. 655-667, Feb. 2012.

Electric Differential for an Electric Vehicle with Four Independent Driven Motors and Four Wheels Steering Ability Using Improved Fictitious Master Synchronization Strategy

M. Moazen, and M. Sabahi*

Faculty of Electrical and Computer Engineering, University of Tabriz, Tabriz, Iran

ABSTRACT

Using an Electric Differential (ED) in electric vehicle has many advantages such as flexibility and direct torque control of the wheels during cornering and risky maneuvers. Despite its reported successes and advantages, the ED has several problems limits its applicability, for instance, an increment of control loops and an increase of computational effort. In this paper, an electric differential for an electric vehicle with four independent driven motors is proposed. The proposed ED is easy-to-implement and hasn't the problems of previous EDs. This ED has been developed for four wheels steering vehicles. The synchronization action is achieved by using an improved fictitious master technique, and the Ackerman principle is used to compute an adaptive desired wheel speed. The proposed ED is simulated and the operation of the system is studied. The simulation results show that ED ensures both reliability and good path tracking.

KEYWORDS: Ackerman principle, Electric differential, Electric vehicle, Fictitious master, Synchronization strategy.

1. INTRODUCTION

In recent decades, transportation investigations have emphasized the development of high efficiency, clean, and safe transportation, therefore Electric Vehicles (EVs) have been typically proposed to replace with conventional vehicles in the future. EVs use energy storage elements, such as batteries, to generate electric energy and transform it into mechanical energy by electrical motors to yield a required driving power.

Now the multi-motor and in-wheel-motor applications have a very attractive field in industrial applications, due to better stability and reducing the traditional mechanical coupling [1,2]. Therefore, they have less tailpipe emission and less fuel consumption.

The use of an Electric Differential (ED) instead of a Mechanical Differential (MD) constitutes a technological advance in vehicle design along the concept of multi-motor applications. The ED is characterized by the features as, no mechanical link between the drive wheels; the separately traction power to each wheel, the applied less power to the inner wheel during a turn and finally the ED act as a differential lock while the wheels of vehicle are driving straight paths. However, despite its long reported advantages, the application of the ED has been limited, mainly due to a number of problems in practical implementations, for instance, an increment of control loops and an increase of computational effort [3]. The fictitious master technique for synchronization of an electric differential for an electric vehicle with two independently driven motors has been presented in [3]. The control strategy has the advantage of being linear and, therefore, easy to implement. However, despite its many advantages, the synchronization strategy has a problem when the vehicle moves straight path.

The ED operation needs to solve two technological problems, wheel synchronization and

*Corresponding author:
M. Sabahi (E-mail: sabahi@tabrizu.ac.ir)

computation of the relative wheel speed as a function of the turn angle. Usually, a synchronization structure such as master–slave, cross coupling, sliding mode, fuzzy or neural network control has been applied to control the relative speed during pathway in the ED [4]. In this reference, to obtain adaptive performance of ED a neural network based control method has been presented. But low reliability of neural network controllers is disadvantageous of the proposed method.

The inner and outer wheel velocity relationship in a corner has usually been described by the use of the Ackerman steering principle. This principle computes the relative speed difference in the wheels by using the data of the turn (steering) angle [5].

Electric differentials, could be classified in two categories. First involves EDs for EVs with two Wheels Drive (2WD) ability [5,6]. In this case, two rear or front wheels of EV are equipped with independent motors and ED system distributes traction power between them, adjusts wheels speed [7-11]. In second one, ED is designed for EV with four Wheels Drive (4WD) ability. In this case, all EV wheels are equipped with electrical motors [12]. In 4WD condition, the traction power divide between four motors, so the required motor sizes are small. Therefore, the chassis level of EV is lower and a part of aerodynamic problem is resolved. Also, EV has better weight distribution, could be helpful for stability of EV. These EDs have been developed for EVs with two Wheels Steering (2WS) [13] or Four Wheels Steering (4WS) ability [14]. However, the complex controller is the main disadvantage of the mentioned references.

Three types of ED have been proposed for in wheel drive EVs in [13], to reduce controller complexity and expensive sensors. But in the proposed methods, because of uncertainty in the estimated speed, can lead the overall performance to become unstable in hard driving conditions. It seems that generally the ED designs with only two in-wheel drive motors, separated from the steering wheels, are studied in the papers.

In the present paper, an electric differential for an electric vehicle with four independent driven motors is proposed. This ED has been developed for four wheels steering vehicles. The method of four wheels steering can steer front wheels and rear wheels at once to least the side slip angle. So this method can increase the yaw response time of vehicle to the steering handle input and rotating performance by decreasing a radius of rotation.

The Ackerman principle is used to compute an adaptive desired wheel speed, and the synchronization action is achieved by using an improved fictitious master technique. The control strategy ensures both reliability and good path tracking for both curve and straight paths, and has the additional advantage of being easy to implement due to its linear nature. So it hasn't the problems of previous EDs.

The proposed ED is simulated and the operation of the system is studied. The simulation results show that ED ensures both reliability and good path tracking.

2. MODELLING AND CONTROL STRATEGY

In this section, the proposed electric differential system for an electric vehicle with 4WD and 4WS ability is described. The four wheels of EV are equipped with four IMs that each of them directly linked to each wheel by virtue of a fixed gear. IMs have the ability to direct control of torque and speed. The kinematic and dynamic models of the system are derived, and the synchronization strategy is presented for the 4WD/4WS vehicle.

2.1. Wheels speed computation

Ackerman principle is used to compute wheels speed. Rudolf Ackerman discovered and defined this principle early in the 19th century. The principle of Ackerman Steering is the relationship between the front inside tire and front outside tire in a corner or curve. The Ackerman steering principle defines the geometry that is applied to all vehicles, whatever they are 2WS or 4WS to enable the correct turning angle of the steering wheels to be generated when negotiating a corner or a curve [15].

To create the proper geometry, the steering arms are angled to turn the inside wheel at a sharper angle than the outside wheel. This allows the inside wheel to follow a smaller radius circle than the outside wheel and prevents scrubbing of the steer tires while turning.

The Ackerman concept is to have all four wheels rolling around a common point during a turn. This can greatly improve cornering ability and performance. Turning of 4WD/4WS vehicle, according to Ackerman principle is shown in Fig. 1.

The ideal turning angles on the front and rear wheels are established by the geometry seen in the Fig. 1, and define the steering angles for the turn.

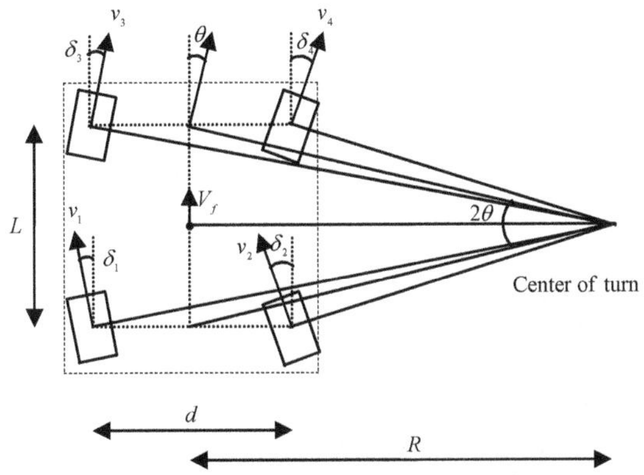

Fig.1. Turning of 4WD/4WS vehicle, according to Ackerman principle [16]

For proper geometry in the turn, the steer angles are given by below equations [16]:

$$\tan(\delta_1) = \frac{L/2}{R + d/2} \tag{1}$$

$$\tan(\delta_2) = \frac{L/2}{R - d/2} \tag{2}$$

$$\tan(\delta_3) = \frac{L/2}{R + d/2} \tag{3}$$

$$\tan(\delta_4) = \frac{L/2}{R - d/2} \tag{4}$$

where d and L are car's width and longitude, respectively, and R is the turn radius which is obtained with (5):

$$R = \frac{L}{2\tan(\theta)} \tag{5}$$

where θ is the steering angle of the vehicle.

The kinematic model of the car when tracking a curve path is given by the Ackerman equations. Such equations describe the relationship between the car angular velocity, ω_f, and the inner and outer wheels velocities using basic trigonometry. So, the linear speed of each wheel for 4WD/4WS vehicle can be expressed in terms of the angular vehicle speed ω_f and the turn radius R (see Fig. 1), *i.e.*:

$$v_1 = \omega_f \sqrt{\left(R + \frac{d}{2}\right)^2 + \frac{L^2}{4}} \tag{6}$$

$$v_2 = \omega_f \sqrt{\left(R - \frac{d}{2}\right)^2 + \frac{L^2}{4}} \tag{7}$$

$$v_3 = v_1 = \omega_f \sqrt{\left(R + \frac{d}{2}\right)^2 + \frac{L^2}{4}} \tag{8}$$

$$v_4 = v_2 = \omega_f \sqrt{\left(R - \frac{d}{2}\right)^2 + \frac{L^2}{4}} \tag{9}$$

Substituting (5) in (6) - (9), the angular speed of each wheel can be determined as follows:

$$\omega_1 = \frac{v_1}{r} = \frac{\omega_f}{r} \sqrt{\left(\frac{L}{2\tan(\theta)} + \frac{d}{2}\right)^2 + \frac{L^2}{4}} \tag{10}$$

$$\omega_2 = \frac{v_2}{r} = \frac{\omega_f}{r} \sqrt{\left(\frac{L}{2\tan(\theta)} - \frac{d}{2}\right)^2 + \frac{L^2}{4}} \tag{11}$$

$$\omega_3 = \frac{v_3}{r} = \frac{\omega_f}{r} \sqrt{\left(\frac{L}{2\tan(\theta)} + \frac{d}{2}\right)^2 + \frac{L^2}{4}} \tag{12}$$

$$\omega_4 = \frac{v_4}{r} = \frac{\omega_f}{r} \sqrt{\left(\frac{L}{2\tan(\theta)} - \frac{d}{2}\right)^2 + \frac{L^2}{4}} \tag{13}$$

where r is the each wheel radius.

The angular velocity of each wheel can be acquired in term of vehicle linear velocity, V_f, with definition of it as follow:

$$V_f = R\omega_f \tag{14}$$

So:

$$\omega_f = \frac{V_f}{R} = V_f \frac{2\tan(\theta)}{L} \tag{15}$$

Substituting (15) in (10) - (13), the angular speed of each wheel in term of the vehicle linear speed can be determined as follows:

$$\omega_1 = \frac{V_f}{r} \sqrt{1 + \frac{2d}{L}\tan(\theta) + \left(1 + \frac{d^2}{L^2}\right)\tan^2(\theta)} \tag{16}$$

$$\omega_2 = \frac{V_f}{r} \sqrt{1 - \frac{2d}{L}\tan(\theta) + \left(1 + \frac{d^2}{L^2}\right)\tan^2(\theta)} \tag{17}$$

$$\omega_3 = \omega_1 = \frac{V_f}{r} \sqrt{1 + \frac{2d}{L}\tan(\theta) + \left(1 + \frac{d^2}{L^2}\right)\tan^2(\theta)} \tag{18}$$

$$\omega_4 = \omega_2 = \frac{V_f}{r} \sqrt{1 - \frac{2d}{L}\tan(\theta) + \left(1 + \frac{d^2}{L^2}\right)\tan^2(\theta)} \tag{19}$$

2.2. Dynamic model and synchronization strategy

By considering a four-wheeled vehicle subjected to the action of four electric motors situated on the all wheels, the equation of motion of the car is given as follows:

$$M\frac{dV_f}{dt} = -Mg f_r \cos(\alpha) - Mg\sin(\alpha) - \frac{1}{2}\rho A C_D V_f^2 + \frac{1}{r}\left(\sum_{i=1}^{4} u_i\right) \tag{20}$$

where M is the vehicle mass, g is the gravity constant, f_r is the rolling friction coefficient, α is the terrain inclination, ρ is the air density, A is the effective area of the aerodynamic resistance of the vehicle, C_D is the aerodynamic drag coefficient, and u_i ($i = 1, 2, 3, 4$) is the produced torque by each motor. Moreover, consider four induction motors with perfect field orientation (e.g., motors under indirect field-oriented control (IFOC) without field weakening), then the actuated wheel dynamics can be shown to be equivalent to:

$$\frac{\omega_i(s)}{u_i(s)} \approx \frac{1}{J_i s + b_i} \tag{21}$$

where s is the differentiator operator, and J_i and b_i are the moment of inertia and friction coefficient of each motor, respectively.

However, since the motors are attached to the vehicle, dynamics of vehicle and motors are coupled, and every perturbation on the vehicle will be reflected to the motors. In this way, we can rewrite the wheel dynamics in the embedded system as:

$$J_i\frac{d\omega_i}{dt} = -b_i\omega_i - T_{Li} + u_i - d_f r\left(Mg f_r\cos(\alpha) + Mg\sin(\alpha) + \frac{1}{2}\rho A C_D V_f^2\right) \tag{22}$$

where T_{Li} shows external perturbations at wheel *i*. Notice that since the vehicle and wheels velocity are related by the Ackerman equation, the whole vehicle–motor system can be described using only four differential equations, which are related to the vehicle plus three wheels dynamics or, alternately, all wheels dynamics, as follows:

The next step in the description of the proposed strategy for the electric differential system is to use the traction model of (22) – (26) to construct the synchronization scheme. In this paper, the used synchronization strategy is Improved Fictitious Master technique.

$$J_1\frac{d\omega_1}{dt} = -b_1\omega_1 - T_{L1} + u_1 - d_f r\left(Mg f_r\cos(\alpha) + Mg\sin(\alpha) + \frac{1}{2}\rho A C_D V_f^2\right) \tag{23}$$

$$J_2\frac{d\omega_2}{dt} = -b_2\omega_2 - T_{L2} + u_2 - d_f r\left(Mg f_r\cos(\alpha) + Mg\sin(\alpha) + \frac{1}{2}\rho A C_D V_f^2\right) \tag{24}$$

$$J_3\frac{d\omega_3}{dt} = -b_3\omega_3 - T_{L3} + u_3 - d_f r\left(Mg f_r\cos(\alpha) + Mg\sin(\alpha) + \frac{1}{2}\rho A C_D V_f^2\right) \tag{25}$$

$$J_4\frac{d\omega_4}{dt} = -b_4\omega_4 - T_{L4} + u_4 - d_f r\left(Mg f_r\cos(\alpha) + Mg\sin(\alpha) + \frac{1}{2}\rho A C_D V_f^2\right) \tag{26}$$

Reference [3] presented fictitious master technique to synchronization of an electric differential for an electric vehicle with two independently driven motors. The control strategy has the advantage of being linear and, therefore, easy to implement. However, despite its many advantages, the synchronization strategy has a problem when the vehicle moves straight path.

In this section, the fictitious master technique for the synchronization of electric differential system for a 4WD/4WS electric vehicle is presented and a basic block diagram of the synchronization controller is shown in Fig. 2. The blocks which are named "motor system 1" to "motor system 4" are related to the traction systems in (22)-(26). With this technique, the reference speed of each wheel is obtained according to Ackerman principle and equations (10)-(13) are used for computation of wheels reference speed.

In Fig. 2, it can be observed that the synchronization scheme is composed of three parts: 1) the fictitious master, 2) a speed controller to each motor (slave system), and 3) a link between them constituted by an average gain K_4 and adaptive references ω_1^*, ω_2^*, ω_3^* and ω_4^*.

Adaptive references $\omega_1^* - \omega_4^*$ are computed based on (10)-(13) using the actual angular velocity of the vehicle (ω_f) rather than the desired reference (ω_f^*). This feature allows the general transient and steady-state response of the overall system to be mainly determined by the fictitious master. In this way, perturbations at the slave stage can be reflected and compensated not only by the slave controller, but also by the virtual master controller, producing new references ω_1^* – ω_4^*; therefore moderated system responses are obtained even under impulsive perturbations. At this point, it is worthy to notice that perturbation functions T_{Li}, represent environmental or non-controller inputs. In addition, T_{Lf} is a function of system states and has the role of reflecting any control change (due to perturbations or saturations) of the slave controllers to the master controller. Summarizing, the fictitious master controls the torque reflected in each motor without using any torque transducer.

The torque load changes are detected and reflected back to each speed controller inside the IM controllers by using each motor signal u_i. Such signals are added and weighted using gain K_4 and then considered like an external torque command to the Fictitious Master. Once the master detects any load disturbance in any motor (T_{Li}), it produces a new speed reference to each motor controller using the Ackerman principle.

In this way, new references ω_1^* - ω_4^* are computed using (10)-(13). The objective of the proposed scheme is to reflect any disturbance in any wheel into all wheel dynamics via references ω_1^* - ω_4^*; therefore, synchronization is maintained without compromising vehicle stability. In the scheme (see Fig. 2), velocity loops in the wheels constitute slave controllers of a fictitious master controller. By using this procedure, speed synchronization during the speed transient and steady state is maintained, even during high-load impacts. In fact, the Fictitious Master transforms the problem of tracking the turn angle θ into a velocity-tracking problem using the dynamic parameterization from a fictitious system. Observe that this dynamic parameterization is not unique and will vary with the master structure. Different parameters on fictitious dynamics will lead to different vehicle desired velocities. Furthermore, the fictitious master establishes the control objective $\omega_f = \omega_f^*$ as a primary control objective, and wheels can be unsynchronized ($\omega_1^* \neq \omega_2^* \neq \omega_3^* \neq \omega_4^*$) to track ω_f^*. In fact, the only case when $\omega_1^* = \omega_2^* = \omega_3^* = \omega_4^*$ is along an unperturbed straight path. As previously stated, the fictitious master parameterizes the vehicle's velocity based on the designer's choice. Also, the general transient and steady-state response of the overall system is mainly determined by the fictitious master; therefore, it is convenient to properly choose system parameters to obtain good performance.

A slow master dynamics constitutes a conservative choice of dynamics, to depart from a stable configuration, opening the possibility for the designer to use a wide window of tuning gains to obtain a given performance. For example, if the virtual moment of inertia J_f is chosen sufficiently large, big peak current demands can be avoided, thus keeping the inverter current demand in a safety range. In this way, it is suggested to choose the following Fictitious Master parameters [3]:

$$J_f \geq \sum_{i=1}^{4} J_i \tag{27}$$

$$b_f \geq \sum_{i=1}^{4} b_i \tag{28}$$

As previously mentioned, the block diagram of Fictitious Master technique for electric differential of an electric vehicle with 4WD ability is shown in Fig. 2. It can be observed that the angular velocity of the vehicle is given to the system as a reference value. When a vehicle moves straight path with different speeds, its angular speed is always zero. So the election of angular velocity of the vehicle as a reference value causes some problems for the system.

If the linear speed of the vehicle is considered as a reference value, the block diagram of Fig. 2 will be changed into Fig. 3. In this status, the reference values of angular speeds in each wheel (ω_1^* - ω_4^*) are obtained by (16)-(19). By comparison of two block diagrams, it can be observed that the block diagram of Fig. 3 has the same structure of Fig. 2. Because of choosing V_f^* as a reference value for the system and using it to calculation of $\omega_1^* - \omega_4^*$ in this improved Fictitious Master technique, the problems of implementation of it when the vehicle moves a straight path has been removed.

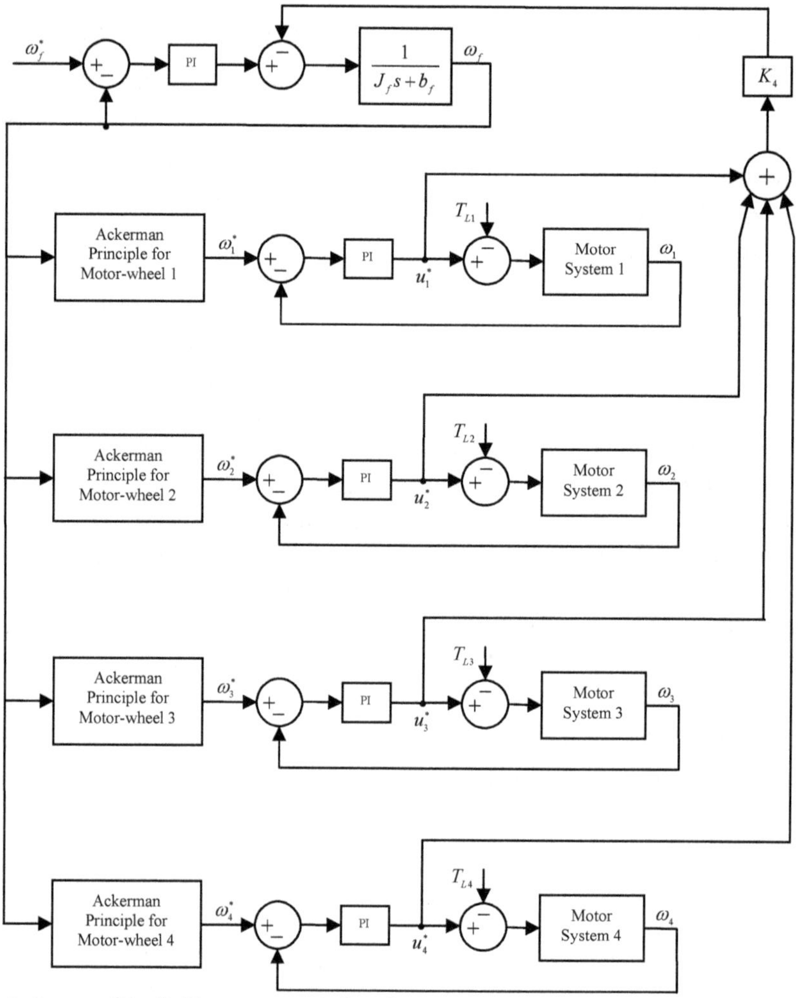

Fig. 2. Block diagram of the fictitious master technique for the synchronization of electric differential system for a 4WD/4WS electric vehicle

3. SIMULATION RESULTS

In this section, performance and robustness of the proposed technique are evaluated during transient and steady state conditions. As previously pointed out, we considered a four driving- wheel EV that places one IM directly linked to each wheel by virtue of a fixed gear. The simulation makes use of rigorous models of vehicle parts to accurately reflect the nonlinear behavior of the overall system. In this way, a complete mechanical model of the vehicle is used to accurately reflect the effect of car load, friction, and aerodynamic forces (parameters can be found in the Appendix).

IFOC is used to move each motor. The IMs work at their constant torque region and provide the full power required by the vehicle. PI compensators are used in DQ controllers. IFOC gains and other parameters can be found in Appendix I. At this point, it is interesting to note that, in contrast to the complexity of the model used; the control structure is simple, constituted by linear controllers, making the implementation task easier.

The transient and steady-state response of the overall system is controlled by the improved

Fictitious Master, which, in this case, is tuned using the biggest moment of inertia of the overall system with the aim to avoid a big peak current demand and to keep the inverter current demand in a safety range. The K_4 gain, which can be seen as an average weight gain, is chosen to be $K_4=1/4$.

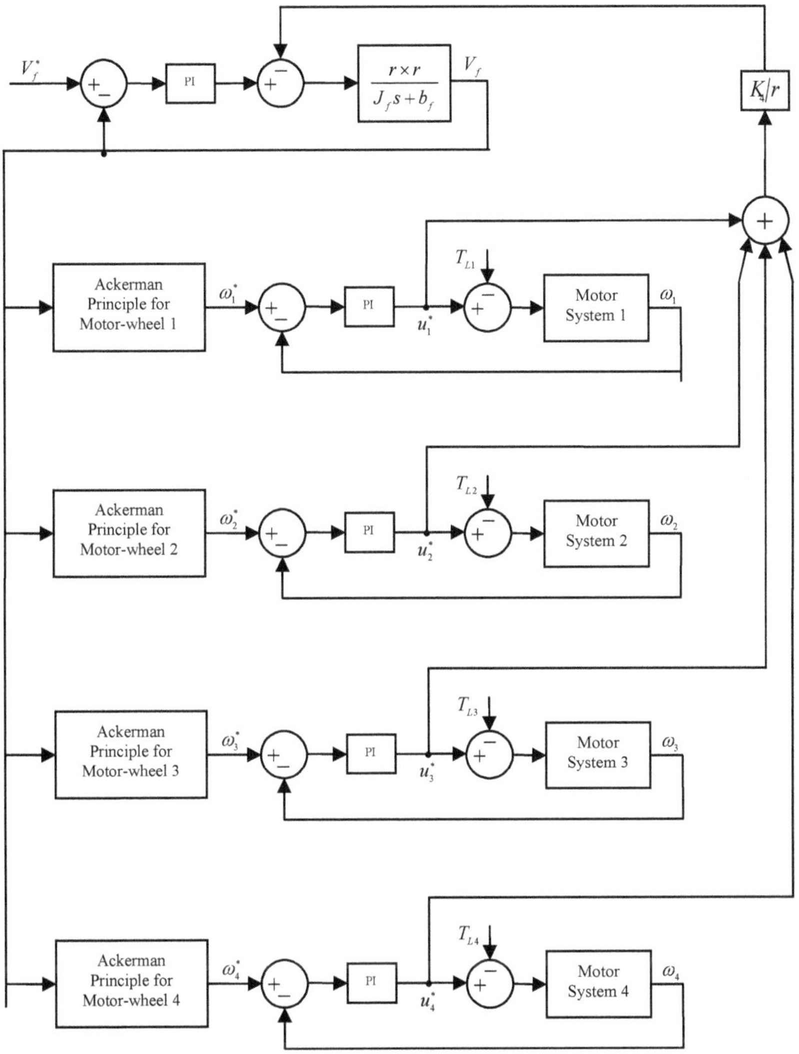

Fig. 3. Block diagram of the improved fictitious master technique for the synchronization of electric differential system for a 4WD/4WS electric vehicle

As a first step, the response of the vehicle during load impacts is evaluated. The system is perturbed while following a straight path with constant speed $V_f^* = 45$ Km/h, with a step-like load change. Fig. 4 shows the applied disturbance torque of motor 1. The simulation results are shown in Figs. 5 and 6. Fig. 6 shows that the produced torque of motor 1 has been increased to provide the applied load torque. Also, it can be seen in Fig. 5 that the speed of this wheel decreases under mentioned condition. If the problem stays in this stage, the synchronization of system has been removed due to the uncoordinated operation of the vehicle wheels. But the electric differential system reflects the disturbance in wheel 1 into all wheels dynamics via producing new references ω_1^* - ω_4^*; then, synchronization is maintained without compromising vehicle stability. So the ED simulates a differential lock while the wheels of vehicle are driving straight paths. Fig. 5

depict that a good degree of speed synchronization during the speed transient and steady state is maintained even during load changes.

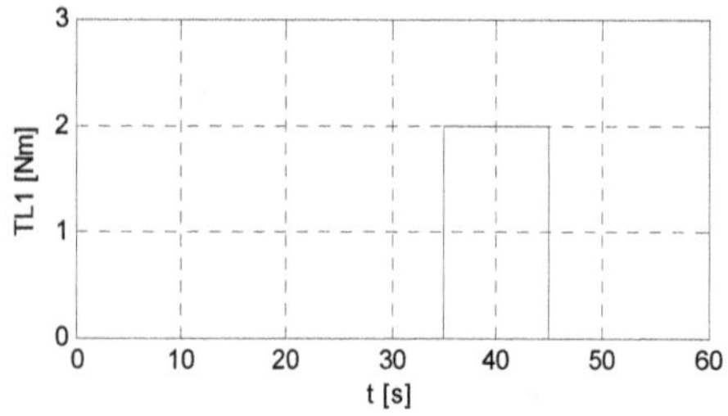

Fig. 4. The applied disturbance torque to motor 1

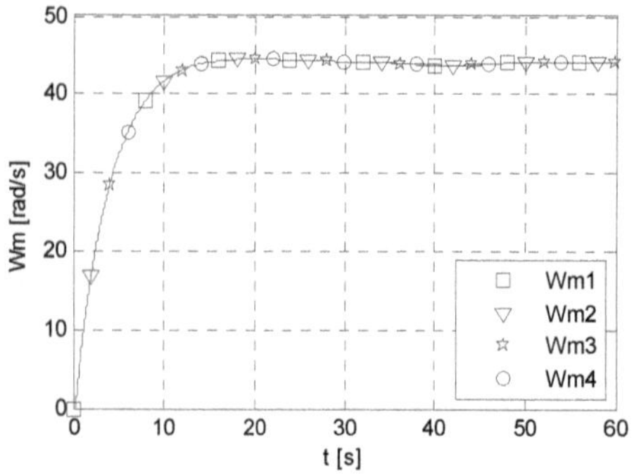

Fig. 5. The angular speeds of vehicle wheels under perturbed condition

On the other hand, to illustrate the performance of the proposed controller while follows curved paths with constant speed, the vehicle is required to follow the curve described by the trapezoidal steering angle in Fig. 7. With this steering angle, the steer angles of the wheels of the vehicle should be such as shown in Fig. 8. In contrast to the case of a straight path, during turns, the wheel velocity is unsynchronized (with respect to each other) for the vehicle to track the virtual reference V_f. That is, during a turn, the outer wheels must be faster than the inner wheels. The simulation results are shown in Figs. 9 and 10. It can be observed that the controller could simulate an electric differential system when the vehicle is turning. The speeds of outer wheels are more than the inner wheels during the turn and the motors produce the proper torque for good path tracking of vehicle.

Fig. 11 shows the curve path that vehicle moves. First, the steering angle of the vehicle is zero and the vehicle moves straight path. With increasing of θ, the vehicle turns and tracks a circular path when θ has its maximum value. With the decreasing of θ, the radius of turning increases and it moves in a straight path when θ becomes zero. The method of 4WS which used in present paper can increase rotating performance by decreasing the radius of rotation. The diameter of turn in recent simulation for the given steering angle is approximately 25m that is obtained from (5) and it can be shown in Fig. 11. If we use the method of 2WS, the diameter of turn in the same steering angle increases to approximately 50 m. It means that by using 4WS method, the vehicle can turn in an easier way.

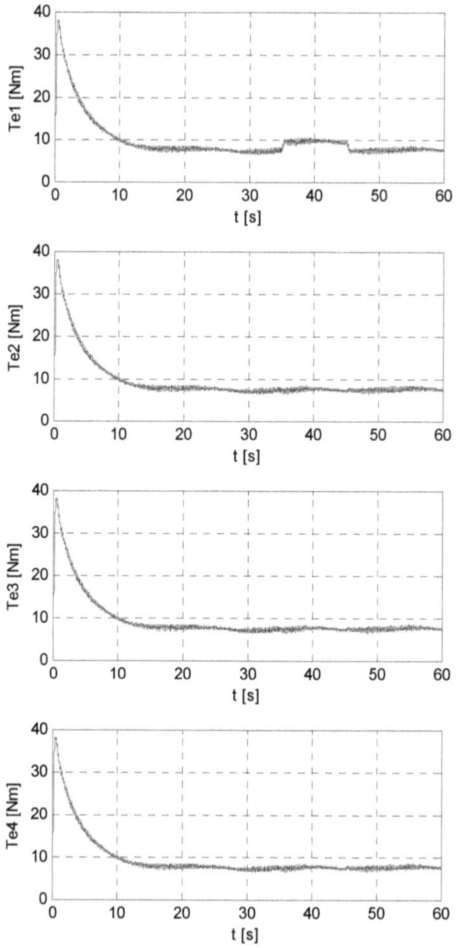

Fig. 6. The torque response of IMs under perturbed condition

It is obvious that by choosing V_f^* as a reference value for the calculation of ω_1^* - ω_4^* in the presented synchronization strategy (improved fictitious master technique), the problems of implementation of ED when the vehicle moves a straight path has been removed. So the system has a good operation even when θ is zero.

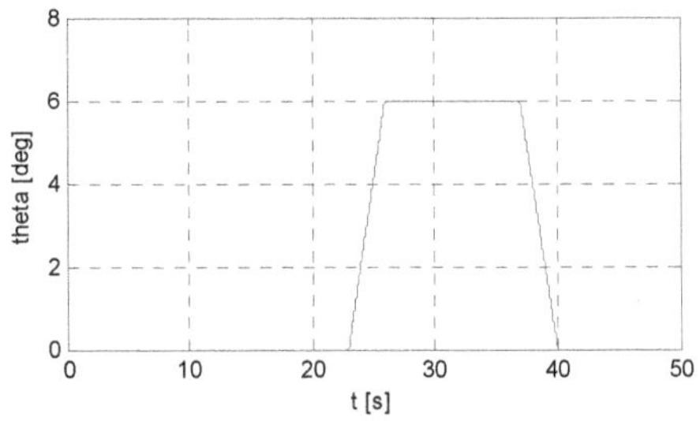

Fig. 7. The steering angle of vehicle

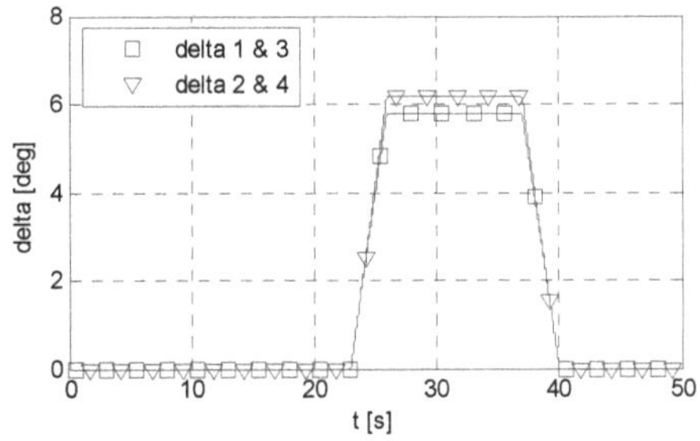

Fig. 8. The steer angles of vehicle wheels

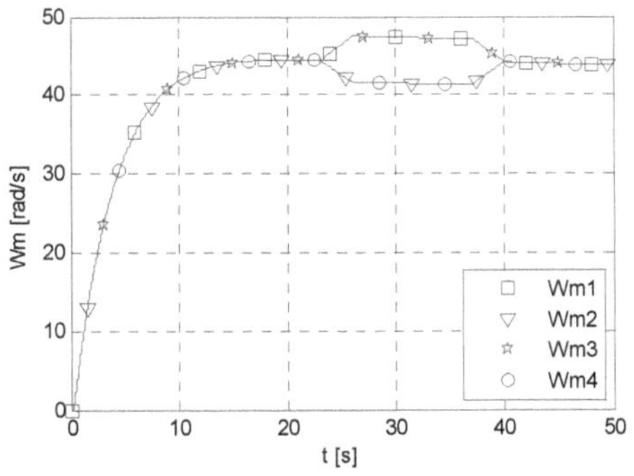

Fig. 9. The angular speeds of vehicle wheels when turning

4. CONCLUSIONS

In this paper, an electric differential for an electric vehicle with four independent driven motors has been proposed. The proposed ED is easy to implement and hasn't the problems of previous EDs. This ED has been developed for four wheels steering vehicles. The method of 4WS can steer both front wheels and rears wheel to minimize the side slip angle. Thus, this method can increase rotating performance by decreasing a radius of rotation.

The synchronization action is achieved by using an improved fictitious master technique, and the Ackerman principle is used to compute an adaptive desired wheel speed. The proposed ED is simulated and the operation of system is studied under perturbed condition, and straight and curved path tracking. The simulation results show that ED ensures both reliability and good path tracking. The study of faulty conditions like motor failure can be performed in the future.

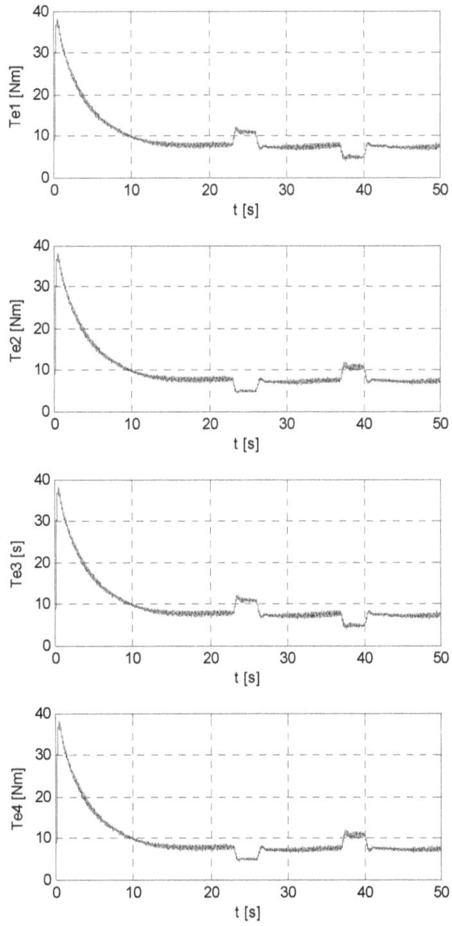

Fig. 10. The torque response of IMs when the vehicle is turning

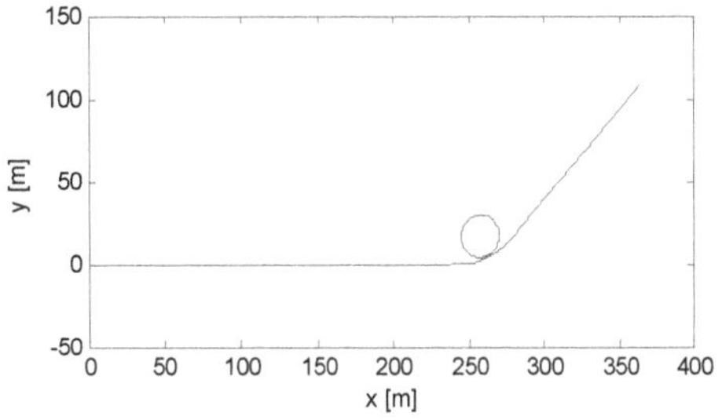

Fig. 11. The curve path that vehicle moves

APPENDIX

For all motors, the following parameters are used: $V_{rated} = 220\text{V}$, $2p = 4$, $L_s = 0.03\text{H}$, $L_r = 0.03\text{H}$, $L_m = 0.44\text{H}$, $R_s = 4\Omega$, $R_r = 2\Omega$, $J = 1.5\text{kg/m}^2$, $b = 0.01\text{Nms/rad}$. The Fictitious Master parameters are: $J_f = 6\text{kg/m}^2$, $b_f = 0.04\text{Nms/rad}$. IFOC controller gains for all motors are: $K_{pd\,IFOC} = 21.779$, $K_{id\,IFOC} = 823.65$, $K_{pq\,IFOC} = 21.779$, $K_{iq\,IFOC} = 823.65$. Other parameters are: $L = 2.7\text{m}$, $d = 1.7\text{m}$, $r = 0.2794\text{m}$, $A = 2\text{m}^2$, $M = 500\text{kg}$, $g = 9.8\text{m/s}^2$, $C_D = 0.25$, $\rho = 1.202\text{kg/m}^3$, $f_r = 0.01$, $d_f = 0.25$.

REFERENCES

[1] K. Zhang, L. Xiong, Z. Yu, Y. Feng, L. Deng," The influence of driving form on stability control in in-wheel-motor EV," *Proceedings of the IEEE International Conference on Vehicular Electronics and Safety*, pp. 57-60, 2013.

[2] J. Gutierrez, J. Romo, M.I. Gonzalez, E. Canibano, J.C. Merino," Control algorithm development for independent wheel torque distribution with 4 In-wheel electric motors," *Proceedings of the European Symposium on Computer Modeling and Simulation*, pp. 257-262, 2011.

[3] F. J. Perez-Pinal, I. Cervantes and A. Emadi," Stability of an electric differential for traction applications," *IEEE Transactions on Vehicular Technology*, vol. 58, no. 7, pp. 3224-3233, 2009.

[4] A. Haddoun, M.E. Benbuzid, D. Diallo, R. Abdessemed, J. Ghouili, and K. Srairi," Modeling, analysis, an neural network control of an EV electrical differential," *IEEE Transactions on Industrial Electronics*, vol. 55, no. 6, pp. 2286 - 2294, June 2008.

[5] B. Tabbache, A. Kheloui, and M.E. Bendouzid, "An adaptive electric differential for electric vehicles motion stabilization," *IEEE Transactions on Vehicular Technology*, vol. 60, no. 1, pp. 104 - 110, 2011.

[6] X. Yuan and J.W. Torque," Distribution strategy for a ront- and rear-wheel-driven electric vehicle," *IEEE Transactions on Vehicular Technology*, vol. 61, no. 8, pp. 3365-3374, 2012.

[7] I. Lachhab and L. Krichen," Torque control development for electric-vehicles motor-wheels using two techniques," *Proceedings of the 14th International Conference on Sciences and Techniques of Automatic Control & Computer Engineering*, pp. 43-46, 2013.

[8] M. Schael, P. Spichartz, M. Rehlander and C. Sourkounis," Electric differential for electric vehicles using doubly-fed induction motors," *Proceedings of the 39th Annual IEEE Conference on Industrial Electronics*, pp. 4618-4623, 2013.

[9] H. Song and X. L. Huang," Study on electric differential control scheme for electric vehicles," *Advanced Materials Research*, vol. 648, pp. 348-352, 2013.

[10] A. Draou," Electronic differential speed control for two in-wheels motors drive vehicle," *Proceedings of the 4th International Conf-erence on Power Engineering, Energy and Electrical Drives*, pp. 764-769, 2013.

[11] S. Hou, Z. Li, T. Wang, L. Pang and Z. Feng," Study on electronic differential control for a mini electric vehicle with dual in-wheel-motor rear drive," *Applied Mechanics and Materials*, vol. 525, pp. 346-350, 2014.

[12] B. Liang, S. Cui and Y. Yu, "Electric differential system based on dual-closed loop slip rate control," *Proceeding of the International Conference on Measurement, Information and Control*, pp. 1056-1061, 2012.

[13] Y. Chen and J. Wang" Design and evaluation on electric differentials for overactuated electric ground vehicles with four independent in-wheel motors," *IEEE Transactions on Vehicular Technology*, vol. 61, no. 4, pp. 1534-1542, 2012.

[14] Y, Zhou, S. Li, X. Zhou and Z. Fang," The control strategy of electronic differential for EV with four in-wheel motors, " *Proceeding of the 2010 Chinese Control and Decision Conference*, pp. 4190-4195, 2010.

[15] M.W. Choi, J.S. Park, B.S. Lee and M.H. Lee," The performance of independent wheels steering vehicle (4WS) applied Ackerman geometry," *Proceedings of the International Conference on Control, Automation and Systems*, pp. 197-202, 2008.

[16] M.W. Choi, J.S. Park, B.S. Lee, M.H. Lee," The performance of independent wheels steering vehicle (4WS) applied Ackerman geometry," *International Conference on Control, Automation and Systems*, pp. 197- 202, 2008.

4

Combined Use of Sensitivity Analysis and Hybrid Wavelet-PSO-ANFIS to Improve Dynamic Performance of DFIG-Based Wind Generation

M. Darabian*, A. Jalilvand, R. Noroozian

Department of Electrical Engineering, University of Zanjan, Zanjan, Iran

ABSTRACT

In the past few decades, increasing growth of wind power plants causes different problems for the power quality in the grid. Normal and transient impacts of these units on the power grid clearly indicate the need to improve the quality of the electricity generated by them in the design of such systems. Improving the efficiency of the large-scale wind system is dependent on the control parameters. The main contribution of this study is to propose a sensitivity analysis approach integrated with a novel hybrid approach combining wavelet transform, particle swarm optimization and an Adaptive-Network-based Fuzzy Inference System (ANFIS) known as Wavelet-ANFIS-PSO to acquire the optimal control of Doubly-Fed Induction Generators (DFIG) based wind generation. In order to mitigate the optimization complexity, sensitivity analysis is offered to identify the Unified Dominate Control Parameters (UDCP) rather than optimization of all parameters. The robustness of the proposed approach in finding optimal parameters, and consequently achieve a high dynamic performance is confirmed on two area power system under different operating conditions.

KEYWORDS: Doubly fed induction generator, Fuzzy logic, Neural networks, Particle Swarm Optimization, Wavelet transform, Sensitivity analysis.

NOMENCLATURE

ψ_{dr}	The d axis rotor flux linkages
ψ_{qr}	The q axis rotor flux linkages
L_{ss}	The stator self-inductance
L_{rr}	The rotor self-inductance
L_m	The mutual inductance
R_r	The rotor resistance
ω_s	The synchronous angle speed
S_r	The rotor slip
X_s	The stator reactance
X'_s	The stator transient reactance
E'_d	The d axis voltage behind the transient reactance
E'_q	The q axis voltage behind the transient reactance
T'_o	The rotor circuit time constant
i_{ds}	The d axis stator current
i_{qs}	The q axis stator current
v_{ds}	The d axis stator terminal voltages
v_{qs}	The q axis stator terminal voltages
v_{dr}	The d axis rotor terminal voltages
v_{qr}	The q axis rotor terminal voltages
ω_r	The generator rotor angle speed
ω_t	The wind turbine angle speed
H_g	The inertia constants of the generator
H_t	The inertia constants of the turbine
θ_{tw}	The shaft twist angle
K_{sh}	The shaft stiffness coefficient
D_{sh}	The damping coefficient
T_{sh}	The shaft torque
T_m	The wind torque
T_{em}	The electromagnetic torque
ρ	The air density
R	The wind turbine blade radius
V_ω	The wind speed
C_f	The blade design constant coefficient
β	The blade pitch angle
λ	The blade tip speed ratio
C_P	The power coefficient
P_s	The stator active power
P_r	The active power at the ac terminal of

Corresponding author :
M. Darabian (E-mail: m.darabian@znu.ac.ir)

	the rotor side converter (RSC)
P_g	The active power at the ac terminal of the grid side converter (GSC)
P_{DC}	The active power of the dc link capacitor
i_{dr}	The d axis rotor currents
i_{qr}	The q axis rotor currents
i_{dg}	The axis currents of the GSC
i_{qg}	The axis currents of the GSC
V_{dg}	The d axis voltage of the GSC
v_{qg}	The q axis voltage of the GSC
V_{DC}	The dc link capacitor voltage
i_{DC}	The current of the dc link capacitor
C	The capacitance of the dc link capacitor
K_{p1}& K_{i1}	The PI gains of the power regulator
K_{p2}& K_{i2}	The PI gains of RSC Current regulator
K_{p3}& K_{i3}	the PI gains of the grid voltage regulator
i_{dr}^{ref}	The current control references for the d axis components of the GSC
i_{qr}^{ref}	The current control references for the q axis components of the GSC
Vs_{ref}	The specified terminal voltage reference
P_{ref}	The active power control reference
K_{pdg}& K_{idg}	the PI gains of the link dc capacitor voltage regulator
K_{pg}& K_{ig}	The PI gains of the grid side converter current regulator
v_{DCref}	The voltage control reference of the dc link capacitor
i_{qgref}	The control reference for the q axis component of the GSC current
K_{p4}& K_{i4}	The PI gains of the WT speed regulator
$\Delta\omega_t$	The deviation of the WT rotating speed

1. INTRODUCTION

Doubly Fed Induction Generator (DFIG) is a popular Wind Turbine (WT) system due to its various abilities such as high-energy efficiency; reduction in mechanical stress on wind turbine; reactive power control; reduction in convertor cost and injection of reactive power in the case of voltage oscillation to regulate terminal voltage [1]. A DFIG system, including induction generator, two mass drive trains, power converters, and feedback controllers, is a multivariable, nonlinear and strongly coupled system. Bifurcation phenomena in such a nonlinear system may occur under certain conditions, leading to oscillatory instability. Hence, practical analysis of DFIG stability will have to involve the bifurcation phenomena. In recent years, a number of researchers studied stability of industrial motor drives with a wealth of nonlinear dynamics according to the bifurcation and chaos theories [2, 3]. Wind speed characteristics of the potential wind farm site as well as governmental support are considered the most important criteria for constructing wind farm projects [4]. In most systems, however, there may be other parameters besides wind speed for optimal development of wind power from the system perspective, whereas there is an urgent need to priorities the construction of wind farm projects throughout the system. Moreover, in the current deregulated environment, there are fewer incentives for transmission system expansion, which causes increased network deficiency, and therefore less reliability [5, 6]. Hence, large-scale wind integration policy should also consider system reliability issues, particularly those related to transmission limitations. Wind power prediction is a primary requirement for efficient large-scale integration of wind generation in power systems and electricity markets [7-10]. The stability analysis and optimal control of wind turbine with DFIG have been studied by many researchers [11-15]. A PSO based approach is presented in [16] to optimize all the control parameters in a DFIG simultaneously in order to improve the performance of the DFIG in the power grid. However, increasing the number of the DFIG in a wind farm will lead to increase the number of the control parameters. The proposed Adaptive Dynamic Programming (ADP) based schemes in [17-20] have presented great success in such a multivariable area. Sensitivity analysis is one of the most significant schemes used in power system to analyze and model [21]. The robust sensitivity analysis and voltage sensitivity analysis are respectively applied to acquire the impact of different DFIG parameters on different critical pairs at different rotor speeds [22, 23]. The identification of a Uniformed Dominate Control Parameters (UDCP) is one of the most challenging problem to mitigate the optimization complexity in large-scale wind farms. For this purpose, this study proposes a novel hybrid approach, combining wavelet transform, particle swarm optimization, and an adaptive-network-based fuzzy Inference system to attain the optimal control of DFIG. The sensitivity analysis is used to distinguish the critical parameters UDCP from all parameters and then the proposed strategy is applied to obtain optimal values to enhance dynamic characteristics. Simulation and comparative analysis reveal the robustness of the proposed method.

This paper is organized as follows: In Sec. 2 the power system model is described. The sensitivity analysis and DFIG Control parameters sensitivity analysis are drawn in Sec. 3 and 4, respectively. In Sec. 5, the UDCP Optimization using hybrid Wavelet-PSO-ANFIS approach is presented. Finally, Sec. 6 and 7 give simulation results and conclusions, respectively.

2. WT WITH DFIG MODEL

In this section the wind farm is presented by one WT with DFIG system. The configuration of the simulated single wind farm infinite bus system is depicted in Fig. 1. The wind turbine uses a DFIG consisting of a wound rotor induction generator and an ac/dc/ac Insulated-Gate Bipolar Transistor (IGBT) based Pulse Width Modulation (PWM) converter. In addition, the stator winding is connected directly to the 60 Hz grid while the rotor is fed at variable frequency through the ac/dc/ac converter [24, 25]. As shown in Fig. 2, the DFIG system utilizes a wound rotor induction generator in which the stator windings are directly connected to the three-phase grid and the rotor windings are fed through three-phase back-to-back bidirectional PWM converters. Also, the back-to-back PWM converter consists of three parts: a Rotor Side Converter (RSC), a Grid Side Converter (GSC) and a dc link capacitor placed between the two converters. For the rotor and grid-side converters, vector control strategy is used to acquire separate control of active and reactive power.

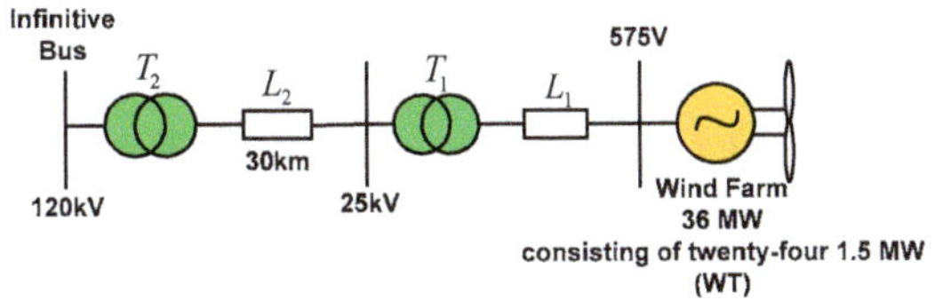

Fig. 1. The diagram of the power system including a DFIG-based wind farm.

2.1 Generator model

According to the voltage and flux-linkage equations of the induction generator [26,27], the differential equations of the stator and rotor circuits of the induction generator with stator and rotor currents as state variables can be given as follows:

$$\frac{X'_s}{\omega_s}\frac{d_{ids}}{dt} = v_{ds} - \left[R_s + \frac{(X_s - X'_s)}{\omega_s T'_0}\right] \times i_{ds} - (1-s_r)E'_d - \frac{L_m}{L_{rr}} v_{dr} \times \left(-\frac{1}{\omega_s T'_0}\right) E'_q - X'_s i_{qs} \quad (1)$$

$$\frac{X'_s}{\omega_s}\frac{d_{iqs}}{dt} = v_{qs} - \left[R_s + \frac{(X_s - X'_s)}{\omega_s T'_0}\right] \times i_{qs} - (1-s_r)E'_q - \frac{L_m}{L_{rr}} v_{qr} \times \left(-\frac{1}{\omega_s T'_0}\right) E'_d - X'_s i_{ds} \quad (2)$$

$$\frac{dE'_d}{dt} = s_r \omega_s E'_q - \omega_s \frac{L_m}{L_{rr}} v_{qr} - \frac{1}{T'_0} \times \left[E'_d + (X_s - X'_s) i_{ds}\right] \quad (3)$$

$$\frac{dE'_q}{dt} = s_r \omega_s E'_d - \omega_s \frac{L_m}{L_{rr}} v_{dr} - \frac{1}{T'_0} \times \left[E'_q + (X_s - X'_s) i_{qs}\right] \quad (4)$$

where,

$$E'_d = -(\omega_s \times \frac{L_m}{L_{rr}} \psi_{qr}),\ E'q = -(\omega_s \times \frac{L_m}{L_{rr}} \psi_{dr})$$

$$X_s = \omega_s \times L_{ss},\ X'_s = \omega_s \times \left[L_{ss} - \left(\frac{L_m^2}{L_{rr}}\right)\right],\ T'_0 = \frac{L_{rr}}{R_r}$$

2.2 Drive train model

A turbine, a low and a high speed shaft, and a gearbox are the main elements of the drive train system which can be expressed by following equations:

$$2H_t \frac{d\omega}{dt} = T_m - T_{sh} \quad (5)$$

$$\frac{d\theta_{t\omega}}{dt} = \omega_t - \omega_r = \omega_t - (1-s_r)\omega_s \quad (6)$$

$$2H_g \frac{d_{sr}}{dt} = -T_{em} - T_{sh} \quad (7)$$

$$T_{sh} = K_{sh}\theta_{t\omega} + D_{sh}\frac{d\theta_{t\omega}}{dt} \quad (8)$$

where

$$T_m = \frac{0.5\rho\pi R^2 C_p V_\omega{}^3}{\omega_t} \quad (9)$$

$$T_{em} = \frac{P_s}{\omega_s} \quad (10)$$

C_p is given by:

$$C_p = \frac{1}{2}\left(\frac{RC_f}{\lambda} - 0.22\beta - 2\right) e^{\frac{-0.225RC_f}{\lambda}} \quad (11)$$

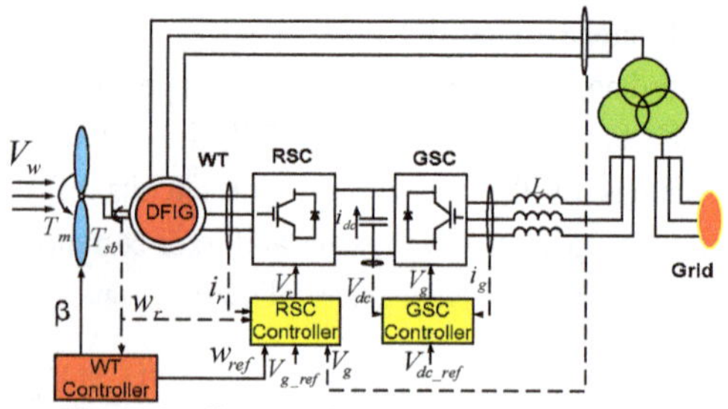

Fig. 2. The diagram of DFIG based wind turbine.

2.3 DC link capacitor model

DC link capacitor is responsible for balancing the active power flow through the back-to-back PWM converter (see Fig. 2) and can be stated as follow:

$$P_r = P_g + P_{DC} \tag{12}$$

where

$$P_r = v_{dr} i_{dr} + v_{qr} i_{qr} \tag{13}$$

$$P_g = v_{dg} i_{dg} + v_{qg} i_{qg} \tag{14}$$

$$P_{DC} = v_{DC} i_{DC} = -C v_{DC} \frac{dv_{dc}}{dt} \tag{15}$$

Substituting (13)-(15) into (12) we have:

$$C v_{DC} \frac{dv_{DC}}{dt} = v_{dg} i_{dg} + v_{qg} i_{qg} - (v_{dr} i_{dr} + v_{qr} i_{qr}) \tag{16}$$

2.4 Rotor Side Controller model

The schematic diagram of RSC is shown in Fig. 3. In this case, the active power and voltage are controlled independently via v_{qr} and v_{dr}, respectively which can be expressed as follows:

$$\frac{dx_1}{dt} = P_{ref} + P_s \tag{17}$$

$$i_{q-ref} = K_{p1}(P_{ref} + P_s) + K_{i1} x_1 \tag{18}$$

$$\frac{dx_2}{dt} = i_{q-ref} - i_{qr} = K_{p1}(P_{ref} + P_s) + K_{i1} x_1 - i_{qr} \tag{19}$$

$$\frac{dx_3}{dt} = v_{s-ref} - v_s \tag{20}$$

$$i_{d-ref} = K_{p3}(v_{s-ref} - v_s) + K_{i3} x_3 \tag{21}$$

$$\frac{dx_4}{dt} = i_{d-ref} - i_{dr} = K_{p3}(v_{s-ref} - v_s) + Ki_3 x_3 - i_{dr} \tag{22}$$

$$v_{qr} = K_{p2}(K_{p1}\Delta P + K_{i1} x_1 - i_{qr}) + Ki_2 x_2 + s_r \omega_s L_m i_{ds} + s_r \omega_s L_{rr} i_{qr} \tag{23}$$

$$v_{dr} = K_{p2}(K_{p3}\Delta U + Ki_3 x_3 - i_{dr}) + Ki_2 x_4 - s_r \omega_s L_m i_{qs} - s_r \omega_s L_{rr} i_{dr} \tag{24}$$

2.5 Grid Side Controller (GSC) model

The schematic diagram of GSC is illustrated in Fig. 4. The GSC has a duty to maintain the dc link voltage and control the terminal reactive power. The reactive power and dc link voltage are controlled independently via *iqg* and *idg*, respectively which can be expressed as follows:

$$\frac{dx_5}{dt} = v_{DC-ref} - v_{DC} \tag{25}$$

$$i_{dg-ref} = -K_{pdg}\Delta v_{DC} + K_{idg} x_5 \tag{26}$$

$$\frac{dx_6}{dt} = i_{dg-ref} - i_{dg} = -K_{pdg}\Delta v_{DC} + Ki_{dg} x_5 - i_{dg} \tag{27}$$

$$\frac{dx_7}{dt} = v_{qg-ref} - iqg \tag{28}$$

$$\Delta v_{dg} = K_{pg}\frac{dx_6}{dt} + K_{ig} x_6 = K_{pg}(-K_{pdg}\Delta v_{DC} + Ki_{dg} x_5 - i_{dg}) + Ki_g x_6 \tag{29}$$

$$\Delta v_{qg} = K_{pg}\frac{dx_7}{dt} + Ki_g x_7 = K_{pg}(i_{qg-ref} - i_{qg}) + Ki_g x_6 \tag{30}$$

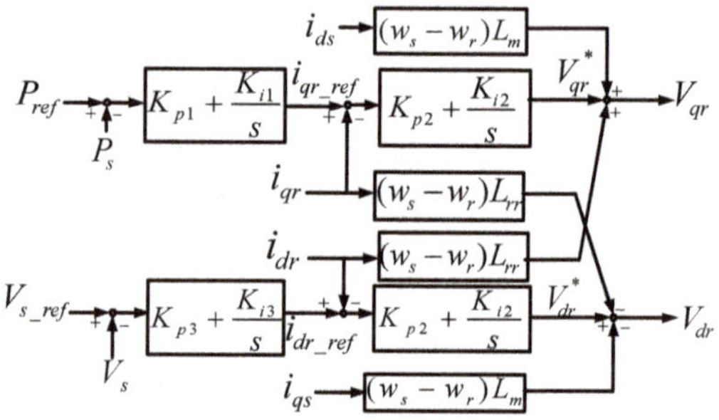
Fig. 3. Schematic diagram of RSC.

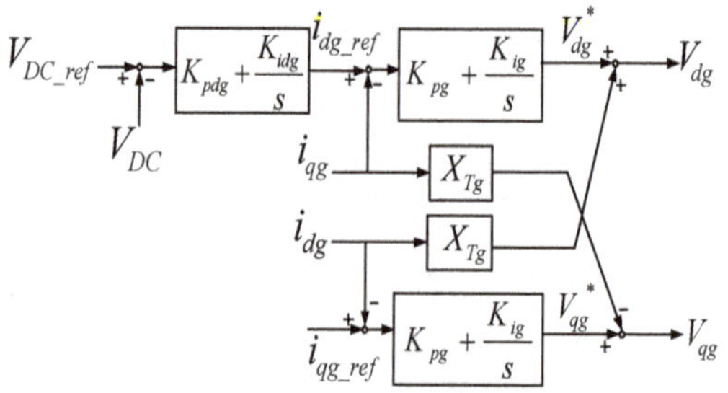
Fig. 4. Schematic diagram of GSC.

2.6 Pitch Controller (PC) model

The schematic diagram of PC is depicted in Fig. 5. The power extraction of WT as well as preventing overrated power production in strong wind are optimized by the pitch angle control which can be modeled as follows:

$$\frac{d\beta}{dt} = K_{p4} - \frac{T_m - T_{sh}}{2H_t} + K_{i4}\Delta\omega_t \tag{31}$$

The control of electrical dynamics can be separated from that of mechanical dynamics due to being much faster than the mechanical dynamics. Therefore, in this paper, we set KP4=KP2 and Ki4=Ki2.

According to above-mentioned, the model of WT with DFIG system can be restated as follows:

$$\dot{X} = f(X,U) \tag{32}$$

Where, X and U are the vectors of the DFIG state variables and input variables defined as follows:

$$X = [\omega_t, \beta, \theta_{t\omega}, s_r, i_{ds}, i_{qs}, E'_d, E'_q, x_1, x_2, x_3, x_4, V_{DC}, x_5, x_6, x_7]^T \tag{33}$$

$$U = [v_{ds}, v_{qs}, i_{dg}, i_{qg}]^T \tag{34}$$

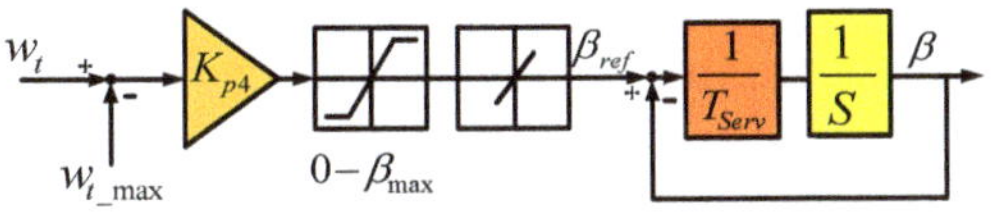

Fig. 5. Schematic diagram of pitch control.

3. SENSITIVITY ANALYSIS

3.1 Trajectory sensitivity analysis

The degree of change of the system when a parameter of the system subjects to small changes and it reflects the derivative relations between the system trajectory and the system parameters is evaluated by the sensitivity analysis which in turn diminish the complexity of simulation [28-29]. A power system applicable sensitivity analysis, is given by the following equation:

$$\begin{cases} \dfrac{dx}{dt} = f(x,y,\theta) \\ 0 = g(x,y,\theta) \\ x(t_0) = x_0 \\ y(t_0) = y_0 \end{cases} \tag{35}$$

Where x is the system state variables, y is the system output variables and θ is the control parameters. The Sensitivity of the control parameters evaluated by the trajectory can be expressed by data from time domain simulations as follows:

$$\frac{\partial y_i(\theta,k)}{\partial \theta_j} = \lim_{\Delta\theta_j \to 0} \frac{y_{i(\theta_1,..\theta_j+\Delta\theta_j,..\theta_m,k)} - y_i(\theta_1,...,\theta_j,..,\theta_m,\theta_k)}{\Delta\theta_j} \tag{36}$$

Where v_i is the trajectory curve of the *ith* variable and corresponds to the *jth* control parameter, m is the total number of control parameters, k is the time instance, and $\Delta\theta_j$ is the change of the parameter of θ. In order to avoid the deflection error and improve the accuracy of the numerical calculation, the system trajectory is needed to calculate twice, by either increment or decrement of a $\Delta\theta_j$:

$$\begin{cases} y_i(\theta_1,...,\theta_j + \Delta\theta_j,...,\theta_m,k) \\ y_i(\theta_1,...,\theta_j - \Delta\theta_j,...,\theta_m,k) \end{cases} \tag{37}$$

Then the relative value of the group of trajectory sensitivity can be calculated as:

$$\frac{\partial\left[\dfrac{y_i(\theta,k)}{y_{i0}}\right]}{\partial\left[\dfrac{\theta_j}{\theta_{jo}}\right]} = \frac{\dfrac{\left[y_i(\theta_1,...,\theta_j+\Delta\theta_j,...,\theta_m,k) - y_i(\theta_1,...,\theta_j-\Delta\theta_j,...,\theta_m,k)\right]}{y_{i0}}}{\dfrac{2\Delta\theta_j}{\theta_{j0}}} \tag{38}$$

where, θ_{jo} is the given value of parameter θ_j and y_{io} is the corresponding steady-state value given θ_{jo}. The mean trajectory sensitivity in order to compare the sensitivity value of these different control parameters can be calculated as follows:

$$A_{ij} = \frac{1}{K}\sum_{k=1}^{k} \left| \frac{\partial[\dfrac{y_i(\theta,k)}{y_{i0}}]}{\partial[\dfrac{\theta_j}{\theta_{j0}}]} \right| \tag{39}$$

where, k is the number of samples on the trajectory sensitivity curves.

3.2 Eigenvalue sensitivity analysis

The power system state equation is obtained by linearizing (35) as follow:

$$\dot{X} = AX \tag{40}$$

Where A is the Jacobian matrix of the system. Then the characteristic equation of the system is:

$$|\lambda I - A| = 0 \tag{41}$$

where, I is an unit matrix with same dimension as A. Therefore, the system eigenvalues $\lambda_1, \lambda_2, ..., \lambda_n$ are obtained via (41). The Jacobin matrix is a function of the control parameter *(A(θ))*. Then each eigenvalue λ_i of $A(\theta)$ will be the function of θ $(\lambda_i(\theta))$. Therefore, the sensitivity of eigenvalue λ_i to the control parameter θ is defined as follows:

$$\frac{\partial \lambda_{i\theta}}{\partial \theta_j} = v_i^T \frac{\partial A(\theta)}{\partial \theta_j} u_j \tag{42}$$

where, u and v are right and left eigenvector, respectively.

4. DFIG CONTROL PARAMETERS SENSITIVITY ANALYSIS

4.1 DFIG control parameters trajectory sensitivity analysis

According to the model of the system shown in Fig.1, a three phase ground fault with the fault resistance 20 Ω in each phase which starts at 15 sec and last for 0.15 sec is located at the Bus-25 KV. The active power trajectory of the DFIG is chosen as the object function in this paper. The active power of DFIG under the fault condition is shown in Fig. 6. The procedure of trajectory sensitivity analysis is as follows:

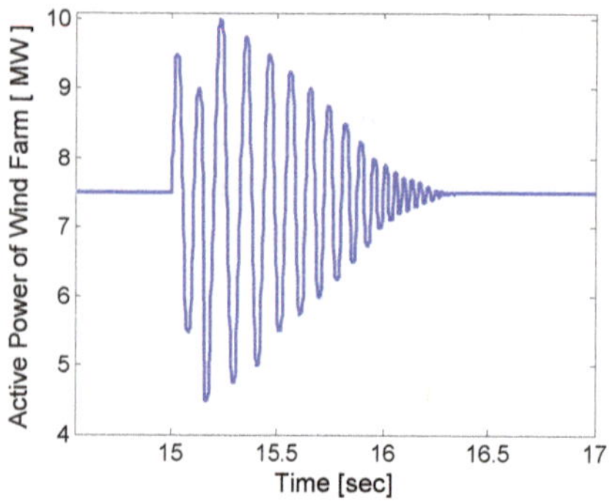

Fig. 6. Active power of WT with DFIG.

a) Consider the parameters for the controller of the WT system are: K_{p1}, K_{i1}, K_{p2}, K_{i2}, K_{p3}, K_{i3}, K_{pdg}, K_{idg}, K_{pg}, K_{ig}. Increase the value of these control parameters 10%, and acquire the trajectory of DFIG active power.

b) Decrease the value of these control parameters 10%, and acquire the trajectory of DFIG active power.

c) Compute the trajectory sensitivity of each control parameter through equation (39).

The sensitivities of the control parameters and their classification are tabulated in Table 1. K_{i3} and K_{pg} with trajectory sensitivity value of 2.1213 and 0.0016 are the largest one and smallest one, respectively (see Table 1). Hence K_{i3} and K_{p1} are assumed as the dominate control parameters for the trajectory sensitivity analysis.

4.2 DFIG control parameters eigenvalue sensitivity analysis

As previously mentioned, the system state vector is defined as follows:

$$X=[\omega_t,\beta,\theta_{t\omega},s_r,i_{ds},i_{qs},E'_d,E'_q,x_1,x_2,x_3,x_4,V_{DC},x_5,x_6,x_7]^T \quad (43)$$

The system eigenvalues and participation factors given by small signal stability analysis are shown in Table 2. Ignoring the stator transients, the state vector can be rewritten as follows:

$$X=[\omega_t,\beta,\theta_{t\omega},s_r,E'_d,E'_q,x_1,x_2,x_3,x_4,V_{DC},x_5,x_6,x_7]^T \quad (44)$$

Table 1. Sensitivity and classification of each control parameters.

Parameters	Sensitivity Value	Order		
		1	2	3
K_{p1}	1.1422	√	-	-
K_{i1}	0.108	-	√	-
K_{p2}	0.5986	-	√	-
K_{i2}	0.0854	-	√	-
K_{p3}	0.4863	-	√	-
K_{i3}	2.1213	√	-	-
K_{pdg}	0.0396	-	-	√
K_{idg}	0.00632	-	-	√
K_{pg}	0.0016	-	-	√
K_{ig}	0.0366	-	-	√

Table 2. System eigenvalue and participation factors analysis.

λ	σ	ω	f	ASV_1	ASV_2
λ_1	-1256	0	0	V_{DC}	-
λ_2	-176.2	0	0	x_5	x_6
$\lambda_{3,4}$	-69.8	112.3	17.8	E'_q	x_3
$\lambda_{5,6}$	-94.7	44.3	7.4	E'_d	x_1
$\lambda_{7,8}$	-2.86	60.4	0.91	$\theta_{t\omega}$	S_r
λ_9	-76.01	0	0	E'_d	x_1
$\lambda_{10,11}$	-0.56	0.69	0.116	ω_t	β
λ_{12}	-15.12	0	0	x_5	x_6
λ_{13}	-24.9	0	0	x_4	-
λ_{14}	-24.35	0	0	x_2	-

The eigenvalue sensitivities of control parameters and the sorted array of their value are given in Table 3 and 4, respectively. As seen in the Tables, K_{pdg} and K_{pg} have the largest eigenvalue sensitivity in response to λ_1 compared to others (with the value of 34.8764 and 12.4459), and consequently are chosen to be the dominate control parameters. Furthermore, K_{i1} and K_{i2} are assumed as a dominate control parameter owing to dominate for most of the eigenvalue ($\lambda_{3,4}$, $\lambda_{5,6}$, $\lambda_{7,8}$, λ_9, $\lambda_{10,11}$, $\lambda_{13,14}$). Hence, the unified dominate control parameters are: K_{i3}, K_{p1}, K_{i1}, K_{i2}, K_{pdg}, K_{pg}.

5. HYBRID WAVELET-PSO-ANFIS APPROACH

The proposed approach is based on the combination of WT, PSO and ANFIS [1]. The WT is used to decompose the wind power series into a set of better-behaved constitutive series. Then, the future values of these constitutive series are estimated using ANFIS. The PSO is used to improve the performance of ANFIS. Three optimization options are used to verify the robustness of the proposed scheme:

a) Optimize all the control parameters (K_{p1}, K_{i1}, K_{p2}, K_{i2}, K_{p3}, K_{i3}, K_{pdg}, K_{idg}, K_{pg}, K_{ig}) simultaneously.

b) Optimize the six UDCP (K_{i3}, K_{p1}, K_{i1}, K_{i2}, K_{pdg}, K_{pg}), and the other control parameters are set as default value.

c) Optimize random six control parameters.

5.1 Wavelet Transform

The WT decomposes the wind power series into a set of better-behaved constitutive series. These constitutive series should be predicted more accurately owing to filtering effect of the WT. WTs could be divided in two categories: Continuous Wavelet Transform (CWT) and Discrete WT (DWT). The CWT $W(a,b)$ of signal $f(x)$ is defined as follow [30]:

$$W(a,b) = \frac{1}{\sqrt{a}} \int_{-\infty}^{+\infty} f(x)\phi\left(\frac{x-b}{a}\right)dx \tag{45}$$

Where, $\Phi(x)$ is a mother wavelet, the scale parameter a controls the spread of the wavelet and translation parameter b determines its central position. DWT is defined as [31]:

$$W(m,n) = 2^{-\left(m/2\right)} \sum_{t=0}^{T-1} f(t)\phi\left(\frac{t-n2^m}{2^m}\right) \tag{46}$$

Where, T is the length of the signal $f(x)$. The scaling and translation parameters are functions of the integer variables m and n ($a=2^m$, $b=n2^m$) is the discrete time index. A multilevel decomposition process can be achieved by successive decomposition of the approximations, where the original signal is broken down into lower resolution components (see Fig. 7).

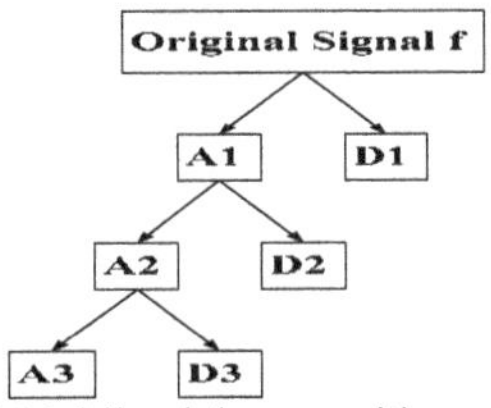

Fig. 7. Multilevel decomposition process.

Table 4. Classify the sensitivity of control parameters.

λ	Control parameters array by eigenvalue sensitivity
λ_1	$K_{pdg}>K_{pg}>K_{p2}>K_{ig}>K_{p1}>K_{p3}>K_{i2}>K_{i1}>K_{i3}>K_{idg}$
λ_2	$K_{ig}> K_{idg}>K_{pg}>K_{pdg}>K_{i2}>K_{p1}>K_{p2}>K_{i1}>K_{i3}>K_{p3}$
$\lambda_{3,4}$	$K_{i2}>K_{p2}>K_{i1}>K_{ig}>K_{p3}>K_{pg}>K_{p1}>K_{i3}>K_{idg}>K_{pdg}$
$\lambda_{5,6}$	$K_{i2}>K_{p2}>K_{i1}>K_{ig}>K_{p1}>K_{pg}> K_{i1}>K_{p3}>K_{idg}>K_{pdg}$
$\lambda_{7,8}$	$K_{i2}>K_{p2}>K_{i1}>K_{p1}>K_{ig}>K_{i3}> K_{pg}>K_{p3}>K_{idg}>K_{pdg}$
λ_9	$K_{i2}>K_{i1}>K_{p2}>K_{ig}>K_{p1}>K_{pg}> K_{i3}>K_{p3}>K_{idg}>K_{pdg}$
$\lambda_{10,11}$	$K_{i2}>K_{i1}>K_{p2}>K_{p1}>K_{ig}>K_{i3}> K_{pg}>K_{idg}>K_{p3}>K_{pdg}$
λ_{12}	$K_{idg}>K_{pdg}>K_{pg}>K_{ig}>K_{i2}>K_{i1}>K_{p2}>K_{p1}>K_{i3}>K_{p3}$
λ_{13}	$K_{i2}>K_{p2}>K_{i3}>K_{i1}>K_{ig}>K_{p3}>K_{p1}>K_{pg}>K_{idg}>K_{pdg}$
λ_{14}	$K_{i2}>K_{p2}>K_{i1}>K_{p1}>K_{i3}>K_{ig}>K_{p3}>K_{pg}>K_{idg}>K_{pdg}$

5.2 Particle Swarm Optimization

The PSO algorithm is based on the biological and sociological behavior of animals searching for their food [32-34]. Consider an optimization problem of D variables. A swarm of N particles is initialized in which each particle is assigned a random position in the D-dimensional hyperspace. Let x denote a particle's position and v denote the particle's flight velocity over a solution space. The best previous position of a particle is *Pbest*. The index of the best

Table 3. Sensitivity of control parameters to each eigenvalue

λ	K_{p1}	K_{i1}	K_{p2}	K_{i2}	K_{p3}	K_{i3}	K_{p1}	K_{idg}	K_{pg}	K_{ig}
λ_1	0.8594	0.0212	7.6912	0.1895	0.3354	0.0092	K_{i1}	0.0062	12.984	1.4163
λ_2	0.0055	0.0108	0.0465	0.0832	0.0062	0.0017	0.2295	0.5496	0.3125	20.026
$\lambda_{3,4}$	0.0281	0.0684	1.1201	2.6795	0.0694	0.1921	$6.74e^{-5}$	0.0016	0.0396	0.0954
$\lambda_{5,6}$	0.1545	0.4762	0.8977	2.598	0.0129	0.0976	0.0026	0.0018	0.1497	0.4752
$\lambda_{7,8}$	$7.85e^{-4}$	0.0046	0.0072	0.0362	$5.96e^{-6}$	$3.25e^{-5}$	$7.81e^{-8}$	$3.54e^{-7}$	$1.22e^{-5}$	$7.65e^{-5}$
λ_9	0.1298	0.5447	0.4203	1.7023	0.0035	0.0125	$6.25e^{-6}$	0.0001	0.0432	0.1729
$\lambda_{10,11}$	$3.91e^{-9}$	$1.24e^{-6}$	$3.55e^{-8}$	$1.16e^{-5}$	$6.2e^{-14}$	$2.12e^{-11}$	$1.11e^{-14}$	$3.28e^{-12}$	$4.94e^{-12}$	$1.15e^{-9}$
λ_{12}	$4.51e^{-8}$	$9.38e^{-7}$	$3.41e^{-7}$	$7.62e^{-6}$	$5.71e^{-10}$	$1.14e^{-8}$	0.0211	0.5496	0.0241	0.0122
λ_{13}	$3.66e^{-4}$	0.0034	0.2212	2.6852	0.0052	0.0061	$6.71e^{-8}$	$1.26e^{-6}$	$7.69e^{-5}$	0.0032
λ_{14}	0.0028	0.0368	0.1821	0.1862	0.0026	0.0024	$1.67e^{-6}$	$2.16e^{-5}$	0.0002	0.0012

particle among all particles in the swarm is *Gbest*. Velocity and position of a particle are updated by the following update rules:

$$v_i(t)=\omega v_i(t-1)+\rho_1\frac{\left(x_{Pbest}-x_i(t)\right)}{\Delta t}+\rho_2\frac{\left(x_{Gbest}-x_i(t)\right)}{\Delta t} \quad (47)$$

$$x_i(t)=x_i(t-1)+v_i(t)\Delta t \quad (48)$$

Where ω is an inertia weight, Δt is the time-step value, ρ_1 and ρ_2 are random variables defined as $\rho_1=r_1c_1$ and $\rho_2=r_2c_2$ with $r_1,r_2\sim U(0,1)$, c_1, c_2 are positive acceleration constants. The time-step is necessary to make the algorithm dimensionally correct. The inertia weight ω is modified according to the following equation:

$$\omega=\omega_{\max}-\frac{\omega_{\max}-\omega_{\min}}{I_{tr\max}}I_{tr} \quad (49)$$

where, ω_{max} and ω_{min} are the initial and final inertia weights, I_{trmax} is the maximum number of iteration, and I_{tr} is the current number of iteration.

5.3 Adaptive-Network-based Fuzzy Inference System (ANFIS)

ANFIS is a class of adaptive multilayer feed-forward networks with the capability of using past samples to forecast the sample ahead [35]. The ANFIS network is comprised of five layers as shown in Fig. 8. Each layer includes several nodes described by the node function. Let Q_i^j denote the output of the *ith* node in layer *j*. In layer 1, every node is an adaptive node with node function

$$Q_i^1=\mu A_i(x), \qquad i=1,2 \quad (50)$$

$$Q_i^1=\mu B_{i-2}(y), \qquad i=3,4 \quad (51)$$

where, *x (or y)* is the input to the *ith* node and A_i (or $B_{i\text{-}2}$) is a linguistic label associated with this node. The membership functions for *A* and *B* are usually described by generalized bell functions.

$$\mu A_i(x)=\frac{1}{1+\left|\dfrac{x-r_i}{p_i}\right|^{2q_i}} \quad (52)$$

where, $\{p_i,q_i,r_i\}$is the parameter set. Parameters in this layer are referred to as premise parameters [35]. In layer 2, each node labeled $\prod$ multiplies incoming signals and sends the product out.

$$Q_i^2=\omega_i=\mu A_i(x)\mu B_i(y), \qquad i=1,2 \quad (53)$$

Each node output represents the firing strength of a rule. In layer 3, each *N* node computes the ratio of the *ith* rules' firing strength to the sum of all rules' firing strengths.

$$Q_i^3=\overline{\omega_i}=\frac{\omega_i}{\omega_1+\omega_2} \quad (54)$$

The outputs of this layer are called normalized firing strengths. In layer 4, each node computes the contribution of the *ith* rule to the overall output:

$$Q_i^4=\overline{\omega_i}z_i=\overline{\omega_i}(a_ix+b_iy+c_i), \quad i=1,2 \quad (55)$$

where, $\overline{\omega_i}$ is the output of layer 3 and $\{a_i,b_i,c_i\}$ is the parameter set. Parameters of this layer are referred to as consequent parameters. In layer 5, the single node $\sum$ computes the final output as the summation of all incoming signals.

$$Q_i^5=\sum_i\overline{\omega_i}z_i=\frac{\sum_i\omega_iz_i}{\sum_i\omega_i} \quad (56)$$

Thus, an adaptive network is functionally equivalent to a Sugeno-type fuzzy inference system.

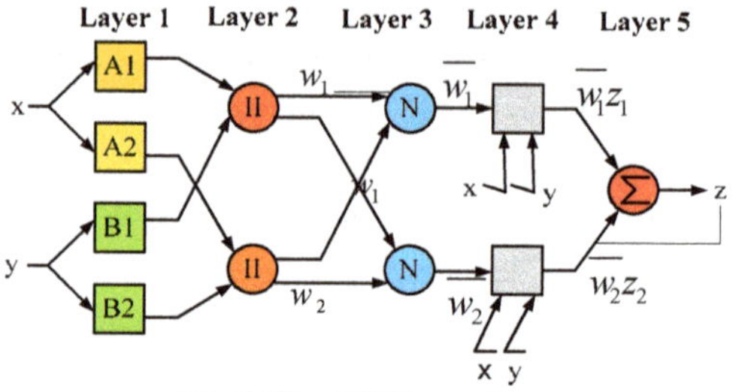

Fig.8. The ANFIS structure.

5.4 Hybrid approach

The structure of the proposed scheme is depicted in Fig. 9 and the implementation of it is described as following steps:

***First step*:** Form a matrix with a set of historical wind power with DFIG data (10 or 6 in this study), arranged in *C* columns of a matrix thereof.

Second step: Select a number of columns of the previous array so that the set of values derived from it represents the real input data. In addition, evaluate the fitness value of the initial group, which is given by:

$$F=\max\{\mathrm{Re}\,al\,(\lambda_l)\} \quad ,l=1,2,...,14 \quad (57)$$

Third step: The WT is used to decompose the input data. The decomposition is made from the choice of basic functions (58) (wavelet family of functions), and the number of levels wanted to split the series. The signal is divided into three levels: namely, a level of approximation (*A*) and details (*D*). The wavelet function used is the Db4 type, which offers a good approach and ability to use a relatively small number of coefficients, making the code faster. Subsequently, in the level of decomposition, the detail series (for high frequencies) obtained is analyzed, so that they make a selection of coefficients in this series. This selection procedure is known as thresholding, because the purpose is to eliminate the coefficients smaller than a given value, with the aim of improving signal quality by removing noise. Finally, there is the process of reconstruction of the series.

Fourth step: Get the signal from the Wavelet, which can be submitted to the entrance of the ANFIS.

Fifth step: Train the ANFIS with the data obtained from the implementation of the previous step. The ANFIS uses a combination of the least-squares method (to determine consequent parameters) and the back propagation gradient descent method (to learn the premise parameters). The training process allows the system to adjust its parameters as inputs/outputs submitted [36]. The knowledge acquired through the learning process is tested by applying new data, called the testing set. The PSO is used to train the parameters within the membership functions of fuzzy inference system which leads to acquire more accurate results.

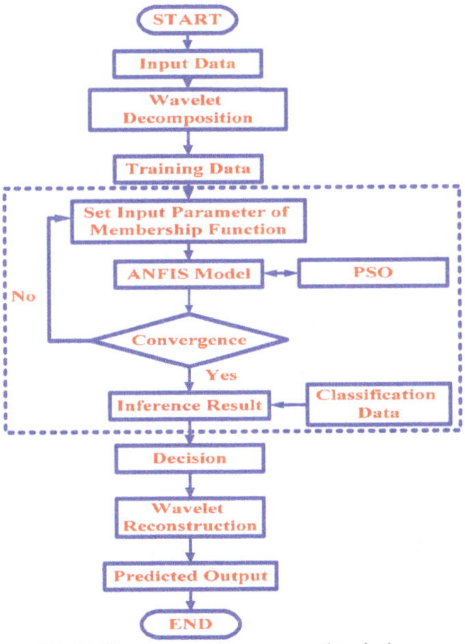

Fig. 9. Structure of the proposed technique.

Sixth step: Create a N-dimension vector, which will be optimized by PSO algorithm, where N is the number of membership functions. The fitness function is defined as the mean squared error.

Table 5. Parameters of PSO.

Parameters	C_1=0.2, C_2=0.2, ω_{min}=0.4, ω_{max}=0.9, Number of Particles=25, Number of Iterations=200

Table 6 shows the result of the three options. In addition, the results of the small signal stability analysis are tabulate in Table 7. As seen in the table, the similar result is achieved through optimizing the UDCP, related to the ten control parameters.

Table 6. Control parameters with and without optimization.

Option	Number of option			
	None	1	2	3
K_{p1}	1	1.1986	1.1992	1.0136
K_{i1}	100	80.032	79.693	98.874
K_{p2}	0.3	0.2541	0.2843	0.2849
K_{i2}	8	4.986	5.023	8.066
K_{p3}	1.25	10.530	1.231	1.1961
K_{i3}	300	220.02	220.82	325.95
K_{pdg}	0.002	0.0123	0.0136	0.0028
K_{idg}	0.05	0.0521	0.046	0.0536
K_{pg}	1	0.7251	0.7526	0.8266
K_{ig}	100	129.56	100.6	98.685

6. SIMULATION RESULTS

In this case, two different power system, single machine infinite-bus system and multi-machine power system are applied to confirm the improvement in dynamic stability by the controller of the WT with DFIG system with the optimized UDCP control parameters.

Case 1: SMIB with small disturbance

The small disturbance is set as the DFIG voltage reference decrease from 1.0 pu to 0.9 pu. The dynamics of the active power, reactive power and dc link voltage are shown in Figs. 10-12, respectively. Each of these figures show four curves with different parameters: the original control parameters optimized all control parameters, optimized random six control parameters and optimized UDCP.

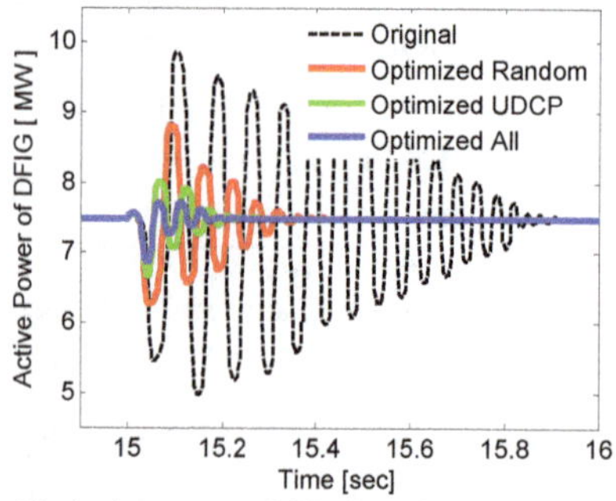

Fig. 10. Active power of SIMB under small disturbance.

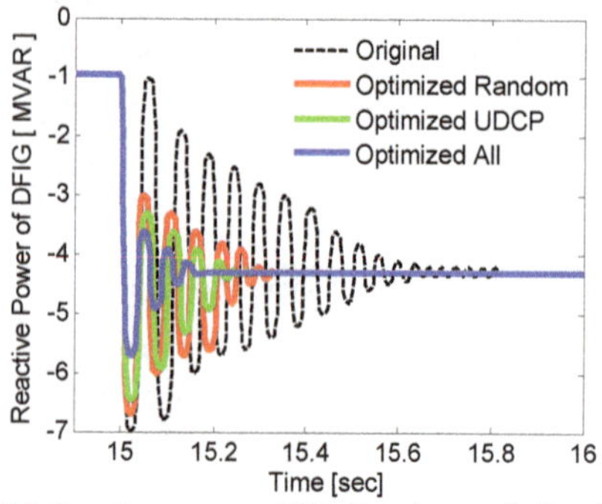

Fig. 11. Reactive power of SMIB under small disturbance.

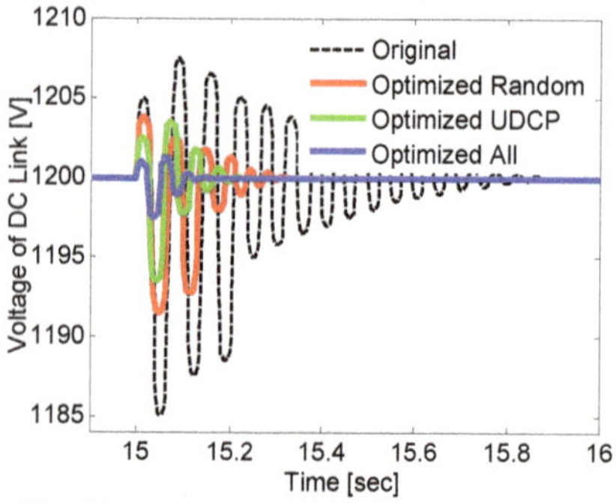

Fig. 12. Voltage of dc link of SMIB under small.

As seen, the dynamic performance of the DFIG base WT using the optimized UDCP is nearly the same as using all the ten optimized control parameters. According to the figures following results are achieved: significant reduction in the magnitudes of the active power, reactive power, voltage sag and overshoot; better oscillation damping with the optimized UDCP than that with_out the optimization. In addition, although the performance of optimized random six control parameters is better than the original one, it cannot compete with the proposed approach with the optimized UDCP.

Case 2: SMIB with large disturbance

In this case, a three-phase ground fault in the middle of the transmission line is applied at t=15 s, and cleared after 0.15 s. Figures 13-15 illustrate the dynamics of the active power, output reactive power and dc link voltage.

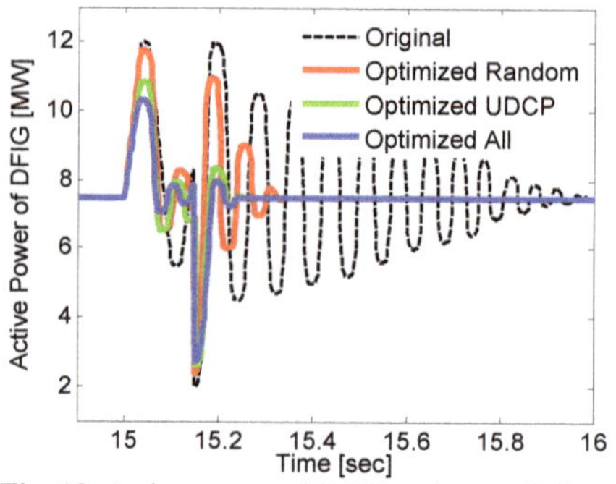

Fig. 13. Active power of SMIB under small disturbance.

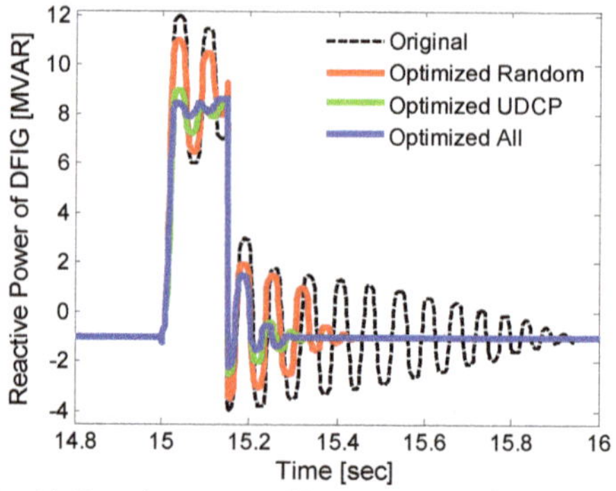

Fig. 14. Reactive power of SMIB under small disturbance.

The oscillation after the disturbance is damped out very quickly as similar as in the small disturbance in case 1. Therefore, the optimized UDCP enhances the fault ride through capability of the WT system.

Table 7. Eigenvalue analysis with different options disturbance.

Option		λ_1	λ_2	$\lambda_{3,4}$	$\lambda_{5,6}$	$\lambda_{7,8}$	λ_9	$\lambda_{10,11}$	λ_{12}	λ_{13}	λ_{14}
1	σ	-66952	-198	-89.53	-111.2	-3.52	-52	-0.46	-4.65	-17.85	-18.22
	ω	0	0	78	57	61.5	0	0.55	0	0	0
2	σ	-66676	-154.2	-98.65	-3.56	-98.41	-47.6	-0.56	-4.22	-17	-17.31
	ω	0	0	98.26	61	54.67	0	0.55	0	0	0
3	σ	-11225	-175.2	-84	-102	-2.56	-84.26	-.55	-25	-26	-17
	ω	0	0	115.2	50.23	63	0	0.53	0	0	0

In addition, as seen in these figures, the performance of optimized random six control parameters is not as good as optimized UDCP.

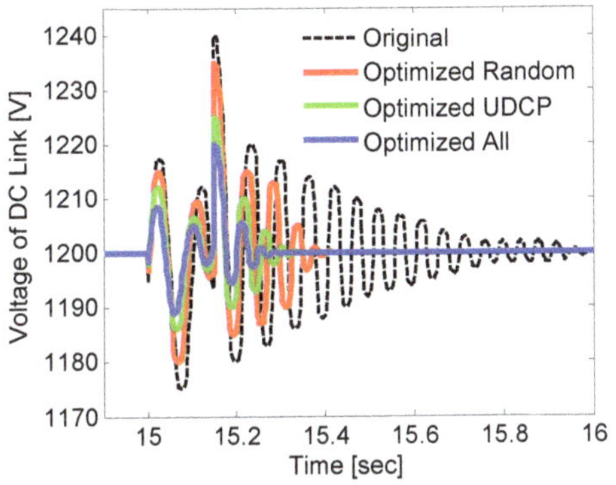

Fig. 15. Voltage of dc link of SIMB under small.

Case 3: Multi-machine power system

A multi-machine power system is also used to verify the strong performance of the proposed scheme.

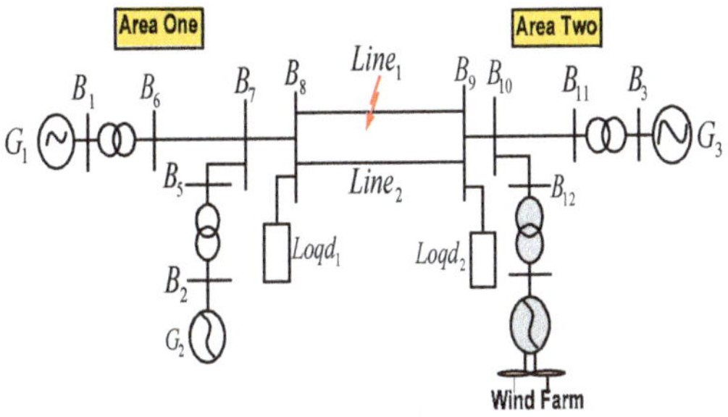

Fig.16. Schematic diagram of the multi-machine power system with DFIG.

The system is divided into two areas, in each of which there are two machines. In addition, the four-machine-two-area system is modified by replacing generator G4 with a DFIG, which is the same as that used in the above SMIB power system. The DFIG capacity is 36 MW as indicated in Section 2, where the other three synchronous machines have the same capacity as the DFIG. Before the fault occurs, the active power transfers from area one to area two is almost 15 MW [37]. Hence, a single-phase ground fault is applied (see Fig. 16) at the t=30 s, and cleared after 0.2 s without tripping the line. The dynamics of the active power, voltage of the DFIG and dc link voltage are shown in Figs. 17–19.

As shown, the dynamic performance of the DFIG based WT using the optimized UDCP is better than the optimized random six control parameters. In addition, damping the oscillation after the disturbance is as quickly as in the SMIB power system. The dynamic responses of generator 3 (G3) will be affected significantly owing to its nearness to the DFIG. The oscillation of the voltage of G3 is damped quickly as shown in Figure 20. Figure 21 shows the convergence of cost function. According to the figure, the convergence of training error of the PSO-ANFIS is meaningfully fast.

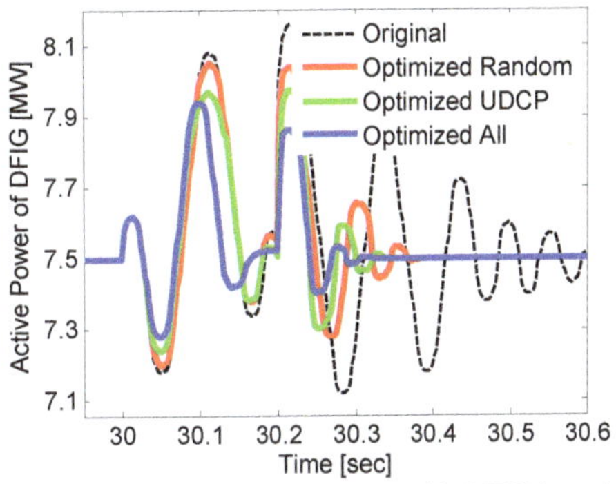

Fig. 17. Active power of the WT with DFIG in multi-machine power system.

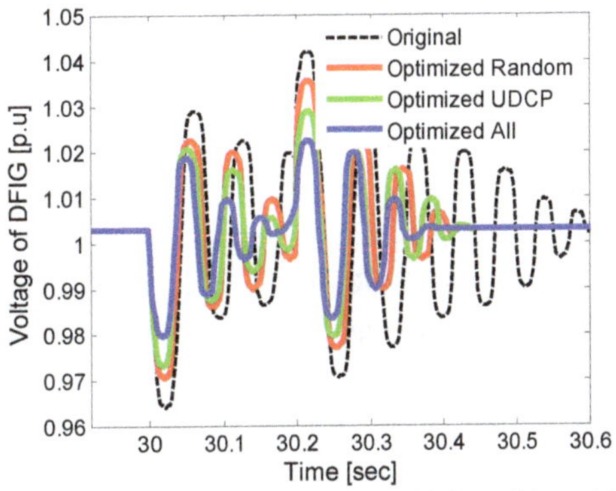

Fig. 18. Voltage of the WT with DFIG in multi-machine power system.

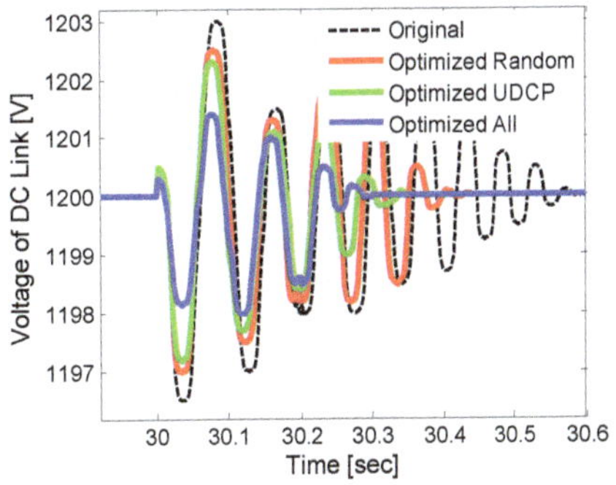

Fig. 19. Voltage of dc link in multi-machine power system.

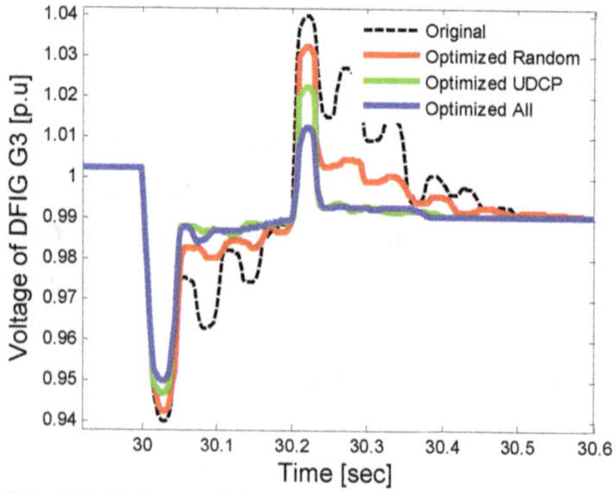

Fig. 20. Voltage of G3 in multi-machine power system.

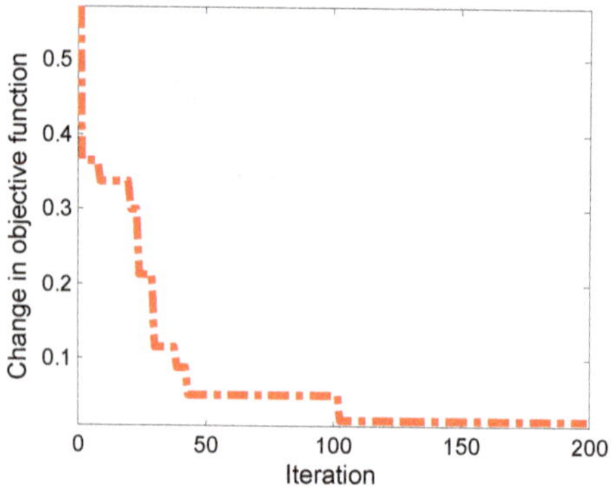

Fig. 21. Training error of the PSO-ANFIS.

7. CONCLUSIONS

A DFIG wind turbine model containing drive train, pitch control, induction generator, back-to-back PWM converters, RSC, GSC and vector-control loops is developed in this paper. A sensitivity analysis based on trajectory sensitivity and eigenvalue sensitivity analysis is performed on the DFIG control parameters to choose the unified dominate control parameters. A novel hybrid approach is applied to find the optimal control parameters in order to acquire the optimal control of the multiple controllers of the wind turbine system. The proposed approach is based on the combination of wavelet transform, particle swarm optimization, and an adaptive-network-based fuzzy inference system. The wavelet transform is used to decompose the wind power series into a set of better-behaved constitutive series. Then, the future values of these constitutive series are estimated using ANFIS. The PSO is employed to improve the performance of ANFIS to obtain a lower error. The main advantages of the proposed approach are concluded as follow:

- Reduction in computational cost due to decrement of selected UDCP parameters via sensitivity analysis;
- Significant reduction in the magnitudes of the active power, reactive power, voltage sag and overshoot, and consequently improving dynamic performance of the WT system;
- Better oscillation damping with the optimized UDCP;
- Reduction in computational time compared with the case of optimizing all control parameters, which leads to improve the optimization efficiency;

REFERENCES

[1] J. P. S. Catalão, H. M. I. Pousinho, and V. M. F. Mendes, "Hybrid wavelet-PSO-ANFIS approach for short-term wind power forecasting in portugal," *IEEE Transaction on Sustainable Energy* , vol. 2, no. 1, pp. 50-59, 2011.

[2] Z. Li, J. B. Park, Y. H. Joo, B. Zhang and G. R.Chen, "Bifurcation and chaos in a permanent-magnet synchronous motor," *IEEE Transaction on Circuits System Fundamental Theory and Application.*, vol. 49, no. 3, pp. 383–387, 2002.

[3] Y. Gao and K. T. Chau, "Hopf bifurcation and chaos in synchronous reluctance motor drives," *IEEE Transaction on Energy Conversion.*, vol. 19, no. 2, pp. 296–302,Jun 2004.

[4] K. E. casey, B.F. Hobbs, D. Shirmohammadi and F.A. Wolak, "Benefit cost analysis of large-scale transmission for renewable generation: principles & California case study," *Proc. Of the IEEE Power Engineering Society General Meeting*, pp. 1–3, June 2007.

[5] G.B. Shrestha and P.A.J. Fonseka, "Congestion-driven transmission expansion in competitive power markets," *IEEE Transaction on Power System.*, vol. 19, no. 3, pp. 1658–1665, 2004.

[6] S. Grijalva and A. M. Visnesky, "The effect of generation on network security: spatial representation, metrics, and policy," *IEEE Transaction on Power System*, vol. 21, no. 3, pp. 1388–1395, 2006.

[7] P. Pinson, S. Lozano, I. Marti, G. N. Kariniotakis, and G. Giebel, "ViLab: A virtual laboratory for

collaborative research on wind power forecasting," *Wind Energy*, vol. 31, no. 2, pp. 117–122, Mar. 2007.

[8] N. Cutler, M. Kay, K. Jacka and T. S. Nielsen, "Detecting, categorizing and forecasting large romps in wind farm power output using meteorological observations and WPPT," *Wind Energy*, vol. 10, no. 5, pp. 453-470, 2007.

[9] H. Aa Nielsen, T. S. Nielsen, H. Madsen, M. J. S. I. Pindado and I. Marti, "Optimal combination of wind power forecasts," *Wind Energy*, vol. 10, no. 5, pp. 471-482, 2007.

[10] P. Pinson, H. A. Nielsen, J. K. Moller, H. Madsen, and G. N. Kariniotakis, "Non-parametric probabilistic forecasts of wind power: Required properties and evaluation," *Wind Energy*, vol. 10, no. 6, pp. 497–516, 2007.

[11] L. Yang, G. Y. Yang, Z. Xu, Z. Y. Dong, K. P. Wong, and X. Ma, "Optimal controller design of a doubly-fed induction generator wind turbine system for small signal stability enhancement," *IET Proceedings on Generation, Transmission and Distribution.*, vol. 5, pp. 579–597, 2010.

[12] W. Qiao, J. Liang, G. K. Venayagamoorthy, and R. G. Harley, "Computational intelligence for control of wind turbine generators," *Proc. Of the IEEE Power Energy Society General Meeting*, pp. 6, 2011.

[13] A. Mendonca and J. A. P. Lopes, "Robust tuning of power system stabilizers to install in wind energy conversion systems," *IET Renewable Power Generation*, vol. 3, no. 4, pp. 465–475, 2009.

[14] N. Kshatriya, U. D. Annakkage, F. M. Hughes, and A. M. Gole, "Optimized partial eigenstructure assignment-based design of a combined PSS and active damping controller for a DFIG," *IEEE Transaction on Power System.*, vol. 25, no. 2, pp. 866–876, 2010.

[15] H. Huang and C. Y. Chung, "Coordinated damping control design for DFIG-based wind generation considering power output variation," *IEEE Transaction on Power Systems*, vol. 27, no. 4, pp. 1916–1925, 2012.

[16] F.Wu, X. P. Zhang, K. Godferey and P. Ju, "Small signal stability analysis and optimal control of a wind turbine with doubly fed induction generator," *IET Proceedings on Generation, Transmission and Distribution*, vol. 1, pp. 751–760, 2007.

[17] P. J. Werbos, "Computational intelligence for the smart grid-history, challenges, and opportunities," *IEEE Computational Intelligence Magazine*, vol.6, no. 3, pp. 14–21, 2011.

[18] G. K. Venayagamoorthy, "Dynamic, stochastic, computational, and scalable technologies for smart grids," *IEEE Computational Intelligence Magazine*, vol. 6, no. 3, pp. 22–35, 2011.

[19] W. Qiao, R. G. Harley and G.K. Venayagamoorthy," Coordinated reactive power control of a large wind farm and a STATCOM using heuristic dynamic programming," *IEEE Transaction on Energy Conversion.*, vol.24, no. 2, pp. 493–503, 2009.

[20] W. Qiao, G. K. Venayagamoorthy and R. G. Harley, "Real-time implementation of a STATCOM on a wind farm equipped with doubly fed induction generators," *IEEE Transaction on Industry Applications*, vol. 45, no. 1, pp. 98–107, Jan./Feb. 2009.

[21] H. L. Xie, P. Ju, J. Luo, Y. Ning, H. Zhu and X.Wang, "Identifiability analysis of load parameters based on sensitivity calculation," *Automation Electric and Power System.*, vol. 33, pp. 17–21, 2009.

[22] L. H. Yang, Z. Xu, J. Østergaard, Z. Y. Dong, K. P .Wong, and X. Ma, "Oscillatory stability and eigenvalue sensitivity analysis of a DFIG wind turbine system," *IEEE Transaction on Energy Conversion*, vol. 26, pp. 328–339, 2011.

[23] R. Aghatehrani and R. Kavasseri, "Reactive power management of a DFIG wind system in micro-grids based on voltage sensitivity analysis," *IEEE Transaction on Sustainable Energy*, vol. 2, pp. 451–458, 2010.

[24] A. D. Hansen, P. Sørensen, F. Iov and F. Blaabjerg, "Control of variable speed wind turbines with doubly-fed induction generators," *Wind Energy.*, vol. 28, no. 4, pp. 411–434, 2004.

[25] A. D. Hansen, P. Sørensen, F. Iov and F. Blaabjerg, "Control of variable speed wind turbines with doubly-fed induction generators," *Wind Energy.*, vol. 28, no. 4, pp. 411–434, 2004.

[26] L. Rouco and J. L. Zamora, "Dynamic patterns and model order reduction in small-signal models of doubly fed induction generators for wind power applications," *Proc. Of the IEEE Power Engineering Society General Meeting*, pp. 1-8, 2006.

[27] P. C. Krause, O. Wasynczuk, and S. D. Sudhoff, "*Analysis of electric machinery and drive systems. piscatawαy,*" NJ: *IEEE Press*, 2002.

[28] P. Juand and D.Q. Ma, "*Identification of power load,*" 2nd ed. Beijing, China: China Electric Power Press, 2008.

[29] Y.Tang, P.Ju, H. He, C. Qin and F. Wu, Optimized control of DFIG-based wind generation using sensitivity analysis and particle swarm optimization", *IEEE Transaction on smart grid*, vol. 4, no. 1, 2013.

[30] N. Amjady and F. Keynia, "Short-term load forecasting of power systems by combination of wavelet transform and neuro-evolutionary algorithm," *Energy*, vol. 34, no. 1, pp. 46–57, 2009.

[31] A. J. R. Reis and A. P. A. da Silva, "Feature extraction via multi-resolution analysis for short-term load forecasting," *IEEE Transaction on Power System*, vol. 20, no. 1, pp. 189–198, 2005.

[32] Z. A. Bashir and M. E. El-Hawary, "Applying wavelets to short-term load forecasting using PSO-based neural networks," *IEEE Transaction on Power Systems*, vol. 24, no. 1, pp. 20–27, 2009.

[33] W. Yu and X. Li, "Fuzzy identification using fuzzy neural networks with stable learning algorithms," *IEEE Transaction on Fuzzy Systems*, vol. 12, pp. 411–420, 2004.

[34] J. S. Heo, K. Y. Lee, and R. Garduno-Ramirez, "Multi-objective control of power plants using particle swarm optimization techniques," *IEEE Transaction on Energy Conversion*, vol. 21, pp. 552–561, 2006.

[35] Z. Yun, Z. Quan, S. Caixin, L. Shaolan, L. Yuming, and S. Yang, "RBF neural network and ANFIS-based short-termload forecasting approach in real-time price environment," *IEEE Transaction on Power System*, vol. 23, no.3, pp. 853-858, 2008.

[36] J. -S. R. Jang, "ANFIS: Adaptive-network-based fuzzy inference system," *IEEE Transaction on Power Systems*, vol. 23, no. 3, pp. 665-685, 1993.

[37] F.Wu, X. P. Zhang, P. Ju, and M. J. H. Sterling, "Decentralized nonlinear control of wind turbine with doubly fed induction generator," *IEEE Transaction on Power Systems*, vol. 23, no. 2, pp. 613-621, 2008.

5

Condition Monitoring Techniques of Power Transformers

Z. Moravej *, S. Bagheri

Faculty of Electrical & Computer Engineering, Semnan University, Semnan, Iran

ABSTRACT

Power transformers provide a vital link between the generation and distribution of produced energy. Such static equipment is subjected to abuse during operation in generation and distribution stations and leads to catastrophic failures. This paper reviewed the techniques in the field of condition monitoring of power transformers in recent years. Transformer monitoring and diagnosis are the effective techniques for preventing the eventual failures and contributing to ensure the plan's reliability. This paper provided a survey on the existing techniques for monitoring, diagnosis, condition evaluation, maintenance, life assessment and possibility of extending the life of the existing assets of power transformers with be appropriate classifications. Thus, this survey could help researchers through providing better techniques for condition monitoring of power transformers.

KEYWORDS: Ageing, Maintenance plans, Condition monitoring techniques, Power transformers.

1. INTRODUCTION

Power transformer condition monitoring is generally considered as one of the most important Condition Monitoring (CM) techniques in power systems. The unscheduled outages of transformers, due to unexpe-cted failures, are catastrophic in many cases. Diagnosis and proper monitoring play a key role in the life expectancy of a power transformer.

There are plenty of proper monitoring methods for evaluating the condition and possible incipient failure of a power transformer. Transformer monitoring, methods are usually too expensive and/or time- consuming to use. However, cost-efficient methods are needed for transformer monitoring and one possibility is to utilize loading and temperature information measured from the network. There are several general and useful refere-nces in this regard [1-3], which contain information on the CM of electrical equipment and mainly power transformers. This paper intended to review the overall literature on this subject by describing the current research situation. However, it should be pointed out that some papers might be missing due to the large number of publications in this area. The main activities of transformer asset management are summarized in Fig. 1. In the next section of this paper, the concept of CM is introduced, its features and functions, are presented, and the related techniques are provided to give a general picture of CM. In the third section, different CM techniques for power transformer are described and summarized. Sec. 4 elaborates on performing maintenance plans, Sec. 5 introduces the end of life assessments and aging, and finally Sec. 6 introduces life cycle cost.

2. GENERAL CONCEPT OF CONDITION MONITORING (CM) AND CONDITION ASSESSMENT (CA)

This Condition monitoring of power transformers has become a reality in recent years. CM is mostly considered for transformer insulation system and winding integrity.

*Corresponding author:
Z. Moravej (E-mail: zmoravej@semnan.ac.ir)

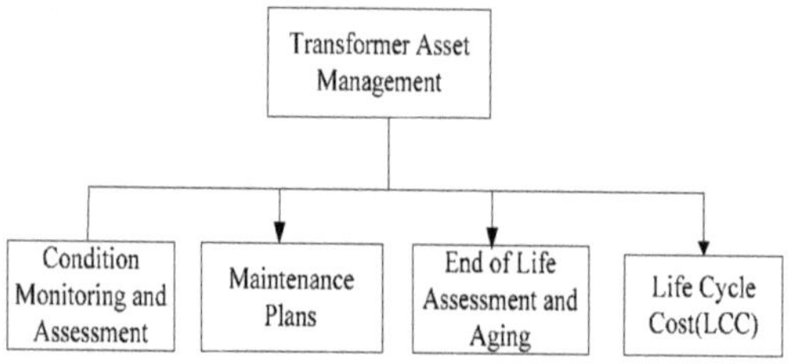

Fig. 1. Transformer asset management activities [3].

It mainly focuses on the detection of incipient faults inside the transformer, which are created from gradual deterioration [3]. The significance of a condition monitoring is in that it allows for the early identification of developing faults such as arcing before resulting in catastrophic failure.

Monitoring and periodic diagnosis together with operational/maintenance strategies would allow for the life extension of assets and ensure the enhanced reliability of the system. CM and life assessment of a costly and critical asset like transformer are aimed to:

-Detect faults at incipient stage and avoid catastrophic faults, and

-Reduce maintenance costs by doing condition-based rather than time-based maintenance.

Monitoring is done in energized as well as de-energized conditions; i.e. off-line and online. Condition assessment [3] means the development of new techniques for analyzing these data to both predict the trends of the monitored transformer and evaluate its current performance. In addition, the improvement of transformer CA renders benefit by giving a decision basis to asset managers. The main target of asset management is to manage transformers to serve longer with reduced lifetime operating cost.

3. DIFFERENT CM TECHNIQUES

In order to have information about the state-of-health of the transformer, the monitored data and the incipient faults detected by the CM system should be analyzed to assess the transformer condition. Transformer CM can be divided into six main categories. Fig. 2 shows the main categories of transformer CM techniques and on-line monitoring method of CM. Based on Fig. 2, each of the CM techniques will be discussed separately.

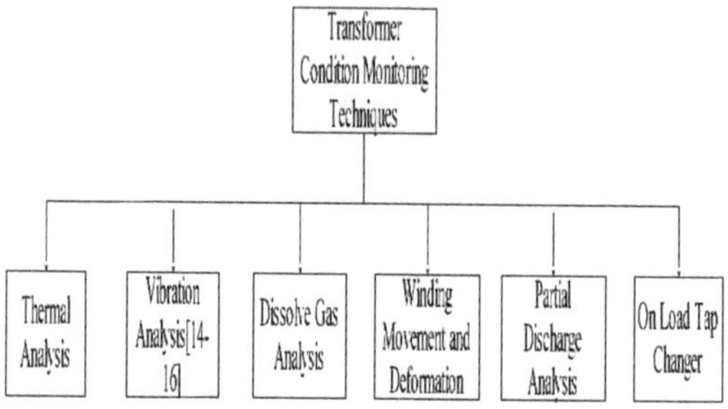

Fig. 2. Transformer condition monitoring techniques.

3.1. Condition monitoring by thermal analysis

Thermal analysis of transformers can provide useful information about their conditions and can be used to detect the inception of any fault [4]. Most of the faults cause change in the thermal behavior of transformers [5]. Abnormal conditions can be detected by analyzing HST (hot spot temperature) or Thermograph [6-12].

The HST is one of the major limiting factors that affect the useful life of the power transformer and its loading. The HST differential is defined by industrial standards (NEMA and ANSI) for each insulation class (type and temperature rating of insulation used on windings). Obviously, the HST is the limiting temperature for a transformer's insulation system [13].

3.2. Condition monitoring by winding vibration analysis

The transformer vibration consists of core vibrations, winding vibrations, on-load tap changer vibrations. The health condition of the core and windings can be assessed using the vibration signature of transformer tank.

Vibration analysis is a very powerful tool for assessing the health of the on load tap changer [14]. Winding vibration is due to the electro dynamic forces caused by the interaction of the current circulating by a winding with the leakage flux. These forces are proportional to the current squared and have components in axial and radial directions. Also, core vibration is caused by magnetostriction and magnetic Forces [15].

Moreover, the mounting place of a measuring transducer on the tank surface of the tested unit does not significantly influence the determined parameters of the registered vibrations and then the analysis of the results obtained in this way [16].

3.3. Condition monitoring by dissolve gas analysis

Dissolved gas analysis is a traditional way for monitoring insulation condition; concentration types and production rates of generated gasses can be used for fault diagnosis. Nevertheless, the concentration of these gases increases in the presence of an abnormality (fault) such as thermal and partial discharge and arcing faults [17]. The combustible gases are produced when insulating oils and cellulose materials are subjected to excessive electrical or thermal stresses. Multiple dissolved gas analysis tests should be taken over time so that the rate of increase of fault gases can be monitored, through which the progress of the fault can be monitored. Fig. 3 shows the classification of dissolve gas analysis.

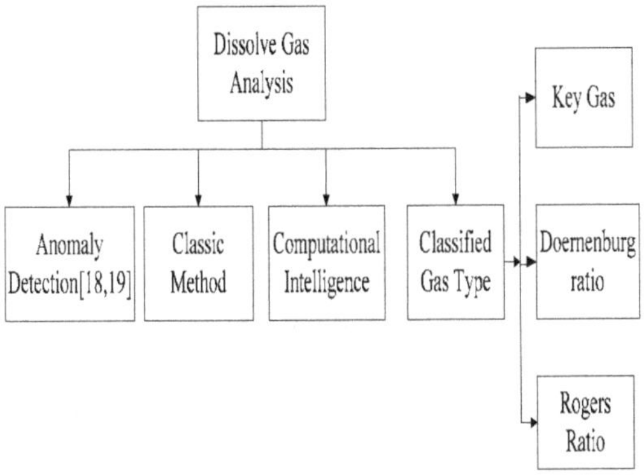

Fig. 3. Classification of dissolved gas analysis activities.

3.3.1. Anomaly detection

Anomaly detection techniques are a way of recognizing changes in plant behaviors. Rather than simply matching patterns of expected faults, a model of behavior specific to each studied transformer can be trained to represent the normal operation of that particular asset. This issue allows for the natural differences between the normal behavior of different transformers and low level of fault behavior can be trained into the model as normal for that unit [18, 19].

3.3.2. Classic methods

In these method, gas ratios and relative proportions of gases are always used to diagnose faults [20, 21]. Currently, DGA is not a science, but an art, subject to variability. IEEE Standard C57.104-2008 describes the key gases, Doernenburg ratios and Rogers ratio method. IEC Standard 60599 introduces the three basic gas ratio methods and the Duval triangle method. All of these classic methods have been generally described in [22-25].

3.3.3. Computational Intelligence (CI)

CI techniques attempt to emulate human and biological reasoning, decision-making, learning and optimization by applying computing techniques that mimic adaptive evolution in living beings. The CI techniques can be either used individually or in combination with other techniques to form complex hybrid methodologies for achieving systems with enhanced capabilities: e.g. a single system can make decisions under uncertainty by fuzzy logic, learn and adapt using artificial neural networks [26, 27], and undergo evolutionary optimization using genetic algorithms, bacterial swarming algorithm, and particle swarm optimization algorithms [28]. Most of the successful attempts for combining techniques have confirmed that fault diagnosis can greatly benefit from CI techniques. Because Condition Based Monitoring (CBM) uses advanced fault diagnostic techniques to identify on-line and off-line incipient faults and to provide real-time transformer conditions, it can also optimize maintenance schedules. CI techniques enable researchers to analyze fault phenomena and use the correlations in data for analyzing faults and diagnosing transformer faults with high accuracy.

3.3.4. Classified gas type

During internal faults, oil produces gases such as H2, CH4 etc. Initially only a hydrogen online monitor was available, but new tools detecting several gases are commercially available with a total-oil monitor recently launched [29-32]. Fig. 4 shows the main categories of classified gas type.

3.3.4.1. Types of faults and generated gases

Stresses due to operation (normal to extreme), ambient conditions, and contamination contribute to the deterioration of the insulation chain and thus shorten the transformer's design life [32, 33]. Table

1 shows the types of faults, generated gases and concentration-generated gases.

3.4. Condition monitoring by winding movement and deformation

When a transformer is subjected to high through fault currents, the windings are subjected to severe mechanical stresses, causing winding movement, deformation, and severe damage in some cases. Such deformities occur due to short circuit forces and sometimes may be caused due to unskilled handling during transportation [34].

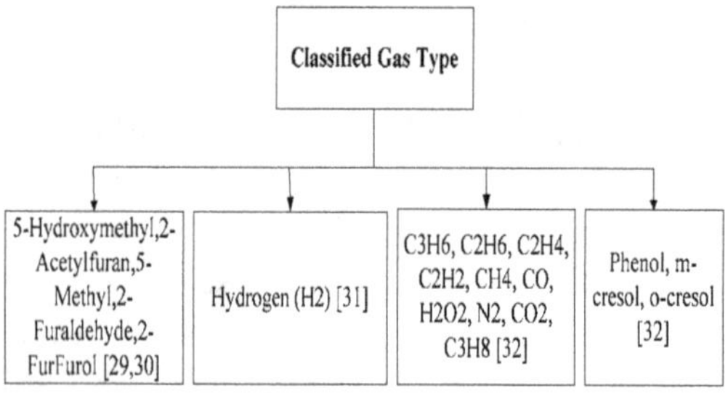

Fig. 4. Classification of gas types.

Table 1. Types of faults and generated gases

Types of Faults	Generated Gases	Concentration (ppm)
Thermal/Oil	C_2H_4, C_2H_6	50-200
Thermal/paper	CO	350-1400
Thermal/paper	CO_2	2500-10000
Electrical/Partial Discharge	H_2, CH_4	120-1000
Electrical/Arcing	H_2, C_2H_2	35-80

Deformation results in relative changes to the internal inductance and capacitance of winding, which can be detected externally by Frequency Response Analysis (FRA) method or transfer function analysis [35, 36]. There are two techniques for FRA measurements on power transformers: Low Voltage Impulse (LVI) based FRA and Sweep Frequency Response Analysis (SFRA), also called Swept Frequency Method (SFM) [37]. Characteristic impedance (Z_C) is more sensitive to changes in the winding structure than SFRA. In addition, the changes in the characteristic impedance signatures take place in different magnitude, ranges whereas the changes in the SFRA all occur in the same magnitude range. The Z_C signature changes most drastically for the radial expansion of the winding which can give the early indication of the winding structure beginning to come apart, especially important toward the end of a transformer's life where the clamps holding the winding in place may begin to fail [38,39]. Fig. 5 shows the classification of winding movement and deformation analysis activities.

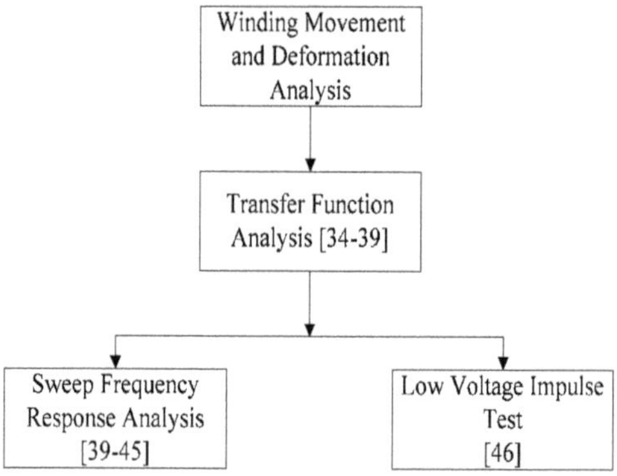

Fig. 5. Classification of winding movement and deformation activities.

3.4.1. Transfer function (TF) analysis

TF method is based on the concept that changes in the windings due to deformation and displacement cause changes in the parameters of transformer (capacitances, inductances,...) and consequently the modification of its TF [35].

Based on a set of TF traces (mainly the amplitude shown over the frequency), an evaluation of the transformers' mechanical condition can be made [35]. Organizations such as IEEE and CIGRE are attempting to develop standards, guidelines and tests for TF method on transformers [37]. For measurements, two test objects have been used: One of them is for the axial displacement modeling and measurements and the other is for the radial deformation tests [38]. All measurements are in the time domain.

In this domain, the test object is excited by low-voltage impulses. The input and output transients are measured and analyzed. Studying the effects of the changes in the parameters of the model on the TFs calculated in the model terminals helps in discovering the correlations between TF variations and fault information. The most important mechanical faults, which are most likely detected using the TF, can be categorized as follows [39-42]:

1) Disc-space variation (DSV)
2) Radial deformation (RD)

3) Axial displacement (AD)

3.4.1.1. Sweep frequency response analysis

Sweep Frequency Response Analysis (SFRA) is an analysis technique for detecting winding displaceme-nt and deformation (among other mechanical and electrical failures) in power and distribution transfor-mers [39]. The SFRA method as a diagnostic tool to detect inters turn winding faults [40, 41]. SFRA involves the injection of sinusoidal signals, one frequency at a time, into one end of the winding [39]. This method has succeeded in detecting mechanical fault conditions in windings where other signatures, such as short circuit impedance and leakage reactance tests [43] have failed to show any difference. Sweep frequency response analysis is diagnosed based on the comparison between two SFRA responses. Any significant difference in low frequency region, shift of existing resonance, creation of new resonance, and change in the shape of plot would potentially indicate mechanical or electrical problem with the winding and core of transformer [41]. It consists of measuring the frequency response of transformer windings over a wide range of frequencies and comparing the results of these measurements using a reference set. examples of fault conditions that can be detected by SFRA are as follows [44, 45]:

Mechanical faults:

i. Winding deformations (including hoop buckling)
ii. Partial collapse of winding
iii. Core displacements
iv. Broken or loosened winding or clamping structure.

Electrical faults:

i. Shorted turns or open circuit winding
ii. Bad ground connection of transformer tank

3.4.1.2. Low voltage impulse (LVI) test

LVI method is adapted from the initial impulse test method. The new technique represents an extension of the traditional low-voltage-impulse method of FRA. Traditional low-voltage-impulse methods have been used to detect mechanical displacements in transformer windings or clamping structures by detecting changes in the transfer function; but, they have been limited in practice due to the sensitivity of the transfer function to differences in the time parameters of the applied impulses as well as sensitivity of the transfer function to variables such as oil condition and temperature. The new technique offers improved repeatability in diagnosing winding movements, and is based on an objective numerical calculation to quantify changes through weighted normalized difference calculation [46].

Low voltage impulse method has the following advantages:

i) Several channels/transfer functions can be measured at the same time; thus, reducing outage time during revision, and
ii) Faster than SFRA. One measurement is usually conducted in one minute.

Low voltage impulse method has the following disadvantages:

i. Fixed frequency resolution resulting in low resolution at low frequencies, which might be a problem for the detection of electrical faults.
ii. Broad band noise cannot be filtered.
iii. Power spectrum of injected signal is frequency-dependent.
iv. Slowly decaying signals are not recorded.
v. Several pieces of equipment are needed.

3.5. Condition monitoring by partial discharge analysis (PD)

Experimental experiences have proved that partial discharges are a major source of insulation failure in power transformers. On the other hand, PD measurements have emerged as an indispensable, non-destructive, sensitive, and powerful diagnostic tool [47]. PD can be detected and measured using Piezo electric Acoustic Emission (AE), Ultra High Frequency (UHF) sensors and optical fiber sensors. PD measurement has been extensively used for the condition assessment of the transformer insulation due to the fact that large numbers of insulation problems start with PD activity.

3.5.1. Piezo electric acoustic emission

Acoustic partial discharge measurement is advantageous in terms of PD source location.

Acoustic sensors mounted on the model experimental transformer tank convert partial discharge acoustic emission (PDAE) signals into electrical signals and are stored on a computer [48].

Acoustic PD detection has following advantages over electrical methods:

i) Acoustic method is Immune to Electromagnetic Interference (EMI); hence, it can be applied for online detection.

ii) Acoustic method can provide an indication of PD source location within a complex system like transformer.

3.5.2. Ultra high frequency

In general, for the CA of a power transformer using UHF PD measuring technique, three steps have to be performed. First of all, detection of any partial discharge activity that indicates harmful insulation defects must be accomplished. After detecting and recording the PD, the analysis of the raw data would provide appropriate information in order to identify the defect in the transformer [49, 50]. The identification of the defect can be done by finding the location of the insulation defect and by comparing its pattern with that of other known defects from a reference database [51].

3.5.3. Fiber optic sensor

Optical fiber transmission is a standard technique for bringing information from high-voltage equipment to monitoring equipment, thus avoiding both hazard to the operator and electromagnetic interference; many examples exist in this regard for both plant management and research. The test of the sensor with real PD shows that it has suitable sensitivity and enough resolution to detect acoustic pressure as low as 1.3-Pa (Pascal). The sensor is found to be able to detect 1-Pa acoustic pressure waves under optimum conditions and to detect partial discharges in small artificial voids in an oil bath [52].

3.6. Condition monitoring by on load tap changer

An On-Load Tap Changer (OLTC) is a part of a power transformer, which allows the transformer output voltage to be regulated at the required levels without interrupting the load current [53]. On-load tap changer by adding or subtracting turns of the windings, regulates the voltage level of the transformer [54]. Thus, the operation of certain on-load tap changers has a significant influence on voltage instability using three objectives of $V_{stability}$, $V_{desired}$ and P_{loss} [55].

The majority of transformer failures are caused by a tap changer fault. Several systems for OLTC online monitoring are currently available [53-55]. However, predicting the online monitoring methods of OLTC can be done by four techniques. Fig. 6 shows the classification of the on load tap changer techniques.

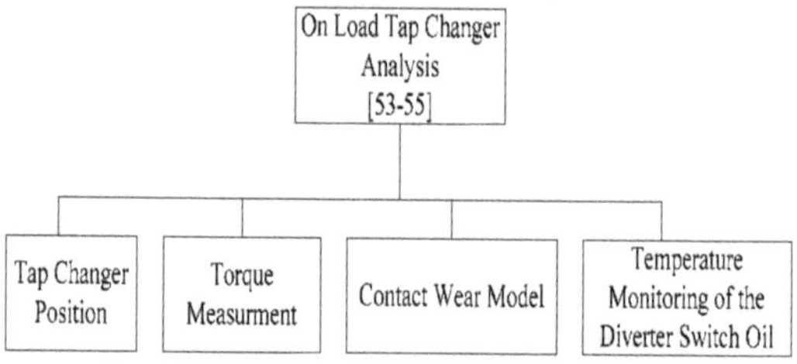

Fig. 6. On load tap changer activities.

4. PERFORMING MAINTENANCE PLANS

Performing maintenance plans is the second transformer asset management activity.

Transformer ageing may also accelerate if the transformer does not undergo proper maintenance and fault diagnosis. Proper fault diagnosis plays a vital role in enhancing the life of a transformer. According to Fig. 7, the maintenance types can be classified into corrective maintenance, preventive maintenance, and reliability centered maintenance.

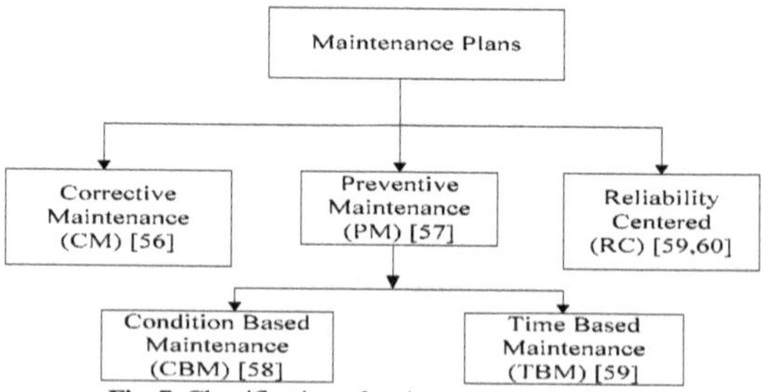

Fig. 7. Classification of maintenance activities.

4.1. Corrective maintenance

Corrective maintenance is designed to perform maintenance activity upon occurrence of failure. The advantages of corrective maintenance are as follow:

i) It saves manpower; *ii*) It spares the system from un-necessary shutdowns; and *iii*) It performs the maintenance only when it is needed, saving unnecessary inspections. In addition, disadvantages of corrective maintenance are:

i) Some transformer failure may be un-repairable if not detected early; and *ii*) Some transformer failure may cause complete shutdown of the production line or the power system for long time [56].

4.2. Preventive maintenance

Preventive maintenance aims to prevent the occurrence of failure. In addition, it aims to guarantee long lifetime of the asset. The preventive maintenance can be classified into condition based maintenance and time based maintenance [57].

4.2.1. Condition based maintenance (CBM)

The main goal is to achieve a cost effective solution through effective asset management. CBM relies on performing maintenance when the CM system detects an incipient fault. Using this technique, the risk of complete failure is reduced.

The advantages of CBM are:

i) Saving costly un-necessary inspections; *ii*) Saving manpower; and *iii*) Reducing the unnecessary shutdowns of the system.

In addition, disadvantages of CBM are:

i) Needing fast data communication and manipulation facilities for successful online monitoring; and *ii*) Being less understood by maintenance engineers and technicians [58].

4.2.2. Time based maintenance (TBM)

TBM is the current maintenance strategy for many industries and utilities. The general meaning of TBM is to perform maintenance in regular intervals. The advantages of the TBM are:

i) It can detect the inception of faults to some extent if the inspection interval is reduced; and *ii*) It increases the lifecycle of the transformer due to regular inspections and maintenance. In addition, its disadvantages are:

i) It is expensive due to regular un-necessary inspections and the large number of the needed maintenance staff; and *ii*) It needs un-necessary shutdowns, which add extra cost to the maintenance activity [59].

4.3. Reliability centered maintenance (RCM)

The fundamental goal of RCM is to preserve the function or operation of a system with a reasonable cost. The general meaning of the RCM maintenance is in optimizing the maintenance plan based on risk analysis [59]. The advantages of RCM are:

i) It guarantees low possibility of occurrence of high-risk failures; *ii*) It saves money paid for unnecessary close timed inspections in case of TBM; and *iii*) It reduces the unnecessary shutdowns for low risk failures. Its disadvantages are:

i) Being less understood by maintenance engineers and technicians; and *ii*) needing for a large amount of data about failure rates, modes, and consequences [60].

5. END OF LIFE ASSESSMENTS AND AGEING

Equipment ageing is a fact of life in power system components. As a piece of equipment ages, it fails more frequently and needs more repair time until reaching its end of life. There are three different concepts of lifetime for power transformers: physical, technical and economic lifetime [59, 60].

5.1. Transformer physical ageing

Transformer life mainly depends on the integrity of its solid insulation (cellulose). The ageing of paper insulation is irreversible [44]. According to Fig. 8, transformers' physical ageing mechanisms can be divided into intransitive and transitive aging.

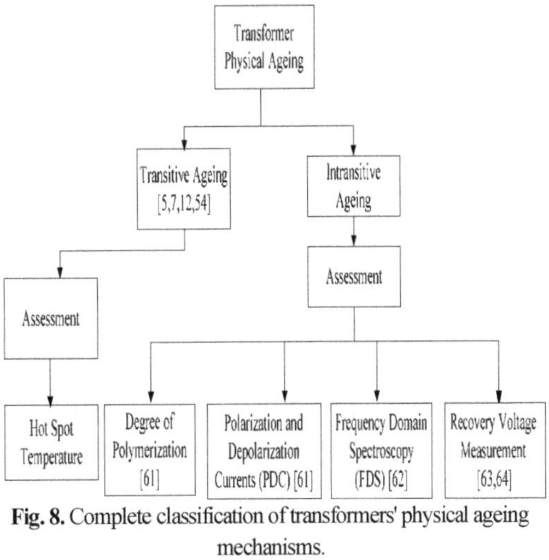

Fig. 8. Complete classification of transformers' physical ageing mechanisms.

5.1.1. Intransitive ageing

The Intransitive Aging is the ability of the solid insulating material to withstand the designed stresses such as electrical, mechanical, and thermal stresses with the passage of time [3].

5.1.1.1. Degree of polymerization (DP)

DP is the main indication of paper health. Paper fibers are composed of cellulose. Glucose monomer molecules are bonded together with glycoside bonds to form cellulose. The average length of the cellulose polymer, measured as the average number of glucose monomers in the polymer chins, is referred to as DP [61]. DP cannot be measured without the opening of transformers. Because of the recent advancement in instrumentation, diagnosing the liquid insulation by Furan detection has been facilitated [61].

5.1.1.2. Polarization/depolarization currents (PDC)

Transformers' polarization/depolarization current measurement method (PDC) is the dielectric response measurement technique in time domain. Owing to the polarization characteristic of the insulation material change with the insulating status, it is possible to realize the non-destructive testing of insulation status using PDC method, which is also a convenient on-site test due to its good noise-proof feature [61].

5.1.1.3. Frequency domain spectroscopy

Frequency Domain Spectroscopy (FDS) measures dielectric loss factor ($tan\delta$) versus frequency. On-line dielectric loss factor $tan\delta$ testing is an extensively used and effective way for insulation aging diagnosis [62]. In Iran, the insulation $tan\delta$ value is being measured under the operating voltage. The structure of this measuring system consists of three parts, of electric bridge method, standard capacitance potentiometer (operating voltage 10kV), and potential transformer (PT). Measurement of the value of $tan\delta$ can reflect a series of defects in insulation and the insulation aging status [62].

5.1.1.4. Recovery voltage measurement

The specific method of Recovery Voltage Measurement called RVM is widely spread in commercial terms [63]. It is used for measuring transformers' oil-paper polarization frequency spectrum [64].

5.1.2. Transitive ageing

It is rapid aging of the asset when subjected to abnormal condition. The abnormal conditions may be overloading, supplying non-sinusoidal loads, or exposure to higher ambient temperature than normal. The main reason for accelerating the end of life of a transformer under the above-mentioned abnormal conditions is its increased HST over the normally accepted values [5, 7], which has an effect on reducing insulation life [12, 54].

5.2. Mechanisms and modeling of economic ageing

This method develops a more systematic approach for determining the life expectancy of transformers. It is based on the economic analysis of the operational characteristics of transformers in conjunction with the technical issues involved in the decision process. The annual lost part of the asset cost is called depreciation cost. The time-based depreciation has two main types: Straight-line depreciation method and accelerated depreciation method. The works [65-68] show these methods.

6. LIFE CYCLE COST

Life Cycle Cost (LCC) refers to all expenses paid in the demonstration, development, production, operation, maintenance, security and post-processing of the equipment in the life cycle [69]. The concept of LCC means to take into account not only the manufacturing cost, but also to consider the operational and disposal costs [70]. Also, LCC analysis structurally deciding and equalizing costs arising within overall life cycle which can be referred to as capital cost and operating cost is a method of analysis essential in economic evaluation [71].

7. CONCLUSIONS

This literature survey is presented a comprehensive overview in the field of fault detection techniques for power transformers, which has evolved rapidly during the last ten years. In addition, advantages and disadvantages of all the techniques and their related applications in detecting transformers' fault were separately explained.

The potential functions of failure prediction and life estimation bring a series of advantages for utility companies: reducing maintenance cost, lengthening equipment's life, enhancing safety of operators, minimizing accident and severity of destruction, as well as improving power quality.

The development of CM for power transformers is now at different stages. Several types of transformer monitoring systems have been already put into practice. However, monitoring and data analysis methods are not satisfactory in terms of special problems such as hot spot temperature, DGA and OLTC analysis.

Research in recent years have clearly shown that advanced signal processing techniques and computational intelligent techniques such as evolutionnary algorithm, analysis of vibroacoustic signals, SFRA and etc are indispensable in developing novel CM systems. Finally, these studies and methods could be effective in terms of timely counteraction to threatening failures.

REFERENCES

[1] W. Tang, and Q. H. Wu, "Condition monitoring and assessment of power transformers using computational intelligence," *International Journal of Electrical Power & Energy Systems*, vol. 33, pp. 1784-1785, 2011.

[2] O. Ristic and V. Mijailovic, "Method for determining optimal power transformers exploitation strategy," *Electric Power Systems Research*, vol. 83, no.1, pp. 255-261, 2012.

[3] M. Hui, T.K. Saha and C. Ekanayake, "Statistical learning techniques and their applications for condition assessment of power transformer," *IEEE Transactions on Dielectrics and Electrical Insulation*, vol. 19, no. 2, pp. 481-489, 2012.

[4] G.J.C. Olivares, A.S. Magdaleno, L.E. Campero, R.P. Escarela and P. S. Georgilakis, "Techno-economic evaluation of reduction of low-voltage bushings diameter in single-phase distribution transformers," *Electric Power Component and Systems*, vol. 39, no. 13, pp. 1388-1402, 2011.

[5] Sh. Taheri, A. Gholami, I. Fofana and H. Taheri, "Modeling and simulation of transformer loading capability and hot spot temperature under harmonic conditions," *Electric Power Systems Research*, vol. 86, pp. 68-75, 2012.

[6] Y. Tamsir, A. Pharmatrisanti, H. Gumilang, B. Cahyono and R. Siregar, "Evaluation condition of transformer based on infrared thermography results," *Proceedings of the IEEE 9th International Conference on the Properties and Applications of Dielectric Materials*, pp. 1055-1058, 2009.

[7] N.S. Beniwal, D.K. Dwivedi and H.O. Gupta, "Creep life assessment of distribution transformers," *Engineering Failure Analysis*, vol. 17, no. 5, pp. 1077-1085, 2011.

[8] M. Lee, A. Abdullah, J.C. Jofriet and D. Patel, "Temperature distribution in foil winding for ventilated dry-type power transformers," *Electric Power Systems Research*, vol. 80, no. 9, pp. 1065-1073, 2010.

[9] B.R. Sathyanarayana, G.T. Heydt and M. L. Dyer, "Distribution transformer life assessment with ambient temperature rise projections," *Electric Power Component Systems*, vol. 37, no. 9, pp. 1005-1013, 2009.

[10] M.A.G. Martins and A.R. Gomes, "Comparative study of the thermal degradation of synthetic and natural esters and mineral oil: effect of oil type in the thermal degradation of Insulating kraft paper," *IEEE Electrical Insulation Magazine*, vol. 28, no. 2, pp. 22-28, 2012.

[11] Ch. Sumi and J. Kuwabara, "Determination of thermal conductivity distribution from internal temperature distribution measurements," *Review of Scientific Instruments*, vol. 77, no. 6, pp. 904-908, 2009.

[12] E.I. Koufakis, C.G. Karagiannopoulos and P.D. Bourkas, "Thermal coefficient measurements of the insulation in distribution transformers of a 20 kV network," *Science Direct Measurement*, vol. 41, no. 1, pp. 10-19, 2008.

[13] A.T. Plesca ,"Thermal analysis of overload protection relays using finite element method," *Indian Journal of Science and Technology*, vol. 6, no. 8, pp. 5120-5125, 2013.

[14] H. Sun, Y. Huang and C. Huang, "Fault diagnosis of power transformers using computational intelligence: a review," *Proceedings of the 2nd International Conference on Advances in Energy Engineering*, vol. 14, pp. 1226-1231, 2012.

[15] B. Garcia, J.C. Burgos and A. Alonso,"Winding deformations detection in power transformers by tank vibrations monitoring," *Electric Power Systems Research*, vol. 74, no. 1, pp. 129-138, 2005.

[16] S. Borucki, "Diagnosis of Technical Condition of Power Transformers Based on the Analysis of Vibroacoustic Signals Measured in Transient Operating Conditions," *IEEE Transactions on Power Delivery*, vol. 27, no.2, pp. 670-676, 2012.

[17] Kh. Bach, S. Souahli and M. Goss, "Power transformer fault diagnosis based on dissolved gas analysis by support vector machine," *Electric Power Systems Research*, vol. 83, no. 1, pp. 73-79, 2012.

[18] J.L. Velasquez-Contreras, M.A. Sanz-Bobi and S.G. Arellano, "General asset management model in the context of an electric utility: application to power transformers," *Electric Power Systems Research*, vol. 81, no. 11, pp. 2015-2037, 2011.

[19] M. Victoria, S.D. Catterson, D.J. McArthur and M. Graham, "Online conditional anomaly detection in multivariate data for transformer monitoring," *IEEE Transactions on Power Delivery*, vol. 25, no. 4, pp. 2556-2564, 2010.

[20] L. Xiaohui, W. Huaren and W. Danning, "DGA interpretation scheme derived from case study," *IEEE Transactions on Power Delivery*, vol. 26, no. 2, pp. 1292-1293, 2011.

[21] M. Daniel, L. Nick, G. Wenyu and O. Yuriy, "Further studies of a vegetable oil filled power transformer," *IEEE Electrical Insulation Magazine*, vol. 27, no. 5,pp. 6-13, 2011.

[22] L. Nick, M. Daniel, G. Wenyu and W. Jaury, "Comparison of dissolved gas-in-oil analysis methods using a dissolved gas-in-oil standard," *IEEE Electrical Insulation Magazine*, vol. 27, no. 5, pp. 29-35, 2011.

[23] M. Daniel, L. Nick, D. Valery and O. Yuriy, "Preliminary results for dissolved gas levels in a vegetable oil–filled power transformer," *IEEE Electrical Insulation Magazine*, vol. 26, no. 5, pp. 41-48, 2010.

[24] H.A. Ivanka and F. Rainer, "Carbon oxides in the interpretation of dissolved gas analysis in transformers and tap changers," *IEEE Electrical Insulation Magazine*, vol. 26, no. 6, pp. 22-26, 2010.

[25] S. Sukhbir and M.N. Bandyopadhyay, "Dissolved gas analysis technique for incipient fault diagnosis in power transformers: a bibliographic survey," *IEEE Electrical Insulation Magazine*, vol. 26, no. 6, pp. 41-46, 2010.

[26] Z. Moravej, D.N. Vishwakarma and S.P. Singh, "Protection and conditions monitoring of power transformer using ANN," *Electric Power Component and Systems*, vol. 30, no. 3, pp. 217-231, 2002.

[27] Z. Moravej, "Evolving Neural Nets for Protection and Condition Monitoring of Power Transformer," *Electric Power Component Systems*, vol. 33, no. 1, pp. 1229-1236, 2005.

[28] M. Darabian, A. Jalilvand and R. Noroozian, "Combined use of sensitivity analysis and hybrid Wavelet-PSOANFIS to Improve Dynamic performance of DFIG-based wind generation," *Journal of Operation and Automation in Power Engineering*, vol. 2, no. 1, pp. 60-73, 2014.

[29] L. Cheim, D. Platts, T. Prevost and X. Shuzhen, "Furan analysis for liquid power transformers," *IEEE Electrical Insulation Magazine*, vol. 28, no. 2, pp. 8-21, 2012.

[30] G. Ma, Ch. Li, Y. Luo, R. Mu and L. Wang, "High sensitive and reliable fiber bragg grating hydrogen sensor for fault detection of power transformer," *Sensors and Actuators B: Chemical*, vol. 169, pp. 195-198, 2012.

[31] L. Yangliu and Zh. Xuezeng, "Estimation of dissolved gas concentrations in transformer oil from membrane," *IEEE Electrical Insulation Magazine*, vol. 27, no. 2, pp. 30-33, 2011.

[32] D. Vrsaljko, V. Haramija and A. Hadzi-Skerlev, "Determination of phenol, m-cresol and o-cresol in transformer oil by HPLC method," *Electric Power Systems Research*, vol. 93, pp. 24-31, 2012.

[33] A. Abu-Siada and S. Islam, "A new approach to identify power transformer criticality and asset management decision based on dissolved gas-in-oil analysis," *IEEE Transactions on Dielectrics and Electrical Insulation*, vol. 19, no. 3, pp.1007-1012, 2012.

[34] K. Pourhossein, G. B. Gharehpetian, E. Rahimpour and B.N. Araabi, "A probabilistic features to determine type and extent of winding mechanical defects in power transformers," *Electric Power Systems Research*, vol. 82, no. 1, pp. 1-10, 2012.

[35] R. Aradhana ,"Application of signal processing techniques for condition monitoring of electrical equipment," Ph.D. dissertation, *Electrical Engineering the Maharaja Sayajirao University*, Barada, India, 2009.

[36] R. Ebrahim, J. Mehdi and T. Stefan, "Mathematical comparison methods to assess transfer functions of transformers to detect different types of mechanical faults," *IEEE Transactions on Power Delivery*, vol.

25, no. 4, pp. 2544-2555, 2010.

[37] A. Singh, F. Castellanos, J.R. Marti and K.D. Srivastava ,"A comparison of trans-admittance and characteristic impedance as metrics for detection of winding displacements in power transformers," *Electric Power Systems Research*, vol. 79, no. 6, pp. 871-877, 2009.

[38] S. Bagheri, R. Effatnejad and A. Salami, "Transformer winding parameter identification based on frequency response analysis using hybrid wavelet transform (WT) and simulated annealing algorithm (SA) and compare with genetic algorithm (GA)," *Indian Journal of Science and Technology*, vol. 7, no. 5, pp. 614-621, 2014.

[39] J.R. Secue and E. Mombello, "Sweep frequency response analysis (SFRA) for the assessment of winding displacements and deformation in power transformers," *Electric Power Systems Research*, vol. 78, no. 6, pp. 1119-1128, 2008.

[40] T.Y. Ji, W.H. Tang and Q.H. Wu, "Detection of power transformer winding deformation and variation of measurement connections using a hybrid winding model," *Electric Power Systems Research*, vol. 87, pp. 39-6, 2012.

[41] K. Pourhossein, G.B. Gharehpetian, E. Rahimpour and B.N. Araabi, "A vector-based approach to discriminate radial deformation and axial displacement of transformer winding and determine defect dxtent," *Electric Power Component and Systems*, vol. 40, no. 6, pp. 597-612, 2012.

[42] E. Kornatowski and S. Banaszak, "Diagnostics of a transformer's active part with complementary FRA and VM measurements," *IEEE Transactions on Power Delivery*, vol. 29, no. 3, pp. 1398-1406, 2014.

[43] L. Satish and S.K. Sahoo, "Locating faults in a transformer winding: an experimental study," *Electric Power Systems Research*, vol. 79, no. 1, pp. 89-97, 2009.

[44] V. Behjat, A. Vahedi, A. Setayeshmehr, H. Borsi and E. Gockenbach, "Sweep frequency response analysis for diagnosis of low level short circuit faults on the windings of power transformers: an experimental study," *Electric Power Systems Research*, vol. 42, no.1, pp. 78-90, 2012.

[45] A. Shintemirov, W.J. Tang, W. H. Tang and Q.H. Wu, " Improved modeling of power transformer winding using bacterial swarming algorithm and frequency response analysis," *Electric Power Systems Research*, vol. 80, no. 9, pp. 1111-1120, 2010.

[46] S.K. Sahoo and L. Satish, "Discriminating changes introduced in the model for the winding of a transformer based on measurements," *Electric Power Systems Research*, vol. 77, no. 7, pp. 851-858, 2007.

[47] M.S. Naderi, T.R. Blackburn, B.T. Phung, M.S. Naderi, R. Ghaemmaghami and A. Nasiri, " Determ-ination of partial discharge propagation and location in transformer windings using a hybrid transformer model," *Electric Power Component and Systems*, vol. 35, no. 6, pp. 607-623, 2007.

[48] P. Kundu, N. K. Kishore and A. K. Sinha, "A non-iterative partial discharge source location method for transformers employing acoustic emission techniques," *Applied Acoustics*, vol. 70, no. 11-12, pp. 1378-1383, 2007.

[49] R. Sarathi, A.J. Reid and M.D. Judd, "Partial discharge study in transformer oil due to particle movement under DC voltage using the UHF technique," *Electric Power Systems Research*, vol. 78, no. 11, pp. 1819-1825, 2008.

[50] M.A. Hejazi, G.B. Gharehpetian, G.R. Moradi, M. Mohammadi and H.A. Alehoseini, "Application of classifiers for on-line monitoring of transformer winding axial displacement by electromagnetic nondestructive testing," *Electric Power Component and Systems*, vol. 39, no. 4, pp. 387-403, 2011.

[51] P. Agoris, P. Cichecki, S. Meijer and J.J. Smit, "Building a transformer defects database for UHF partial discharge diagnostics," *Proceedings of the IEEE conference on Lausanne on Power Tech*, pp. 2070-2075, 2007.

[52] I. Bua-Nunez, J.E. Posada-Roman, J. Rubio-Serrano and J.A. Garcia-Souto, "Instrumentation system for location of partial discharges using acoustic detection with piezoelectric transducers and optical fiber sensors," *IEEE Transactions on Instrumentation and Measurement*, vol. 63, no. 5, pp. 1002-1013, 2014.

[53] E. R. Trujillo, E.L. Jácome, J.L. San-Románb and V. Díaz, "Characterizing the diverter switch of a load tap changer in a transformer using wavelet and modal analysis," *Engineering Structures*, vol. 32, no. 10, pp. 3011-3017, 2010.

[54] M. Sefidgaran, M. Mirzaie and A. Ebrahimzadeh, "Reliability model of the power transformer with ONAF cooling," *International Journal of Electrical Power & Energy Systems*, vol. 35, no. 1, pp. 97-104, 2012.

[55] G. Yesuratnam and D. Thukaram, "Optimum reactive power dispatch and Identification of critical on-load tap changing (OLTC) transformers," *Electric Power Component and System*, vol. 35, no. 6, pp. 655-674, 2007.

[56] Ch.Y. Lee, H.Ch. Chang and Ch. Liu, "Emergency dispatch strategy considering remaining lives of transformers," *IEEE Transactions on Power Systems*, vol. 22, no. 4, pp. 2066-2073, 2007.

[57] V. Mijailovic, "Method for effects evaluation of some forms of power transformers preventive maintenance," *Electric Power Systems Research*, vol. 78, no. 5, pp. 765-776, 2008.

[58] X. Zhang and E. Gockenbach, "Age-dependent maintenance strategies of medium-voltage circuit-breakers and transformers," *Electric Power Systems Research*, vol. 81, no. 8, pp. 1709-1714, 2011.

[59] M.A. Martins, M. Fialho, J. Martins, M. Soares, M. Cristina, R. Castro Lopes and H.M.R. Campelo, "Power transformer end-of-life assessment pracana case study," *IEEE Electrical Insulation Magazine*, vol. 27, no. 6, pp. 15-26, 2011.

[60] C. Yang, H. Shi and T. Liu , "Aging and life assessment of large and medium-sized power transformers in nuclear power plants," *Proceedings of the Second International Confe-rence* on *Intelligent System Design and Engineering Application*, pp. 768-772, 2012.

[61] T. Leibfried, M. Jaya, N. Majer, M. Schafer, M. Stach and S. Voss," Postmortem investigation of power transformers rofile of degree of polymerization and correlation with furan foncentration in the oil," *IEEE Transactions on Power Delivery*, vol. 28, no. 2, pp. 886-893, 2013.

[62] H. Shuai, L. Qingmin, L. Chengrong and Y. Jiangyan, "Electrical and mechanical properties of the oil-paper insulation under stress of the hot spot temperature," *IEEE Transactions on Dielectrics and Electrical Insulation*, vol. 21, no. 1, pp. 179-185, 2014.

[63] V. Aschenbrenner and T. UEik, "Using of parameters of RVM measurement for qualitative appreciation of power transformers insulation state," *Proceedings of the IEEE/PES Transmission and Distribution Conference and Exhibition.*, pp. 1829-1833, 2002.

[64] H. ITO, H. Kajino, Y. Yamagata and K. Kamei, "Study on transient recovery voltages for transformer-limited faults," *IEEE Transactions on Power Delivery*, vol. 29, no. 5, pp. 2375-2384, 2014.

[65] G. Ji, W. Wu, B. Zhang and H. Sun, "A renewal-process-based component outage model considering the effects of aging and maintenance," *International Journal of Electrical Power & Energy Systems*, vol. 44, no. 1, pp. 52-59, 2013.

[66] W.H. Xing, Y.Q. Ping and Z.Q. Ming, "Artificial neural network for transformer insulation aging diagnosis," *Proceedings of the Third International Conference on Electric Utility Deregulation and Restructuring and Power Technologies*, pp. 2233-2238, 2008.

[67] A.E.B. Abu-Elanien, M.M.A. Salam and R. Bartnikas, "A techno-economic method for replacing transformers," *IEEE Transactions on Power Delivery*, vol. 26, no. 2, pp. 817-829, 2011.

[68] M. Yazdani-Asrami, M. Mirzaie and A. S. Akmal, "Investigation on impact of current harmonic contents on the distribution transformer losses and remaining life," *Proceedings of the IEEE International Conference on Power and Energy*, pp. 689-694, 2010.

[69] L.Li , W. Huang, X.L. Ma, J.B. Ge and L.J. Yang, "Investment decision-making of power distribution transformers transformation based on life cycle cost theory," *Proceedings of the China International Conference on Electricity Distribution*, pp. 1-6, 2010.

[70] I. Jeromin, G. Balzer, J. Backes and R. Huber, "Life cycle cost analysis of transmission and distribution systems," *Proceedings of the IEEE conference in Bucharest on Power Tech*, pp. 1-6, 2009.

S.H. Lee, A.K. Lee and J.O. Kim, "Determining economic life cycle for power transformer based on life cycle cost analysis," *Proceedings of the IEEE International Power Modulator and High Voltage Conference*, pp. 604-607, 2012.

6

Optimal Reconfiguration and Capacitor Allocation in Radial Distribution Systems Using the Hybrid Shuffled Frog Leaping Algorithm in the Fuzzy Framework

M. Sedighizadeh[1,*], M.M. Mahmoodi[2]

[1]Faculty of Electrical and Computer Engineering, Shahid Beheshti University, Tehran, Iran

[2]Department of Electrical Engineering, College of Engineering, Saveh Branch, Islamic Azad University, Saveh, Iran

ABSTRACT

In distribution systems, network reconfiguration and capacitor placement are commonly used to diminish power losses and keep voltage profiles within acceptable limits. In this paper, the Hybrid Shuffled Frog Leaping Algorithm (HSFLA) has been used to optimize the balanced and unbalanced radial distribution systems using a network reconfiguration and capacitor placement. High accuracy and fast convergence are the major advantages of the proposed approach regarding the result of solving the multi-objective reconfiguration and capacitor placement in a fuzzy framework. These objectives are minimizing the total network real power losses and buses voltage violation, and balancing the load in the feeders. Each objective is transferred into fuzzy domain using its membership function and fuzzified separately. Then, the overall fuzzy satisfaction function is formed and considered as a fitness function. The value of this function has to be maximized to gain the optimal solution. In the literature review, several reconfiguration and capacitor placement methods which had already been implemented separately have been investigated, but there are few studies which simultaneously apply these two methods. The proposed algorithm has been implemented in three IEEE test systems (two balanced and one unbalanced systems).The numerical results obtained by the simulation carried out in this study show that the HSFLA algorithm improves the performance much more than other meta-heuristic algorithms.

KEYWORDS: Artificial intelligence, Distribution systems, Multi-objective optimization, Optimal reconfiguration and capacitor placement, SFLA

1. INTRODUCTION

It is a common application to use capacitors in power systems in order to compensate for reactive power losses as well as to provide a good voltage profile by preventing occurrence of under- or over-voltages. An issue of exploiting maximum advantage of compensation effect of capacitors is the size and location of these components. On the basis of switches used in power systems there are two types of these devices called normally closed switches (sectionalizing switches) and normally open switches (tie switches), which by applying either, topology of system may be changed. The change happens when altering the status of these switches from open or closed, and by this way, feeder is reconfigured due to the change in topology and configuration of distribution systems. Regarding this, the need to optimally reconfigure network and find the optimum placement of capacitors have raised and separately been investigated in many papers. For solving the aforementioned problem associated with reconfiguration of feeder and finding the optimum placement for capacitors, many different methods have been used with various objective functions and optimization theories.

Recently there have been so many algorithms developed for different goals including power loss

*Corresponding author:
M. Sedighizadeh (m_sedighi@sbu.ac.ir)

reduction and major utilization factors using reconfiguration of distribution systems, most of which based on artificial intelligence methods and heuristic techniques. Examples of studies focused on reconfiguration of network can be mentioned as following: in [1], a new meta-heuristics fireworks algorithm was proposed to optimize the radial distribution network while satisfying the operating constraints. Ref. [2] presents a step-by-step heuristic algorithm for the reconfiguration of radial electrical distribution systems, aiming at power loss minimization, based on a dynamic switches set approach, which is updated due to topological changes in the electrical network and to avoid the premature convergence of the algorithm in suboptimal solutions. A method to improve the power quality and reliability of distribution systems by employing optimal network reconfiguration was presented in [3], which was applied independently to a system in a specified period to minimize the number of propagated voltage sags and other reliability indexes. The quantum-inspired binary firefly algorithm is used to find the optimal NR.

In [4] a modified Tabu Search (MTS) algorithm is used to reconfigure distribution systems so that active power losses are globally minimized with turning on/off sectionalizing switches. TS algorithm is introduced with some modifications such as using a tabu list with variable size according to the system size. A salient feature of the MTS method is that it can quickly provide a global optimal or near-optimal solution to the network reconfiguration problem. A methodology for the reconfiguration of radial electrical distribution systems based on the bio-inspired meta-heuristic artificial immune system to minimize energy losses was presented in [5], in which radiality and connectivity constraints were considered as well as different load levels for planning the system operation. In [6] an efficient hybrid big bang–big crunch optimization algorithm to solve the multi-objective reconfiguration of balanced and unbalanced distribution systems in a fuzzy framework was rpresented. The objectives considered were the minimization of total real power losses, the minimization of buses voltage deviation, and load balancing in the feeders. In [7] allocation of power losses to consumers connected to radial distribution networks before and after network reconfiguration in a deregulated environment was reported. The network reconfiguration algorithm is based on the fuzzy multi-objective approach and the max-min principle was adopted for the multi-objective optimization in a fuzzy framework. Multiple objectives were considered for real-power loss reduction in which nodes voltage deviation is kept within a range, and an absolute value of branch currents is not allowed to exceed their rated capacities. An adapted ant colony optimization for the reconfiguration of radial distribution systems with minimizing real power loss was used in [8] that conventional ant colony optimization was adapted by the graph theory to always create feasible radial topologies during the whole evolutionary process which avoids tedious mesh check and hence reduces the computational burden. In [9] size and location of FACTs devices in a power system are calculated and a Dedicated Improved Particle Swarm Optimization (DIPSO) algorithm was developed for decreasing the overall costs of power generation and maximizing of profit.

There are many articles that have presented wide researches on the capacitor placement problem for reduction of losses in power distribution systems. For instance, an approach based on fuzzy method was proposed in [10]. For determining the location, size and number of capacitor banks in distribution systems a mixed integer LP model was reported in [11]. For loss reduction in [12] a two stage method was used for formulation and the optimal operation status of the devices by applying a genetic algorithm. To solve the problem of capacitor placement, Ref. [13] was applied a single objective probabilistic optimal allocation, and in [14] for optimal placement of capacitors in order to reduce harmonic distortion, a honey bee foraging approach was used. In [15] a hybrid optimization algorithm for the optimal placement of shunt capacitor banks in radial distribution networks was used in the presence of different voltage-dependent load models, which the algorithm was based on the combination of genetic algorithm and binary particle swarm optimization algorithm. Optimal capacitor allocation and sizing using big bang big crunch optimization algorithm is represented in [16].

Papers, which work with both capacitor placement and network reconfiguration at the same time, are reviewed now. Zhang et al. in [17] treated capacitor placement and reconfiguration by using Improved Adaptive Genetic Algorithm (IAGA) and a simplified branch exchange algorithm, respectively. Farahani et al. in [18] solved the reconfiguration problem by using simple branch exchange method and the outcome was that loops selection sequence is an affecting factor which has effects on network loss as well as optimal configuration and also proposed a new algorithm for combining improved method of reconfiguration and capacitor placement, in which for optimizing the location and size of capacitors and sequence of loops selection, discrete genetic algorithm (GA) was used. Chung-Fu Chang in [19] worked on ant colony search algorithm and used it as a solver for the problems of feeder reconfiguration optimization and capacitor placement simultaneo-usly. Montoya et al. in [20] by using a minimum spanning tree algorithm determined the minimum losses optimum configuration in reconfiguration problem and utilized GA to obtain the greatest savings through the problem of optimal capacitor problem. Guimara˜es et al. in [21] used a modified dedicated approach based on GA. Development and implementation of this algorithm was successful, as well, it has low computations and was capable of obtaining appropriate configurations. In [22] based on a new Improved Binary PSO (IBPSO) algorithm, some suggestions for planning priority associated with problems of capacitor placement and reconfiguration in distribution systems are being investigated. This suggested method applies a new structure in order to obtain an optimization for the aforementioned problem.

The proposed method is to use an efficient Hybrid Shuffled Frog Leaping Algorithm (SFLA) [23] associated with fuzzy objective function to get a proper solution for the problems of feeder reconfiguration and capacitor placements at the same time. The objective functions which have been considered in this paper are the minimization of total real power losses and bus's voltage violation as well as load balancing in the feeders. One of the commonly used methods to increase loading capability of system, dwindling real power losses and reducing voltage drops is load balancing. The main objectives considered in this paper are to obtain maximum reduction of loss, present an in-limits-maintained voltage profile, and have the current in each branch maintained within the capacity limits of the branch. The first step, is to use trapezoidal fuzzy membership function in order to transfer the objectives to fuzzy domain and fuzzify them separately. The second step is to develop the overall fuzzy satisfaction function. We will see in the next step that this function is considered as an overall fitness function and the value of it will increase until it reaches to maximum value. The test system for the suggested method is a balanced 33 and 94-bus and an unbalanced 25-bus distribution system. In comparison with PSO and IPSO and other various algorithms, the suggested HSFLA has a better efficiency which is verified by numerical results.

2. FUZZY MULTI-OBJECTIVE FORMULATION

Since the objective functions have different dimensions, for easier comparison a fuzzy multi-objective approach is used. In fuzzy domain, a membership function is defined for each objective which represents the degree of fuzzy satisfaction of the objective. The membership value of each objective is a real number between 0 and 1 and in this section is determined by using the trapezoidal fuzzy membership function. In this paper, power losses minimization, minimizing the buses voltage deviation and load balancing in the feeders are considered as the objectives and fuzzified as explained below.

2.1 Membership function for the real power loss (λP_i)

Mathematically, the real power loss in the network can be formulated as follows:

$$P_{loss} = \sum_{i=1}^{N_{br}} R_i \frac{P_i^2 + Q_i^2}{|U_m|^2} \tag{1}$$

The voltage magnitude at each bus must remain within its permissible intervals. On the other hand, the current of each branch must satisfy the branch current limitations. Therefore:

$$U_{min} \leq |U_m| \leq U_{max} \quad (2)$$

$$|I_i| \leq I_{i,max} \quad (3)$$

Where R_i, P_i and Qi are, the branch resistance, real and reactive power flows through branch i respectively, and U_m is the voltage at bus m and N_{br} is the total number of branches in the system. U_{min} and U_{max} are minimum and maximum allowable voltages, respectively, which are considered as $U_{min} = 0.95$ and $U_{max} = 1.05$. The following index `for the power loss minimization is defined as follows [6]:

$$XP_i = \frac{P_{lossi}}{P_{loss0}} \quad (4)$$

where, P_{loss0} represents the initial real power loss before reconfiguration and capacitor placement of the network and P_{lossi} represents the real power loss after reconfiguration and capacitor placement in ith radial system.

The degree of fuzzy satisfaction of power loss objective function can be determined using the membership function as defined in fuzzy domain. The membership function is expressed as follows:

$$\lambda P_i = \begin{cases} 1 & XP_i < XP_{min} \\ \dfrac{XP_{max} - XP_i}{XP_{max} - XP_{min}} & XP_{min} < XP_i < XP_{max} \\ 0 & XP_i > XP_{max} \end{cases} \quad (5)$$

where, XP_{min} and XP_{max} are the lower and upper limits of XP_i index, respectively. To determine the XP_{min} and XP_{max}, the best and the worst system configuration for real power losses is considered. P_{lossi} for the best system configuration is minimum value of the power loss and for the worst system configuration is assumed to be equal with the power loss of the initial configuration.

2.2 Membership function for maximum bus voltage violation (λU_i)

For the purpose of minimizing the bus voltage deviation, the index of XUi is defined as follows:

$$XU_i = max\left(\left|1 - U_{min}\right| and \left|1 - U_{max}\right|\right) \quad (6)$$

where, U_{min} and U_{max} are the minimum and maximum values of bus voltage respectively. Membership function of maximum bus voltage deviation index is formulated as follows[6]:

$$\lambda U_i = \begin{cases} 1 & XU_i \leq XU_{\min} \\ \dfrac{XU_{\max} - XU_i}{XU_{\max} - XU_{\min}} & \mathrm{XU}_{\min} < XU_i < XU_{\max} \\ 0 & \mathrm{XU}_i \geq XU_{\max} \end{cases} \quad (7)$$

where, XU_{min} and XU_{max} are the lower and upper limits of XU_i index, respectively. To determine the XU_{min} and XU_{max}, the best and the worst system configuration is considered for minimum and maximum bus voltage deviation, respectively.

2.3 Membership function for load balancing index (LBI) (λI_i)

For the purpose of load balancing, first an appropriate parameter is defined, indicating what portion of the branches has been loaded. This portion is defined as the line usage index for the ith branch, calculated as follows [24]:

$$Line\,Usage\,Index = \frac{I_i}{I_i^{max}} \quad (8)$$

where, I_i^{max} is the maximum current capacity of the ith branch of the system. For all the branches of the system LBI index is calculated as follows:

$$Y = \left[\frac{I_1}{I_1^{max}} \frac{I_2}{I_2^{max}} \frac{I_3}{I_3^{max}} \quad \cdots\cdots \quad \frac{I_N}{I_{N_{br}}^{max}} \right] \quad (9)$$

$$LBI = Var(Y) \quad (10)$$

where, Var represents the variance operation. However, the smaller value of the LBI index indicates that the load balancing has been conducted more efficiently. In the next stage, the index of XB_i for load balancing is defined as:

$$XB_i = \frac{LBI_i}{LBI_0} \quad (11)$$

where, LBI_0 is the load balancing before network reconfiguration and capacitor placement, calculated in initial power flow for each case study, and LBI_i is the load balancing of the ith radial system after reconfiguration and capacitor placement.

Membership function of feeder load balancing index is formulated as follows:

$$\lambda I_i = \begin{cases} 1 & XB_i \leq XB_{min} \\ \dfrac{XB_{max} - XB_i}{XB_{max} - XB_{min}} & XB_{min} < XB_i < XB_{max} \\ 0 & XB_i \geq XB_{max} \end{cases} \quad (12)$$

where, XB_{min} and XB_{max} are the lower and upper limits of XB_i, respectively. To determine the XB_{min} and XB_{max}, the best and the worst system configuration is considered for feeder load balancing.

In the proposed algorithm, the worst system configuration is considered to be the initial configuration of system before reconfiguration and capacitor placement, and the best system configuration after reconfiguration and capacitor placement is obtained by optimizing each objective separately.

2.4 Degree of overall fuzzy satisfaction (λO_i)

The idea of multi objective function is proposed for the following purposes:

- Finding the best and most compatible system configuration satisfying every objectives.
- Satisfying operational limits such as voltage and current constraints and also preventing load islanding.

In this paper, a new operator named "max-geometric mean" is utilized to determine the degree of overall fuzzy satisfaction in the proposed method. This operator is expressed as follows [25]:

$$\lambda O_i = \left(\lambda P_i \times \lambda U_i \times \lambda I_i \right)^{\frac{1}{3}} \quad (13)$$

where, λO_i in the HSFLA is considered as the fitness function, maximized during the optimization process to obtain the best compatible configuration. This operator has several advantages. For instance, if any membership function of each objective reaches the value of zero, λO_i is assigned a value of zero. Furthermore, this function provides correct information as about how to make this algorithm achieving an ideal state, namely a value of 1.

3. SHUFFLED FROG LEAPING ALGORITHM

3.1. Original algorithm

The first thing to do in SFLA is to randomly create initial population of F frogs. Then it is necessary to sort the population of F frogs in increasing performance level and separate them into m memeplexes each of which containing n frogs (i.e. $F=m \times n$); in this sorting the first frog goes to the first memeplex, the second frog goes to the second memeplex, the mth frog goes to the mth memeplex, and the $(m+1)$th frog goes back to the first [8]. After the previous is done it is time to evaluate each memeplex. In this step, the best frog is a sample from which each frog in the memeplex by learning from it, leaps toward the location which is the optimum. The new position, the worst frog has in the memeplex, is calculated as represented below:

$$x\_worst^{k+1} = x\_worst^k + r^k(x\_sbest - x\_worst^k) \quad (14)$$

where, x_worst is the position of the worst frog in the memeplex, x_sbest is the position of the best frog in the memeplex, r is a random number between 0 and 1, and k is the iteration number of the memeplex [26].

In case that, this process introduces a better answer (frog), the older frog is being replaced. Otherwise, x_sbest is replaced by x_gbest in Eq. (14), and the way we calculate the new position is as below:

$$x\_worst^{k+1} = x\_worst^k + r^k(x\_gbest - x\_worst^k) \quad (15)$$

In a case that, there is no improvement observed, the old frog is replaced by random frog [27].

3.2. Proposed SFLA based hybrid algorithm

The basis of this new method is identification of drawbacks of the basic SFLA, which was initially used on various functions and to mention a critical ones of them can be explained in this way that because some memeplexes have been wasted in local minima, the effective frogs eliminated from the solving procedure. To prevent this as much as possible, enhancement of the guiding particle in each memeplex is necessary; in the SFLA this guiding article in each memeplex is x_sbest. So in the suggested method the movement of the frog which has the best position is determined through the search space toward the position which the global best frog has, is given by Eq. (16)

$$x\_best^{k+1} = x\_best^k + r^k (x\_gbest - x\_best^k) \quad (16)$$

4. IMPLEMENTATION OF THE HSFLA

In the proposed algorithm, the number of switch's which should be opened to maintain a feasible radial configuration and the capacitors that should be placed in candidate buses are considered as control variables. So control variables are integer numbers, and the number of those is the sum of the number of tie switches and the number of buses that candidate for capacitor placement, is expressed as follows:

$$N_{cv} = N_L + N_{bus} \quad (17)$$

where, N_{cv} is the number of control variables, N_L is the number of tie switches and N_{bus} is the number of network buses that candidate for capacitor placement. The number of tie switches is obtained as follows:

$$N_L = N_{br} - N_{bus} + 1 \quad (18)$$

where, N_{br} is the total number of network branches.

For example, in 33-bus system shown in Fig. 2 the number of tie switches is 5 and the number of buses for capacitor placement is 32 (the bus zero is slack bus and is ignored for capacitor placement). So the total number of control variables is 37. Each candidate solution or individual has 37 sections.

In the first step, loop and capacitor vectors should be defined. In the proposed algorithm each loop vector consists of switches that form a loop in network. In other words, the number of loop vectors is equal to the number of fundamental loops or tie switches. In 33-bus system the number of fundamental loop is five, and so the number of loop vectors is five too.

$$Loop\ vectors1 = [s_2 s_3 s_4 s_5 s_6\ s_7 s_{33} s_{20} s_{18} s_{19}]$$

$$Loop\ vectors2 = [s_8 s_9 s_{10} s_{11} s_{35}\ s_{21} s_{33}]$$

$$Loop\ vectors3 = [s_9 s_{10} s_{11} s_{12} s_{13} s_{14} s_{34}]$$

$$Loop\ vectors4 = [s_{22} s_{23} s_{24} s_{37} s_{28}\ s_{27} s_{26} s_{25} s_5 s_4 s_3]$$

$$Loop\ vectors5 = [s_{25} s_{26} s_{27} s_{28} s_{29}\ s_{30} s_{31} s_{32} s_{36} s_{17} s_{16} s_{15} s_{34} s_8 s_7 s_6]$$

To define the capacitor vectors for one bus, six types of capacitors 300, 600, 900, 1200, 1500 and 1800 kVar are used. In this paper is assumed that for each bus of system a capacitor is selected and placed from capacitor vectors as follows:

$$Capacitor\ vector = [0\,300\,600\,900\,1200\,1500\,1800]$$

This capacitor vector is repeated for all buses that should be candidate for capacitor placement.

For the initialization of each individual, one switch is randomly chosen from each loop vector to be opened and one capacitor is also chosen from each capacitor vector to be allocated. The HSFLA is applied to the problem of the multi-objective network reconfiguration and capacitor placement as follows:

Step 1: Defining the input data. In this step, the input data are defined including the initial network configuration, line impedance, the total number of fundamental loops and capacitor vectors for each bus, the number of switches in each loop, the number of population ($P = n \times m$), the number of memeplexes (m), number of frogs in each memeplex (n), and the number of iterations (G).

Step 2: Generating the initial population. For the initialization of each individual (frog), one switch from each fundamental loop or loop vector to be opened and one capacitor from capacitor vector to be placed is randomly chosen.

Step 3: Checking the radiality of the network and all loads being in service for each individual. To check whether radiality is maintained as well as to make sure that all loads are in serviceso as to prevent load islanding, the graph theory can be used. If in a tree the vertices those degree is equal to 1 along all edges connected to them are removed and this procedure is repeated, finally, all vertices will be deleted. If the network graph is not the tree, it means that the network is not radial or that at least one load has been isolated. In this state, the value of fitness function is considered to be zero.

Step 4: Performing the load flow. By allocating capacitors that are determined by each individual in candidate buses a direct approach proposed in [28] is used for load flow solution. The value of the fitness function (λO_i) is calculated using the results of distribution load flow for each radial structure (for each individual or frog).

Step 5: Sort the population P in descending order of their fitness and then divide P into m memeplexes;

for each memeplex, determine the best and worst frogs;

Step 6: Improve the worst frog position using Eqs. (14) and (15).

Step 7: Improve the best frog in each memeplex toward the global best using Eq. (16)

Step 8: Combine the evolved memeplexes;

Step 9: Repeating steps 3-8 until a termination criterion is satisfied. In this paper, the termination criterion is considered to be the number of iterations. Furthermore, if the maximal iteration number is satisfied, the algorithm is terminated.

Fig. 1. shows the flowchart of the proposed algorithm.

5. SIMULATION RESULTS

To demonstrate the performance of the proposed algorithm, three case study systems consisting of two balanced distribution systems (33-bus system and 94-bus system) and one unbalanced distribution systems (25-bus system) are investigated and numerical results are compared with another algorithm such as PSO and IPSO. These methods have been implemented using MATLAB software.

5.1. Case study 1

Baran and Wu [29] distribution test system is used as first example with 3 feeders which is shown in Fig. 2. The system consists 32 sectionalizing switches (normally closed switches), and 5 tie switches (normally open switches) and 37 branches. The total real and reactive power loads on the system are 3715 kW and 2300 kVra, respectively.

The initial power loss is 202.677 kW and minimum bus voltage is 0.913 p.u. To optimize the multi objective fitness function, in this simulation, number of frogs in each memeplex is 5 and number of memeplexes is 6, so the number of population is set $P=m\times n=30$. Maximum iteration to achieve the convergence was set $G=50$. In the first step, the objective functions, including loss reduction, minimization of voltage violation and load balancing, are separately optimized. The results for these three objectives are respectively shown in Tables. 1, 2 and 3. Results obtained by optimizing the multi-objective fitness function for case study 1 are shown in Table. 4. The results indicated for all three objectives and also multi objectives are the best results obtained after 50 instances of running the proposed method and other algorithms.

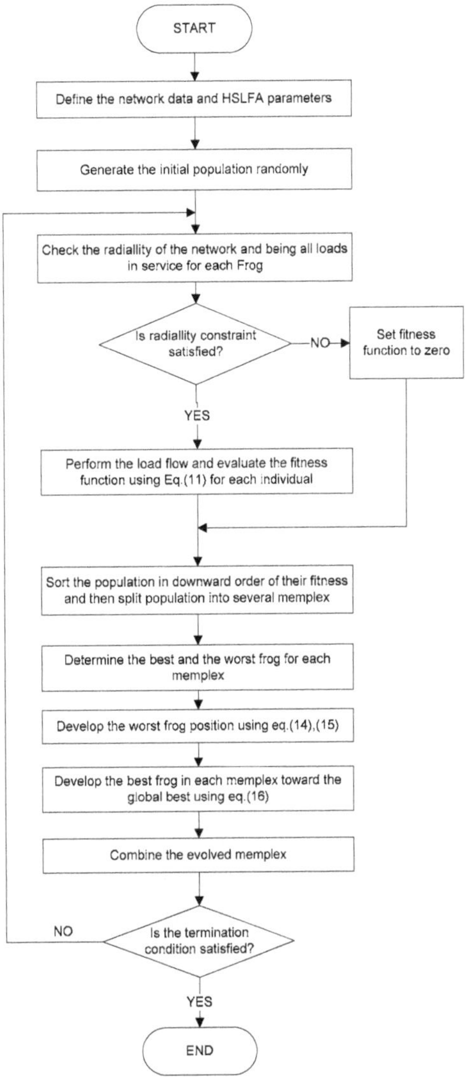

Fig. 1. Flowchart of the proposed HSFLA

As demonstrated in Table. 1, it is observed that the loss reduction ratio obtained by the HSFLA is more than the PSO, IPSO, IBPSO and ACO algorithms. Thus, the proposed method has a higher performance compared to the other methods. It can be seen from Table. 2 that when the only optimization objective is improving the voltage profile, the proposed algorithm by minimum voltage drop of 0.98441544 is not as appropriate as PSO and IPSO algorithms. On the other hand the total used capacitance is equal by the ones used in IPSO method but their arrangement became more distributed.

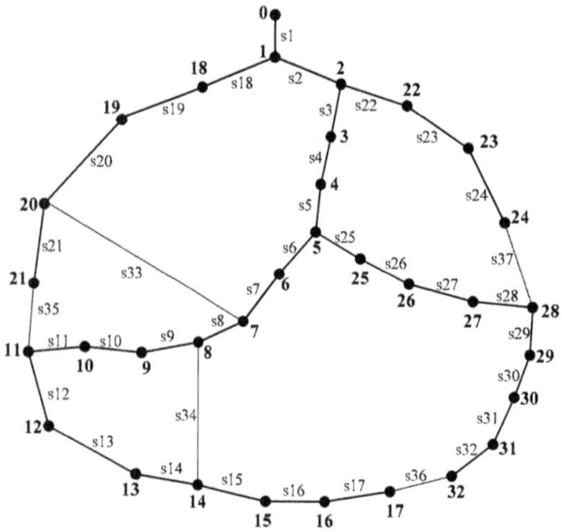

Fig. 2. Baran and Wu distribution test system (33-bus)

By considering Table. 3 which shows simulations for a load balancing of a single objective case, it is shown that LBI index is 0.039968 for proposed algorithm does not provide the best result, but is close to PSO and IPSO results. But the weakness of this method is its capacitance (3300 KVAR) versus 2700 KVAR of IPSO algorithm. Table. 4 shows result of multi-objective simulations, it can be seen that for all three objectives, the proposed algorithm has better results than PSO algorithm and very close to IPSO results, but it used less capacitors than IPSO. Figures 3 and 4 show the voltage and branches current profiles before and after optimal reconfiguration and capacitor placement, respectively. As shown in these figures, the voltage and current branches profile is obviously improved by using the HSFLA algorithm.

Fig. 5 indicates the convergence characteristic of the HSFLA for the multi-objective function for case study 1.It is shown that after 19 iterations HSLFA algorithm reaches to full convergence and fitness function value at approximately 0.83 remains constant.

5.2. Case study 2

The second example is a practical distribution network of the Taiwan Power Company [31]. It is a three-phase, 11.4-kV system which consists of 94-bus, 96 branches, 11 feeders, 83 sectionalizing switches (normally close switches), and 13 tie switches (normally open switches).Fig. 6 shows a diagram of this system which has a total load of 28,350 kW and 20,700 kVAr. Details of the data of this example can be found in [31]. The initial power loss is 531.99 kW and minimum bus voltage is 0.9285 p.u. In this simulation to optimize the multi objective fitness function, number of each cycle frogs and number of memeplexes are considered as 5 and 5, respectively. So, the number of population was $P = m \times n = 25$. By considering the statistical nature of this algorithm, the results indicated for all the three objectives and also the multi objective function are the best results obtained after 50 times running the proposed method (*G*=50). The optimal solutions for minimization of total real power losses, the minimization of buses voltage violation, and load balancing and optimal solution for the multi-objective function are illustrated in Table. 5 and 6. The optimal solution for the minimization of total real power losses using the HSFLA and SA, GA and ACSA is shown in Table. 7.

Table 1. Results obtained by optimizing the real power losses for case study 1

Methods	Power losses (kW)	Loss reduction (%)	Minimum voltage (p.u.) at bus#17	LBI	Open switches	Capacitor located at (buses)
Initial state	202.677	----------------	0.9130905	0.1575671	33-34-35-36-37	------------------
HSFLA	92.5768	54.32	0.95858645	0.0448181	7-11-14-37-32	300(2-4-10-11-18-24-28-29-30)
PSO	95.38	52.93	0.9635100	0.046994	7-10-14-37-36	300(9-10-31) 600(6-29)
IPSO	98.834	51.23	0.965607	0.0400872	11-28-33-34-36	300(5-13-32) 1200(28)
IBPSO[22]	93.061	54.08	0.9585	0.0433806	7-9-14-32-37	300 (11-24-32) 600 (6-29)
ACO[30]	95.79	52.73	0.9656	0.0469611	7-9-14-32-37	450(28) 600 (20-29)

Table 2. Results obtained by optimizing the voltage violation of the buses for case study 1

Methods	Power losses (kW)	Loss reduction (%)	Minimum voltage(p.u.) at bus#17	LBI	Open switches	Capacitor located at (buses)
Initial state	202.677	------------------	0.9130905	0.1575671	33-34-35-36-37	------------------------
HSFLA	187.3621	7.55	0.98441544	0.0946267	6-35-13-37-17	300 (1-2-12-16-17-18-19) 600(13-24) 900 (24) 1200 (30)
PSO	103.1509	49.105	0.96942101	0.0432883	7-11-34-28-36	300(9-14-19-25) 600 (28) 900 (31)
IPSO	183.073	9.67	0.98617336	0.10167143	7-9-34-37-36	300 (1-9-14 -15-20-22-32) 1200 (23) 1500 (28) 600 (29)

Table 3. Results obtained by optimizing the load balancing for case study 1

Methods	Power losses (kW)	Loss reduction (%)	Minimum voltage (p.u.) at bus#17	LBI	Open switches	Capacitor located at (buses)
Initial state	202.677	------------------	0.9130905	0.1575671	33-34-35-36-37	--------------------------
HSFLA	127.472	37.10	0.96323607	0.039968	7-35-34-37-32	300 (5-19-23) 600 (16-18) 1200 (31)
PSO	149.534	26.22	0.9703912	0.0282468	7-35-34-37-32	300 (7) 600 (29-31) 900 (32)
IPSO	135.541	33.12	0.9601967	0.030369	33-11-34-28-36	300 (25-26-27) 900 (16-32)

Table 4. Results obtained by optimizing the multi-objective fitness function for case study 1

Methods	Power losses (kW)	Loss reduction (%)	Minimum voltage (p.u.) at bus#17	LBI	Open switches	Capacitor located at (buses)
Initial state	202.677	------------------	0.9130905	0.1575671	33-34-35-36-37	------------------
HSFLA	98.44	51.45	0.95418297	0.0464747	7-10-14-37-32	300 (10-12-26) 600(3) 900 (29)
PSO	100.05	50.63	0.9616666	0.046695	7-11-34-37-36	300 (16-25-30-32) 600(1-5)
IPSO	101.11	50.11	0.9706953	0.04698	7-10-14-37-36	300 (11-17-25) 600(28-32) 900 (2)

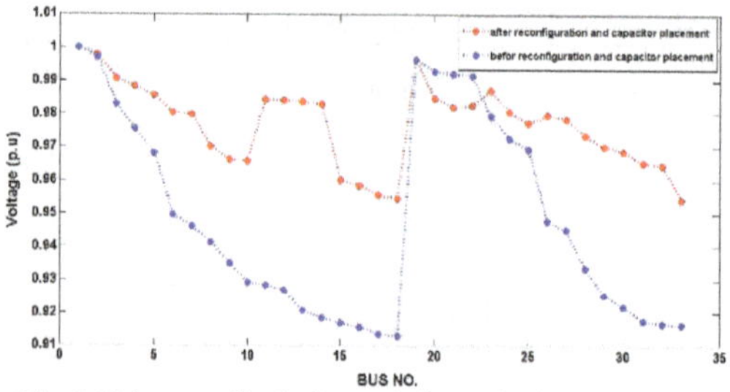

Fig. 3. Voltage profiles before and after optimal reconfiguration and capacitor placement in 33-bus system

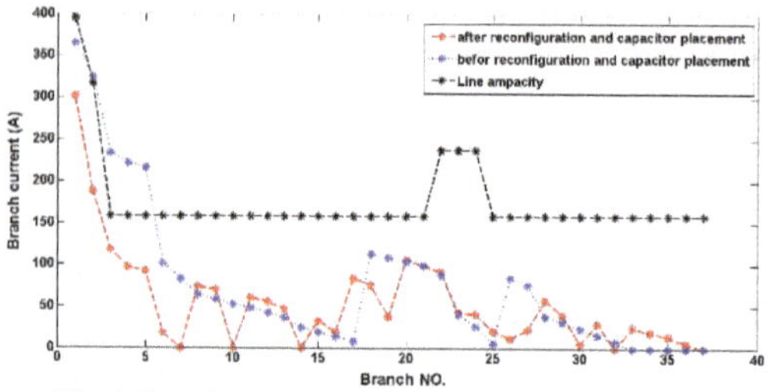

Fig. 4. Branches current profiles before and after optimal reconfiguration and capacitor placement in 33-bus system

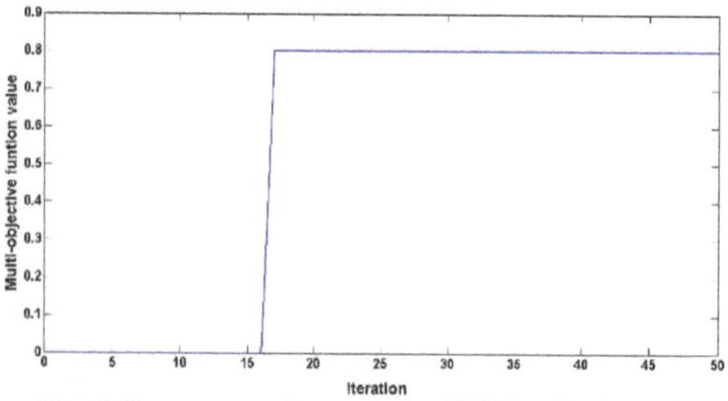

Fig. 5. Convergence characteristic of HSFLA for the multi-objective function for case study 1

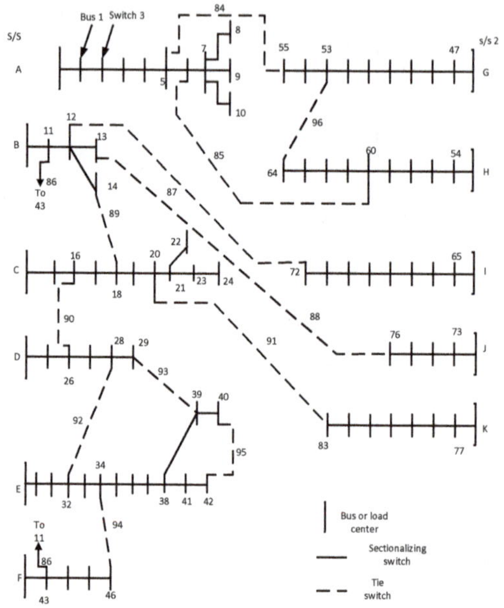

Fig. 6. 94-bus system

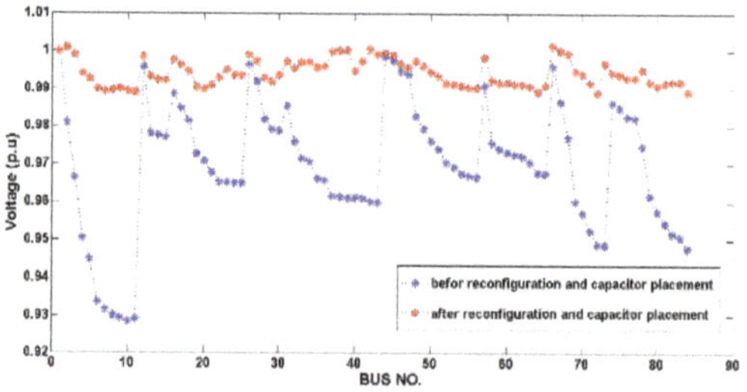

Fig. 7. Voltage profiles before and after optimal reconfiguration and capacitor placement in 94-bus system

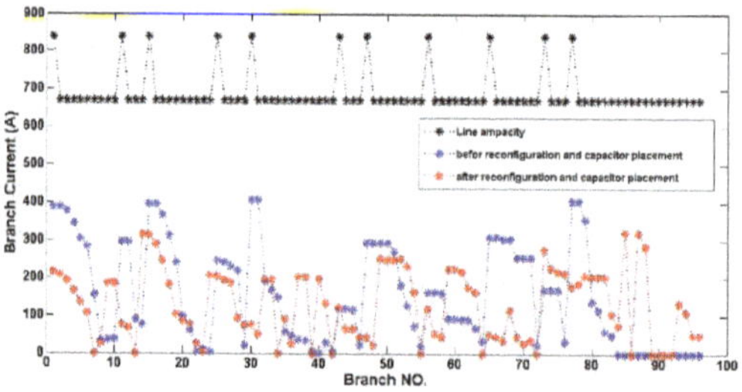

Fig. 8. Branches current profiles before and after optimal reconfiguration and capacitor placement in 94-bus system

As can be seen from Table. 6,the proposed method have better performance compared to the SA algorithm, but GA and ACSA have better perform-ance compared to the HSFLA algorithm. Figures 7 and 8 show the voltage and branches current profiles before and after optimal reconfiguration and capac-itor placement and convergence characteristic of the HSFLA for Case study 2, respectively. As shown in these figures, the voltage and current branches profile are obviously improved by using the HSFLA algorithm.

5.3. Case study 3

The third case study is a 25-bus unbalanced distribution 4.16-kV system consisting of 24 sectio-nalizing switches (normally close switches) and 3 tie switches (normally open switches). Details for the line and load data of the system can be found in [32]. This system is shown in Fig. 9. The initial power loss is 150.13 kW and minimum bus voltage in phase

a, b and c is 0.9284, 0.9284 and 0.9366 p.u respectively. In this simulation to optimize the multi objective fitness function, number of each cycle frogs and number of memeplexes are considered as

4 and 5, respectively. So, the number of population is $P = m \times n = 20$. The optimal solutions for only minimizing the total real power losses, only minimizing the buses voltage violation, only load balancing and the optimal solution for the multi-objective function are presented in Table. 8. By considering the statistical nature of this algorithm, the results indicated for all the three objectives and also the multi-objective function are the best results obtained after 50 instances of running the proposed method. Fig. 10 shows the voltage profiles in phase a, b and phase c of case study 3 before and after optimal reconfiguration and capacitor placement.

Fig. 11 shows the convergence characteristic of the HSFLA for case study 3. As shown in Fig. 10, fitness function after 35 iterations converges to 0.97 and the voltages profile is obviously improved using the HSFLA algorithm in each phase.

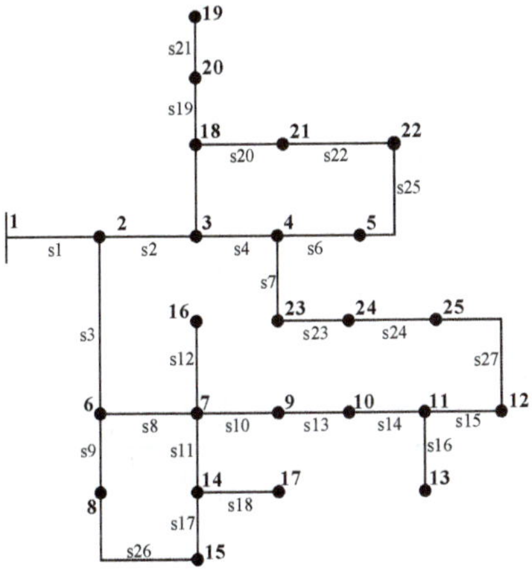

Fig. 9. 25 bus unbalanced distribution system

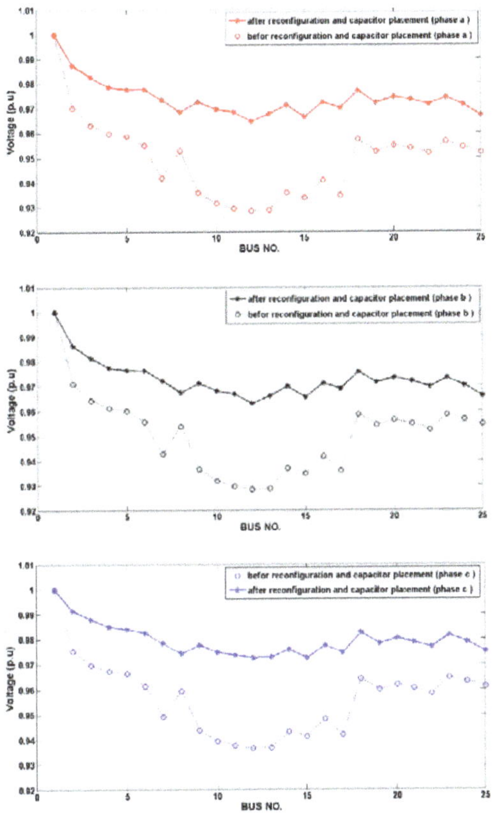

Fig. 10. Voltage profiles before and after optimal reconfiguration and capacitor placement in 25-bus system

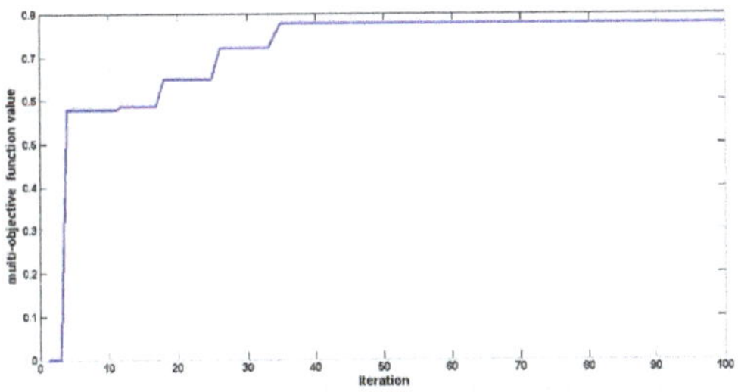

Fig. 11. Convergence characteristic of HSFLA for the multi-objective function for case study 3

Table 5. Results obtained by the HSFLA algorithm for case study 2

Item	Initial state	Only optimizing real power losses	Only optimizing voltage violation	Only optimizing load balancing	optimizing the multi-objective fitness function
Power losses(kW)	531.99	296.47	491.82	474.06	317.836
Loss reduction(%)	---------------	43.6	7.55	10.89	40.25
Minimum voltage (p.u.) at bus#72	0.9285191	0.9850667	0.992976	0.9921594	0.9890495
LBI	0.0329944	0.0180701	0.03202964	0.0101664	0.0141092
Open switches	84-85-86-87-88-89-90-91-92-93-94-95-96	55-7-86-72-13-89-90-83-92-39-34-95-63	55-4-86-87-76-89-90-91-28-39-94-40-64	54-7-86-72-13-89-90-91-92-93-34-40-61	55-7-86-72-13-89-90-91-92-39-34-42-64

Table 6. Results obtained for optimal size and location of capacitors by the HSFLA algorithm for case study 2

Bus		Capacitor (KVAr)		Capacitor (KVAr)		Capacitor (KVAr)		Capacitor (KVAr)	
Slack(0)	43	0	0	0	0	0	600	0	0
1	44	300	300	300	0	600	600	300	300
2	45	0	600	300	300	0	0	900	300
3	46	0	0	0	900	0	0	0	0
4	47	300	0	600	900	300	600	300	300
5	48	300	300	0	300	900	300	300	300
6	49	600	300	1200	300	300	0	300	300
7	50	0	0	600	300	300	300	0	0
8	51	300	300	300	300	900	300	600	0
9	52	300	600	300	0	600	0	300	300
10	53	300	300	300	0	900	300	0	600
11	54	0	300	1800	300	900	900	0	0
12	55	300	300	900	0	300	300	600	300
13	56	600	0	600	0	1500	900	900	0
14	57	900	0	300	600	1500	0	0	300
15	58	300	900	300	0	600	0	300	0
16	59	300	0	300	0	300	300	600	600
17	60	300	0	1500	300	0	0	300	0
18	61	600	0	300	0	0	600	0	300
19	62	300	0	0	0	0	600	300	300
20	63	900	0	600	600	600	600	300	0
21	64	600	300	300	1200	300	300	0	600
22	65	0	0	600	300	300	1500	900	1200
23	66	0	300	0	300	300	600	300	300
24	67	0	0	300	300	1500	300	0	600
25	68	0	0	300	900	300	300	0	300
26	69	300	600	600	300	300	1200	300	600
27	70	600	300	300	900	600	300	300	600
28	71	600	900	1800	300	900	300	0	300
29	72	300	300	600	600	1200	1200	300	900
30	73	600	0	600	900	1800	900	0	0
31	74	900	300	300	300	300	300	900	0
32	75	0	600	600	300	900	300	900	600
33	76	600	300	0	600	1500	600	300	300
34	77	600	0	600	300	300	600	600	300
35	78	0	1200	600	600	0	0	0	300
36	79	0	300	0	300	0	600	300	600
37	80	0	300	0	1200	300	600	0	300
38	81	300	300	0	900	300	0	300	900
39	82	0	300	300	300	300	300	0	600
40	83	0	0	300	0	1500	1200	300	0
41		300		0		0		300	
42		0		300		300		600	

Table 7. Results obtained by optimizing real power losses with HSFLA algorithm along in comparison with SA, GA and ACO

Item		Power losses (KW)	Minimum voltage (p.u) at bus#72	LBI
Original configuration		531.99	0.9285191	0.0329944
HSFLA	Best	296.47	0.9850667	0.0180701
	Worst	303.7	0.98378407	0.01724226
	Average	300.08	0.9844253	0.01765618
	Average Loss reduction	43.6	----------------	--------------
SA [19]	Best	309.12	-------------------	-----------
	Worst	315.86	-------------------	-----------
	Average	312.30	-------------------	-----------
	Average Loss reduction	41.3	-------------------	-----------
GA [19]	Best	295.39	-------------------	-----------
	Worst	299.13	-------------------	-----------
	Average	297.75	-------------------	-----------

	Average Loss reduction	44.03	-------------------	----------
ACO [19]	Best	295.12	-------------------	----------
	Worst	299.46	-------------------	----------
	Average	296.89	-------------------	----------
	Average Loss reduction	44.19	-------------------	----------

Table 8. Results obtained by the HSFLA algorithm for case study 3

Item	Initial state	Only optimizing real power losses	Only optimizing voltage violation	Only optimizing load balancing	fitness function
Power losses (kW)	150.13	91.28	146.377	149.973	94.179
Loss reduction (%)	-----------------	39.2	2.5	0.104	37.29
Minimum voltage Phase a (p.u.) at bus #12	0.9284107	0.9640415	0.9877222	0.9740478	0.964586
Minimum voltage Phase b (p.u.)at bus #12	0.9283703	0.9626266	0.985857	0.9694966	0.963076
Minimum voltage Phase c (p.u.)at bus #12	0.9365706	0.9695176	0.9932599	0.9804585	0.9725021
LBI	0.1009584	0.0454020	0.0735533	0.0328862	0.0455454
Open switches	25-26-27	22- 17-15	20-17-15	5-11-13	25-17-15
Capacitor (KVAr) (Bus)	------------	300(3-4-7)	300 (5-8-11-12-14)	300 (10-16-17-19)	300 (2-3-9)

6. CONCLUSIONS

An HSFLA optimization algorithm as an efficient algorithm for multi-objective reconfiguration and capacitor placement of balanced and unbalanced distribution systems in a fuzzy framework has been introduced in this paper. An important property of the proposed approach is introduced for solving the problem of multi-objective reconfiguration and capacitor placement problem in the fuzzy framework. The minimization of total network real power losses and buses voltage violation as well as balancing the load in the feeders, are the major objectives of this approach. To obtain the optimal solution for the multi-objective fitness function; first, each objective is transferred into the fuzzy domain using the membership function and then the resulting overall fuzzy satisfaction function is considered as a fitness function, which is maximized during the optimization process. The proposed method has been successfully tested in three case studies (consisting of two balanced and one unbalanced system). In case study 1, the HSFLA has achieved better performance compared to other algorithms. In case study 2, the HSFLA has obtained a better performance compared to the SA, and shown a performance almost similar to that of the GA and ACO. As it can be seen from simulation results, the proposed algorithm is an effective method for finding the optimal solution. It is also a powerful method for solving optimization problems in the fuzzy framework for balanced and unbalanced distribution networks.

REFERENCES

[1] A. M. Imran and M. Kowsalya, "A new power system reconfiguration scheme for power loss minimization and voltage profile enhancement using fireworks algorithm", *International Journal of Electrical Power & Energy Systems*, vol. 62, pp. 312-322, 2014.

[2] E.J.D. Oliveira, G.J. Rosseti, L.W.d. Oliveira, F. V. Gomes and W. Peres, "New algorithm for reconfiguration and operating procedures in electric distribution systems", *International Journal of Electrical Power & Energy Systems*, vol. 57 , pp. 129-134, 2014.

[3] H. Shareef, A.A. Ibrahim, N. Salman, A. Mohamed and W. Ling Ai, "Power quality and reliability enhancement in distribution systems via optimum network reconfiguration by using quantum firefly algorithm", *International Journal of Electrical Power & Energy Systems*, vol. 58, pp. 160-169, 2014.

[4] A. Abdelaziz, F. Mohamed, S. Mekhamer and M. Badr, "Distribution system reconfiguration using a modified tabu search algorithm" *Electric*

Power Systems Research, vol. 80, no. 8, pp. 943-953, 2010.

[5] L.W.d. Oliveira, E.J.d. Oliveira, F.V. Gomes, I.C. Silva Jr, A.L.M. Marcato and P. V.C. Resende, "Artificial Immune Systems applied to the reconfiguration of electrical power distribution networks for energy loss minimization", *International Journal of Electrical Power & Energy Systems*, vol. 56, pp. 64-74, 2014.

[6] M. Sedighizadeh, S. Ahmadi and M. Sarvi, "An efficient hybrid big bang–big crunch algorithm for multi-objective reconfiguration of balanced and unbalanced distribution systems in fuzzy framework", *Electric Power Components and Systems*, vol. 41, no. 1, pp. 75-99, 2013.

[7] J.S. Savier and D. Das, "Impact of network reconfiguration on loss allocation of radial distribution systems," *IEEE Transactions on Power Delivery*, vol. 22, pp. 2473-2480, 2007.

[8] S. Anil, N. Gupta and K. R. Niazi, "Adapted ant colony optimization for efficient reconfiguration of balanced and unbalanced distribution systems for loss minimization," *Swarm and Evolutionary Computation*, vol. 1, no. 3, pp. 129-137, 2011.

[9] H. Shayeghi and M. Ghasemi, "FACTS devices allocation using a novel dedicated improved PSO for optimal techno-economic operation of power system", *Journal of Operation and Automation in Power Engineering*, vol. 1, no. 2, pp. 124-135, 2013.

[10] H.A. Ramadan, M.A.A. Wahab, A.H.M. El-Sayed and M. M. Hamada, "A fuzzy-based approach for optimal allocation and sizing of capacitor banks", *Electric Power Systems Research*, vol.106, pp. 232-240, 2014.

[11] J.F. Franco, M.J. Rider, M. Lavorato and R. Romero," A mixed-integer LP model for the optimal allocation of voltage regulators and capacitors in radial distribution systems" *International Journal of Electrical Power & Energy Systems*, vol. 48, pp. 123-130, 2013.

[12] A.R. Abul'Wafa, "Optimal capacitor allocation in radial distribution systems for loss reduction: A two stage method", *Electric Power Systems Research*, vol. 95, pp. 168-174, 2013.

[13] G. Carpinelli, C. Noce, D. Proto, A. Russo and P. Varilone, "Single-objective probabilistic optimal allocation of capacitors in unbalanced distribution systems", *Electric Power Systems Research*, vol. 87, pp. 47-57, 2012.

[14] M. Sedighizadeh and F. Kalimdast, "A honey bee foraging approach to optimal capacitor placement with harmonic distortion consideration", *International Review of Electrical Engineering*, vol. 7, no. 1, pp. 3592-3599, 2012.

[15] R. Baghipour and M. Hosseini, "A hybrid algorithm for optimal location and sizing of capacitors in the presence of different load models in distribution network," *Journal of Operation and Automation in Power Engineering*, vol. 2, no. 1, pp. 10-21, 2014.

[16] M. Sedighizadeh and D. Arzaghi-haris, "Optimal allocation and sizing of capacitors to minimize the distribution line loss and to improve the voltage profile using big bang-big crunch optimization", *International Review of Electrical Engineering*, vol. 6 , no. 4 , pp. 2013-2019, 2011.

[17] D. Zhang, Z. Fu and L. Zhang, "Joint optimization for power loss reduction in distribution systems", *IEEE Transactions on Power Systems*, vol. 23, no. 1, pp. 161-169, 2008.

[18] V. Farahani, B. Vahidi and H.A. Abyaneh, "Reconfiguration and capacitor placement simultaneously for energy loss reduction based on an improved reconfiguration method", *IEEE Transactions on Power Systems*, vol. 27, no. 2, pp. 587-595, 2012.

[19] C.F. Chang, "Reconfiguration and Capacitor Placement for Loss Reduction of Distribution Systems by Ant Colony Search Algorithm", *IEEE Transactions on Power Systems*, vol. 23, no. 4, pp. 1747-1755, 2008.

[20] D.P. Montoya and J.M. Ramirez, "Reconfiguration and optimal capacitor placement for losses reduction", *Proceedings of the IEEE/PES, Transmission and Distribution: Latin America Conference and Exposition*, pp. 1-6, 2012.

[21] M.A.N. Guimaraes, C.A. Castro and R. Romero, "Distribution systems operation optimization through reconfiguration and capacitor allocation by a dedicated genetic algorithm", *IET Proceedings on Generation, Transmission & Distribution*, vol. 4, no. 11, pp. 1213-1222, 2010.

[22] M. Sedighizadeh, M. Dakhem, M.Sarvi and H. H. Kordkheili, "Optimal reconfiguration and capacitor placement for power loss reduction of distribution system using improved binary

particle swarm optimization", *International Journal of Energy and Environmental Engineering*, vol. 5, no. 1, pp. 1-11, 2014.

[23] M. Sedighizadeh, M. Sarvi. And E. Naderi, "Multi objective optimal power flow with FACTs devices using shuffled frog leaping algorithm", *International Review of Electrical Engineering*, vol. 6 ,no. 4, pp. 1794-1801, 2011.

[24] A. Saffar, R. Hooshmand, and A. Khodabakh-shian, "A new fuzzy optimal reconfiguration of distribution systems for loss reduction and load balan-cing using ant colony search-based algorithm," *Applied Soft Computing*, vol. 11, no. 5, pp. 4021-4028, 2011.

[25] N. Gupta, A. Swarnkar, K. R. Niazi, and R. C.Bansal, "Multi-objective reconfiguration of distribu-tion systems using adaptive genetic algorithm in fuzzy framework," *IET Proceedings on Generation, transmission & Distribution*, vol. 4, no. 12, pp. 1288-1298, 2010.

[26] J. Luo and M.C. Xia Li, "A Novel Hybrid Algorithm for Global Optimization Based on EO and SFLA", *Proceedings of the 4th Conference on Industrial Electronics and Applications*, Xi'an, China, pp. 1935-1939, 2009.

[27] T. Niknam and E. Azad Farsani, "A hybridself-adaptive particle swarm optimization and modified shuffled frog leaping algorithm for distribution feeder reconfiguration", *Engineering Applications of Artificial Intelligence*, pp. 1340-1349, 2010.

[28] J.H. Teng, "A direct approach for distribution system load flow solutions," *IEEE Transactions on Power Delivery*, vol. 18, no. 3, pp. 882-887, 2003.

[29] M.E. Baran, and F.F. Wu, "Network reconf-iguration in distribution systems for loss reduction and load balancing," *IEEE Transactions on Power Delivery*, vol. 4, no. 2, pp. 1401-1407, 1989.

[30] M. Kasaeiand and M. Gandomkar, "Loss reduction in distribution system with simultaneous using of capacitor placement and reconfiguration by a colony algorithm" *Proceedings of the 24th International Power System Conference*, Chengdu, China, pp. 1-4, 2009.

[31] C.T. Su and C.S. Lee, "Network reconfiguration of distribution systems using improved mixed-integer hybrid differential evolution," *IEEE Transactions on Power Delivery*, vol. 18, no. 3, pp. 1022-1027, 2003.

[32] G. Vulasala, S. Sirigiri, and R. Thiruveedula, "Feeder reconfiguration for loss reduction in unbalanced distribution system using genetic algorithm", *International Journal of Electronics and Electrical Engineering*, vol. 3, no. 12, pp. 754-762, 2009.

7

Combined Economic Dispatch and Reliability in Power System by Using PSO-SIF Algorithm

N. Ghorbani[1], E. Babaei[2,*]

[1]Eastern Azarbayjan Electric Power Distribution Company, Tabriz, Iran
[2]Faculty of Electrical and Computer Engineering, University of Tabriz, Tabriz, Iran

ABSTRACT
Reliability investigation has always been one of the most important issues in power systems planning. The outages rate in power system reflects the fact that more attentions should be paid on reliability indices to supply consumers with uninterrupted power. Using reliability indices in economic dispatch problem may lead to the system load demand with high reliability and low probability of power's outage rate. In this paper, the Economic Dispatch (ED) problem is optimized using the reliability indices. That is, ED problem and system reliability are proposed as Combined Economic Dispatch and Reliability (CEDR) problem. In CEDR problem, it is tried to utilize generating units in a way that we have high reliability in supplying the system load demand as well as the minimum fuel costs. Due to multi-objective and non-convex characteristics of this problem, Particle Swarm Optimization with Smart Inertia Factor (PSO-SIF) is used to solve the problem. In this research, the ED of power plants is successfully implemented in two systems with 6 and 26 generating units considering emission and system reliability.

KEYWORDS: Economic dispatch, Reliability, Particle swarm optimization, Smart inertia, Non-convex.

1. INTRODUCTION

Power supplying with high quality and uninterrupted to consumers is one of the main tasks of the power networks. The rate of supplying consumers' power demand with minimum outage is measured by reliability concept. Reliability is always one of the major aims in power systems [1] and is one of the most important factors in power systems planning, design, maintenance, and operation [2]. The reliability of a system is generally represented by its indices. Recent outages in power systems depict that the reliability indices should be more under attention in supplying consumers with uninterrupted power. The reliability indices play an important role in power system planning. In [3-5], the reliability parameters such as Loss of Load Probability (LOLP), Expected Energy Not Supplied (EENS) and Forced Outage Rate (FOR) are defined and explained. The concept of reliability can be investigated in three generation, transmission and distribution sections. In this paper, the reliability parameters are evaluated in the power generation section considering the Economic Dispatch (ED) problem.

The ED aims in thermal plants to minimize the plants fuel costs. This is accomplished in a system by determining the output power of the plants in a way that the total network power is supplied with the minimum cost amount and constraints satisfaction. For simplicity, the cost function of each power plant is specified by a quadratic function [6]. The mathematical approaches require some information about the derivation of the cost function. Unfortunately, the input-output characteristics of generation units are non-convex due to the prohibited operating zones, valve-point loadings and etc. The practical ED problem, considering constraints should optimize the non-convex problem which cannot be directly solved by the mathematical methods [7]. Hence, some advanced techniques such as Particle

*Corresponding author:
E. Babaei (E-mail: e-babaei@tabrizu.ac.ir)

Swarm Optimization (PSO) and its improvement versions have been developed to optimize the economic dispatch problem.

In [8], a new hybrid particle swarm optimization algorithm was proposed and applied successfully to solve the dynamic economic dispatch problem with valve-point effects. The obtained results revealed the ability of this new version of PSO in solving ED problem. In [9], a hybrid Bacterial Foreign Algorithm and PSO (BFA-PSO) algorithm was reported for solving the economic load dispatch problem with valve-point loading effects. This method combines the advantages of both the bacterial foraging algorithm and PSO by incorporating the best bacterium in velocity in order to reduce the randomness and increase the swarming effect. In [10], the $\theta - PSO$ algorithm is proposed to solve non-convex ED problem considering practical constraints. The results show the ability of this improved version of PSO for solving ED problem with high constraints.

Particle swarm optimization with time varying acceleration coefficients is another improved version of PSO which proposed to solve multi-objective heat and power economic dispatch problem [11]. The obtain-ed results of this paper demonstrate the its superiority in solving non-convex and constrained combined heat and power economic dispatch problem.

PSO with Smart Inertia Factor (PSO-SIF) is another new and robust version of PSO implemented successfully in ED problems [12]. The obtained results of this paper prove the robustness and effectiveness of this method and show that it could be used as a reliable tool for solving optimization problems.

In order to optimize the multi-object function of this paper which aims to decrease the fuel cost of power plants along and greenhouse gases (GHGs) emission costs with system reliability enhancement, the PSO-SIF algorithm is applied.

PSO is a population-based search algorithm and searches in parallel using a group of particles. Kennedy and Oberhart presented the PSO algorithm based on the analysis of the behavior of birds and fishes [13]. In PSO, each particle tries to decide considering its previous experiences and that of its neighbors. The simple concept, easy implementtation, relative robustness to control parameters and computational efficiency are some of the advantages of the PSO algorithm [14-15].

In PSO, once the iteration increases, inertia weight and consequently the velocity of the particles will reduce. The concept of inertia weight was introduced in order to balance the local and global search. A high inertia weight during initial part of search ensures global exploration, while a low value leads to the end facilitated global convergence. Thus, if the algorithm is not able to find the optimum points in the initial iterations and with high inertia weight, it will not discover global points near the optimal point [12]. To overcome the problem of search area of PSO algorithm with increasing the iteration number, the present article puts forward a new method in which the value of inertia coefficient, unlike classic PSO, is smart and is not same for all the population.

The objective function of the proposed problem consists of plants fuel cost, emission costs and EENS. In order to investigate the functionality of the proposed method, the economic dispatching of the plants is accomplished on two systems with 6 and 26 units, aiming to decrease the system fuel cost, emission cost, and increase the system reliability.

2. FORMULATION OF THE PROBLEM

2.1. Objective function in proposed problem

In solving the Economic Emission Dispatch (EED) problem with reliability, it is aimed to decrease the plants fuel and emission cost, and at the same time increase the system reliability by applying it in solution process. Thus, the objective function of the problem is consists of three independent functions. The variables of the problem are the generated powers of plants defined as follows:

$[P_G] = [P_1, P_2, ... P_n]^T$

minimizing: $F = [F_{FC}, F_{GHG}, EENS]$ (1)

Subjected to: $h(P_i) = 0$ *and* $g(P_i) \leq 0$

where n is the number of the last generator and P_i is the real power generated by the ith generator. The parameter $h(P_i)$ is the equality constraint and $g(P_i)$ is the problem's inequality constraint. F is the multivariable objective function that should be minimized. The parameter F_{FC} is the fuel cost of the units and F_{GHG} shows the greenhouse gases emission costs. In the next, these functions are separately investigated before combining them in the objective function.

2.2. Economic dispatch formulation

Aim of ED problem is minimizing the cost function of the system considering the system constraints. The more details have been presented in [12, 16]. Generally, the simplified fuel cost function of each generation unit is as follows:

$$F_{FC} = \sum_{i=1}^{n} F_i(P_i) \tag{2}$$

$$F_i(P_i) = a_i + b_i P_i + c_i P_i^2 \tag{3}$$

where, F_{FC} is the total generation cost, F_i is the cost function of the ith generator, a_i , b_i and c_i are the cost coefficients of the ith generator, P_i is the output power of the ith generator and n is the last generator number.

In order to balance the power, an equality constraint should be satisfied. The total generated power should be the same as the total load demand (P_{Load}) as follows:

$$\sum_{i=1}^{n} P_i = P_{load} \tag{4}$$

The output power of each generator should correspond to the following inequality constraint:

$$P_{i,\min} \le P_i \le P_{i,\max} \tag{5}$$

where $P_{i,min}$ and $P_{i,max}$ are the minimum and maximum power amounts of ith plant, respectively.

The generating units with multi-steam valve create more variations in plant cost function. Since the existence of steam valves leads to ripple in plants characteristics, the cost function would have a more nonlinear formula. Therefore, the cost function (3) should be replaced by the following cost function:

$$F_i(P_i) = a_i + b_i P_i + c_i P_i^2 + |e_i \times \sin(f_i \times (P_{i,\min} - P_i))| \tag{6}$$

where e_i and f_i are the coefficients of generator i reflecting valve-point loading [12].

2.3. Emission formulation

It is aimed to decrease the released emission of fossil fuel of power plants. The emission from each unit depends on the power generated by that unit and can be modelled as the sum of a quadratic function [17], which is given by Eq. (7):

$$F_{GHG} = \sum_{i=1}^{n} h.EM_i(p_i) \tag{7}$$

$$EM_i(p_i) = ef_i(f_i + g_i p_i + h_i p_i^2) \tag{8}$$

Where $EM_i(p_i)$ is the GHGs emissions of thermal generator i; ef the fuel emission factor of GHGs for thermal generator; f_i,g_i, and h_i the fuel consumption coefficients of thermal unit; h is the given GHGs emission price which is determined by regulations and markets. The GHGs is CO_2 emission in this paper.

2.4. Reliability formulation

The target in the proposed problem is choosing the optimal generator power in such a way that the fuel cost and EENS of system reduce. The probability of any generation unit to be downed is equal to its *FOR* value.

In solving the CEDR problem, there are some generation units with different FOR value which each of the generation units produce a part of the power that is required by system. In calculating the amount of systems EENS our aim is creating a relationship between each unit's *FOR* value and amount of the production power of that unit. In a way that the units which has lower FOR value and consequently has more reliable quality participate more in producing the power required by system. In this way, we can compute the EENS of each unit that depends on the value of FOR and production power of each unit by using the following equations [4]:

$$EENS_i = FOR_i \times T \times P_i \quad (MWh) \tag{9}$$

$$EENS = \sum_{i=1}^{n} EENS_i \quad (MWh) \tag{10}$$

where n is the number of the last unit, T is the evaluation time interval in terms of hour and P_i is the ith unit's power generation capacity in terms of MW. As it is shown in Eq. (9) in a constant value of EENS, more power will be produced by unit which has lower FOR. Eqs .(9) and (10) have been used to compute EENS in power market and deregulated systems [18].

2.5. Combination of ED, emission and reliability in objective function

The objective function of the proposed problem consists of three independent functions. Since the ED and emission cost, and EENS are in terms of (\$/h) and MWh, respectively, and because the optimum values of functions are numbers with different range of values and the algorithm would not be able to similarly optimize all functions in the objective function, it is necessary to express each function in per unit form to enable the objective function to search optimum powers of plants in per unit. Another advantage of per unit form falls in the fact that it would be easy to indicate what percentage of each function is applied by the problem optimization. The objective function of the evaluated problem is as follows:

$$\min imize(F = \gamma \times F_{FC,pu} + \eta \times F_{GHG,pu} + \mu \times EENC_{pu}) \quad (pu) \tag{11}$$

where, $F_{FC,pu}$ is the fuel cost of the units in per unit based on its maximum value and is:

$$F_{FC,pu} = \frac{F_{FC}}{F_{FC,\max}} \quad (pu) \tag{12}$$

where the followings are valid:

$$F_{FC,\max} = \sum_{i=1}^{n} (a_i + b_i P_{i,\max} + C_i P_{i,\max}^2) \quad (\$/h) \tag{13}$$

$F_{GHG,pu}$ is the emission cost of the units in per-unit based on its maximum value and equals:

$$F_{GHG,pu} = \frac{F_{GHG}}{F_{GHG,\max}} \quad (pu) \tag{14}$$

$$F_{GHG,\max} = \sum_{i=1}^{n} h \times ef_i (f_i + g_i p_{i,\max} + h_i p_{i,\max}^2) \tag{15}$$

In Eq. (11), $EENS_{pu}$ is the per unit form of EENS based on its maximum value and equals to the following:

$$EENS_{pu} = \frac{EENS}{EENS_{\max}} \quad (pu) \tag{16}$$

where, the followings are valid:

$$EENS_{\max} = \sum_{i=1}^{n} FOR_i \times T \times P_{i,\max} \quad (MWh) \tag{17}$$

The parameters γ, η and μ are constants related to the influence percentage of each economic dispatch, emission and system reliability on objective function and it is necessary to be initialized in a way that the sum of these parameters be equal to one.

3. PARTICLE SWARM OPTIMIZATION

3.1. A review on PSO algorithm

Kennedy and Oberhart suggested the PSO algorithm based on individuals (particles or ingredients) behavior in a population. Its base refers to the Zoology and models of subjects' manner within a group. It seems that the group members share information between each other, which leads to group efficiency increasing. In this algorithm, each particle represents a solution for the problem. Here, each particle moves toward the optimum value considering three factors. These factors are current velocity, previous experiences and neighbors' experiences [19].

In a n-dimensional search space, the position and the velocity of the ith particle are determined by $X_i = (X_{i1}, X_{i2}, \ldots X_{in})$ and $V_i = (V_{i1}, V_{i2}, \ldots V_{in})$ vectors, respectively. $(P_{best} = X_{i1}^P, X_{i2}^P, \cdots, X_{in}^P)$ and $(G_{best} = X_{i1}^G, X_{i2}^G, \cdots, X_{in}^G)$ represent the best position of the ith particle and its neighbor respectively. The corrected velocity and the position of each particle at the end of any iteration are given:

$$V_i^{k+1} = \omega V_i^k + c_1.r_1.(P_{best}^k - X_i^k) + c_2.r_2.(G_{best}^k - X_i^k) \tag{18}$$

$$X_i^{k+1} = X_i^k + V_i^{k+1} \tag{19}$$

where, V_i^k is the velocity of the ith particle in the kth iteration, ω represents the weight inertia factor, and c_1 and c_2 are the acceleration coefficients. The parameters r_1 and r_2 are random numbers within [0 1] and X_i^k shows the position of the ith particle in the kth iteration.

During the updating process of the velocity, the values of parameters such as ω should be determined in a progressive form. Generally, in order to increase the convergence feature, the weight inertia (ω) is updated in a way that it linearly decreases and in each iteration has same weight for all population [12].

3.2. A review on PSO-SIF algorithm

In the PSO-SIF, each population has its own inertia factor changing with the feedback from best obtained cost in the range [0.3, 0.9]. In this state, decline of the inertia factor and the search space of algorithm are prevented by increasing the iterations.

In the proposed algorithm, the minimum inertia factor is selected to be 0.3, resulting in a situation in which the populations have the costs near the optimum global cost, searching over an optimal point with lower velocities.

In the proposed algorithm, the smart inertia factor is determined by Eq. (20):

$$\omega_j = \frac{0.6 \times (\lambda_j - 1)}{\delta_m} + 0.3 \qquad (20)$$

$$\lambda_j = \frac{\cos t_j}{\cos t_{gbest}} \qquad (21)$$

$$\delta_m = \delta_1 - (\frac{iter}{iter_{\max}}) \times \delta_2 \qquad (22)$$

where, $cost_j$ is j^{th} population cost, $cost_{gbest}$ refers to the best group cost; λ_j is j^{th} population cost ratio to the cost of the best group solution; and δ_m refers to cost variation percent of jth population from the best group solution rate. *iter* is program iteration number; $iter_{max}$ refers to the most number of program iteration; and δ_1 , δ_2 are the adjustment parameters of this algorithm.

The program implementation process through the PSO-SIF technique is summarized as follows:

Step 1: Algorithm initialization,

Step 2: Randomly initial population and particle's initial velocity generation,

Step 3: CEDR problem cost calculation and costs sorting and selecting P_{best} and G_{best}.

Step 5: Calculation of ω_j for each population according to Eq. (20),

Step 4: Updating particles velocity according to Eqs. (18) and (19),

Step 5: Correcting the new positions of the particles to satisfy the constraints of the problem,

Step 6: Go to the third step until the problem's ending criterion was not satisfied,

Step 7: Extracting the best cost's values of each function from the per unit form after program implementation ending; The best values correspond-ding to the best cost amount, which is the best position of particles (G_{best}) or the best arrangement of units power generation depicted initially are applied in Eqs. (2) and (7) to calculate the optimum system fuel and emission cost in terms of ($/h). In continuous, it is applied in Eq. (10) to obtain system optimum EENS in terms of (MWh).

4. NUMERICAL EXPERIMENTATIONS

All the programs are developed and simulated using MATLAB version 7.01. The system configuration is Pentium IV processor with 3.2 GHz speed and 2 GB RAM. In all experimentations, the ED is considered for just one hour. For each case study of the problem, thirty separate experimentations are conducted to be able to compare the solution quality and convergence features. The initial population size and iteration number are 100 and 1000, respectively and, c_1, c_2 are considered as 2.0. The objective function's penalty factor in per unit form is 0.07 and 100 in non-per-unit form.

The proposed method is applied on two systems: 6-unit system considering ED, emission and reliability level and 26-unit system considering ED and reliability level.

4.1. Six units system

Tests are carried out on 6 generating units system with equality and inequality constraints and valve-point effects. The system total load is 1200 MW. The fuel cost coefficients, generator limits and emission factors are reported in [17].

The experimentations are conducted in three separate sections as CEDR problem solving considering different influence percentages of reliability, CEDR problem solving considering different outage rate of power plants and CEDR problem solving considering emission cost.

4.1.1. Solving CEDR problem considering different influence percentages of reliability

Six independent experimentations cases 1 to 6 are conducted considering different influence percentta-ges of each independent function on the objective function to investigate the accurately optimized problem.

In order to calculate the system reliability, it is assumed that the *FOR* values of units in different cases (1 to 6) are shown in Table 1-(A).

It is tried to optimize the economic dispatching in six different experimentations applying different perce-nttages of system reliability. The experimentations are detailed as follows:

1) It is aimed to reduce the unit's fuel costs consid-ering reliability without level. The coefficients

related to the influence percentage of each function in objective function are equal to $\gamma = 1$ and $\mu = 1$.

2) It is aimed to decrease the fuel cost of plants applying 20% reliability level influence in the object-tive function.
3) It is aimed to decrease the fuel cost of plants applying 40% reliability level influence in objective function.
4) It is aimed to decrease the fuel cost of plants applying 50% reliability level influence in objective function.
5) It is aimed to decrease the fuel cost of plants applying 60% reliability level influence in objective function.
6) It is aimed to increase the reliability level without considering fuel cost.

The results of the experimentations are illustrated in Table 2. The parameter TP in Table 2 depicts the total power amount of the system and *F* represents fitness values in objective function in per unit form. The results of the first experiment are shown in the second column of Table 2. In this experimentation, the ED was accomplished to decrease the optimized system's fuel costs without considering system reliability. The system fuel cost in per unit form is 0.615701 pu, which is the minimum among the other case studies. In this case, the EENS is 46.6279 MW, which is the most and is the worst case in comparison with the other case studies.

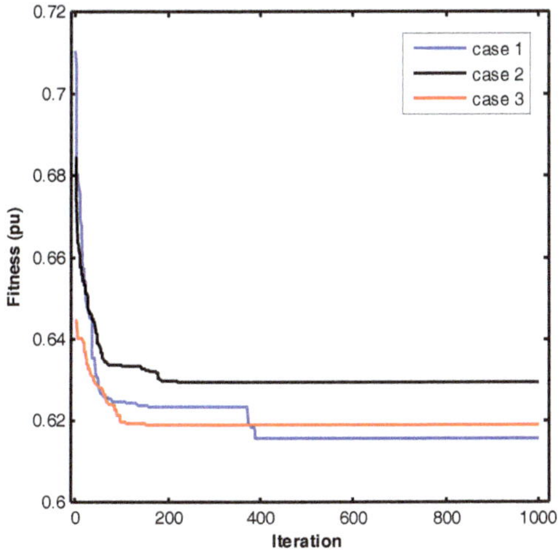

Fig. 1. Convergence characteristics of the PSO-SIFs for test system 1

Comparing the results of the second experiment-tation is shown in the third column of Table 2 depicts that the system EENS value decreases by 5.8503 MWh and reaches to 40.7776 MWh in comparison with the previous case and increases the system reliability influence percentage up to 20% in objective function. It is obvious, as the reliability influence percentage in system objective function increases, the EENS value decreases proportionally and as the fuel cost influence percentage in objective function decreases, its value increases proporti-onally. This is accomplished in a way that as the system reliability influence percentage increases by 40% and the influence percentage of the units fuel cost decreases by 40% in the fourth case study, the EENS value decreases in comparison with the first case by 6.9984 MWh. The notable point in the search algorithm with the objective function in per unit form is its ability to simply detect the best cost with an accuracy equals to the case in which the objective function is not considered in per unit form. In Fig. 1, the convergence characteristics of the CEDR problem' objective function in per-unit form is presented (case studies 1, 2 and 3) optimized through the PSO-SIF algorithm.

4.1.2. Solving EDR Problem Considering Different Outage Rate of Power units

In order to investigate the influence of units outage rate on the amount of power delivered to the system six experimentations are conducted. In all case studies, the aim is to decrease the units' fuel cost considering 50% system reliability influence on objective function. The FOR values for B-G experimentations are shown in Table 1 and are detailed as follows:

A) The outage rates of *A* stat mode are presented in column 2 of Table 1. These values are applied in all six case studies.
B) The *FOR* value of unit 1 is increased by 25% in comparison with *A* case. In other words, $FOR_{B,1} = 1.25 \times FOR_{A,1} = 0.05$.
C) The *FOR* value of unit 2 is decreased by 57% in comparison with *A* case.
D) The *FOR* value of unit 3 is decreased by 40% in comparison with *A* case.

Table 1. Different forced outage rate values applied in six units system

Case study	A	B	C	D	E	F	G
Unit 1	0.04	0.05	0.04	0.04	0.04	0.04	0.04
Unit 2	0.035	0.035	0.02	0.035	0.035	0.035	0.035
Unit 3	0.05	0.05	0.05	0.02	0.05	0.05	0.05
Unit 4	0.02	0.02	0.02	0.02	0.03	0.02	0.02
Unit 5	0.03	0.03	0.03	0.03	0.03	0.05	0.03
Unit 6	0.04	0.04	0.04	0.04	0.04	0.04	0.02

Table 2. Results of ED in a system with six-generation applying different reliability percentages

Unit Output (MW)	$\gamma = 1.0$ $\mu = 0.0$	$\gamma = 0.8$ $\mu = 0.2$	$\gamma = 0.6$ $\mu = 0.4$	$\gamma = 0.5$ $\mu = 0.5$	$\gamma = 0.4$ $\mu = 0.6$	$\gamma = 0.0$, $\mu = 1.0$
Unit 1	94.7998	94.8074	94.8044	94.7998	94.8000	20.0000
Unit 2	100.0000	26.9860	99.9883	20.0027	99.9865	20.0000
Unit 3	568.7989	419.2013	344.3994	269.5996	120.0013	120.0000
Unit 4	259.5996	508.9365	510.6424	515.9981	510.8098	519.9996
Unit 5	136.8015	110.0686	110.1652	259.5996	334.3979	480.0004
Unit 6	40.0000	40.0000	40.0000	40.0000	40.0000	40.0000
F_{FC}	29491.4289	31191.78	31616.91	32862.8637	34217.1802	35725.7273
EENS	46.6279	40.7776	39.6295	37.6800	35.1399	33.9000
TP	1200	1200	1200	1200	1200	1200
F	0.615701	0.627417	0.618371	0.607280	0.603604	0.622536
RT	1.5765	1.6724	1.6765	1.6761	1.6801	1.3765

*F_{FC} :Fuel Cost [$/h], TP: Total Power [MW], F: Fitness [pu], TC: Total Cost [$/h], RT: Run Time [sec.]

Table 3. CEDR problem results considering different FOR of power units

Unit Output (MW)	A	B	C	D	E	F	G
Unit 1	**94.7998**	**57.3999**	20.0000	94.7998	94.7993	94.7998	20.0001
Unit 2	**20.0027**	20.0021	**100.00**	20.0004	100.0000	100.0000	99.9992
Unit 3	**269.5996**	269.5996	269.5996	**575.8662**	269.5996	344.3994	269.5996
Unit 4	**515.9981**	366.3990	409.1993	359.3328	**409.20000**	510.7999	508.9324
Unit 5	**259.5996**	446.5992	361.2009	110.0007	259.5993	**110.0000**	133.2403
Unit 6	**40.0000**	40.0000	40.0000	40.0000	66.8030	40.0007	**168.2282**
F_{FC}	32862.8637	33531.4483	33435.4890	29716.8443	32565.4949	31616.8925	33706.4957
EENS	37.6800	39.3760	36.9000	28.0960	43.5080	41.82799	35.32038
TP	1200	1200	1200	1200	1200	1200	1200
F	0.607280	0.621957	0.613348	0.573768	0.624306	0.587282	0.614260

Table 4. CEDR problem results considering emission cost

Unit Output (MW)	Minimization of F_C	Minimization of EC	Minimization of $EENS$	Minimization of F_C, EC (TC)	Minimization of F_C, E, $EENS$
Unit 1	94.7998	20.0000	20.0000	20.0006	20.0000
Unit 2	100.0000	20.0000	20.0000	20.0000	20.0000
Unit 3	568.7989	120.0000	120.0000	344.3995	269.5996
Unit 4	259.5996	520.0000	519.9996	515.9995	508.9324
Unit 5	136.8015	479.9995	480.0004	259.6001	341.4675
Unit 6	40.0000	40.0005	40.0000	40.0002	40.0005
F_{FC}	29491.4289	35725.7406	35725.7273	32853.5682	33733.8192
EC	20227.6538	15064.2739	15064.2988	16456.5568	15971.7365
TC	49719.0828	50790.0145	50790.0261	49310.1250	49705.5557
EENS	46.6279	33.9001	33.9000	38.4280	37.0026
TP	1200	1200	1200	1200	1200
F	0.653968	0.475455	0.622536	0.640650	0.6821548

*EC :Emission Cost [$/h], TC: Total Cost [MW],

E) The *FOR* value of unit 4 is increased by 50% in comparison with *A* case.
F) The *FOR* value of unit 5 is increased by 66% in comparison with *A* case.
G) The *FOR* value of unit 6 is decreased by 50% in comparison with *A* case.

The parameter $FOR_{B,1}$ indicates the FOR value of unit 1 in case B, with value equal to 0.05 and is shown in Table 1. The results obtained from A-G experimentations are shown in Table 3.

The experimentations aim to investigate the influence of different values of FOR on each unit's delivered power amount, system fuel cost, and reliability. In experiment *B*, the outage rate of unit 1 is increased by 25% in comparison with case *A*. As a result, the reliability of the unit 1 is decreased. As shown in Table 3, as the outage rate of unit 1 in state *B* increased in comparison with case *A* of the same unit, the amount of delivered power in constant load of 1200 MW decreases from 94.7998 MW to 57.3999 MW. This shows the existence of linear relation between *FOR* and consequently the unit 1 reliability and the power amount delivered to the system. Therefore the total EENS of the system increases from 37.6800 MW in *A* case study to 39.3760 MW in *B* case. This depicts the influence of generated power of a unit on total system EENS.

In case study *C*, the outage rate of unit 2 is decreased by 57% in comparison with that of the case *A*. The decreasing of outage rate results in considerable increasing in unit 2 reliability and generated power amount. As it is shown in Table 3, the generated power of unit 2 increases from 20.0027 MW in case *A* to 100 MW in case *C*. Here, the amount of EENS is decreased in comparison with case *A* as expected.

4.1.3. Solving CEDR problem considering emission cost

Tests are conducted on a system with six generation units considering fuel cost, emission cost and reliability level. The system data for emission is presented in [17]. The aim is minimization of fuel cost, emission cost and the EENS of system.

The simulation results using PSO-SIF are presented in Table 4. In Table 5, the results of solving CEDR problem considering emission cost through PSO-SIF are compared with that of PSO and PSO-TVAC methods. In this comparison, per-unit coding is used to combine the proposed multi-objective problem and offer a single objective function. For PSO, c_1 and c_2 are set to 2.0. The weighting inertia coefficients for both PSO and PSO-TVAC were considered a varying number in the range [0.3, 0.9] and the initial populat-ion size and iteration number are 100 and 1000, respectively. The adjustable parameters for PSO-TVAC algorithm were chosen as: $C_{1f} = C_{2f} = 2.5$ and $C_{1i} = C_{2i} = 0.5$ [11].

Table 5. Comparison of results of each method for CEDR problem considering emission cost

Unit (MW)	Minimization of F_C , ED , $EENS$		
	PSO	PSO-TVAC	PSO-SIF
Unit 1	94.8007	20.4902	20.0000
Unit 2	100.00	20.0180	20.0000
Unit 3	269.5996	269.4790	269.5996
Unit 4	510.7996	445.5142	508.9324
Unit 5	184.7999	404.2935	341.4675
Unit 6	40.0000	40.2070	40.0005
F_{FC}	32427.5813	33926.2194	33733.8192
EC	18174.6878	15980.4931	15971.7365
TC	**50602.2692**	**49906.7125**	**49705.5557**
EENS	**38.1320**	**37.6415**	**37.0026**
TP	1200	1200	1200
F	0.6491588	0.644692	0.6821548

As it is obvious from Table 5, the minimum total cost (fuel and emission cost) obtained using PSO-SIF is 49705.5557 ($/h) and the related EENS] is 37.0026 (Mw/h) that are lower than both total cost and system's EENS obtained through PSO and PSO-TVAC approaches which shows the superiority of the proposed PSO-SIF method over the mentioned techniques.

4.2. 26-unit test system

Tests are conducted on a system with 26 units considering fuel cost and reliability level functions. The system total load is 2430 MW and the generation units' data are available in [20]. Reliability data are presented in Table 6 and adapted from [21]. Test are conducted in three separate parts of fuel cost minimization, EENS minimization, and simultaneous cost and EENS level minimization, the results of which by PSO-SIF method are shown in Table 7.

Table 6. Forced outage rate values applied in 26 units system

Units	FOR	Units	FOR	Units	FOR
Unit 1	0.12	Unit 10	0.02	Unit 19	0.02
Unit 2	0.12	Unit 11	0.02	Unit 20	0.02
Unit 3	0.08	Unit 12	0.04	Unit 21	0.02
Unit 4	0.04	Unit 13	0.04	Unit 22	0.02
Unit 5	0.04	Unit 14	0.04	Unit 23	0.1
Unit 6	0.04	Unit 15	0.05	Unit 24	0.1
Unit 7	0.04	Unit 16	0.05	Unit 25	0.1
Unit 8	0.02	Unit 17	0.05	Unit 26	0.1
Unit 9	0.02	Unit 18	0.02		

Table 7. CEDR problem results in 26 units system

UNIT (MW)	Minimization of F_c	Minimization of *EENS*	Minimization of F_c and *EENS*
Unit 1	399.9995	100.0003	288.8665
Unit 2	399.9981	100.0000	284.5642
Unit 3	350.0000	338.7472	349.9972
Unit 4	155.0000	155.0000	154.9999
Unit 5	155.0000	155.0000	154.9997
Unit 6	154.9998	155.0000	154.9999
Unit 7	155.0000	155.0000	154.9999
Unit 8	75.9998	76.0000	75.9971
Unit 9	75.9992	76.0000	75.9961
Unit 10	75.9999	76.0000	75.9998
Unit 11	75.9975	76.0000	75.9995
Unit 12	47.7311	100.0000	99.9982
Unit 13	40.4191	100.0000	99.9985
Unit 14	33.0057	100.0000	99.9994
Unit 15	68.9500	197.0000	68.9513
Unit 16	68.9500	197.0000	69.0070
Unit 17	68.9500	197.0000	68.9500
Unit 18	2.4000	12.0000	11.9763
Unit 19	2.4000	12.0000	11.9606
Unit 20	2.4000	12.0000	11.8017
Unit 21	2.4000	12.0000	11.9391
Unit 22	2.4000	12.0000	11.9969
Unit 23	4.0000	4.1195	4.0000
Unit 24	4.0000	4.0155	4.0005
Unit 25	4.0000	4.0491	4.0000
Unit 26	4.0000	4.0681	4.0000
TP	2430.0000	2430.0000	2430.0000
F_{FC}	33630.0528	42212.3306	36269.9568
EENS	171.9084	126.3550	152.8301
F	0.674581	0.614477	0.734283
TC	1.9702	1.8207	1.9904

In the PSO-SIF algorithm, selecting optimal values for δ_1 and δ_2 is important and plays a key role in quality of optimization process [12]. The results of solving CEDR problem on 26-unit system in terms of various values for δ_1 and δ_2 after 30 independent testes are given in Table 8 and the optimal values for δ_1 and δ_2 are determined as 0.05 and 0.04, respe-ctively.

Table 8. Determination of δ_1 and δ_2 for PSO-SIF in 26 units system

Case	δ_1	δ_2	Minimum F (pu)	Average F(pu)
1	0.1	0.08	0.811464	0.830021
2	0.09	0.07	0.7910099	0.809265
3	0.08	0.065	0.761201	0.778234
4	0.07	0.055	0.739401	0.742561
5	0.06	0.05	0.735021	0.736198
6	**0.05**	**0.04**	**0.734283**	**0.734285**
7	0.04	0.03	0.735114	0.736601
8	0.03	0.024	0.745243	0.750458
9	0.02	0.016	0.759852	0.772213
10	0.01	0.008	0.774485	0.800049

5. CONCLUSIONS

In this paper, the reliability indices are used to optimize the ED problem and it is solved by considering units reliability. In the proposed problem, it is tried in economic dispatching problem to utilize units with higher reliability in addition to lower fuel cost.

The objective function of the proposed problem consists of three independent fuel cost, emission cost and reliability functions. In order to combine these functions in the objective function, this paper proposed a per-unit coding in a way that each function is converted to per unit form based on their maximum amounts.

The results obtained from the experimentations of the first section depict the efficiency of the per uniting several functions in objective function and show the possibility of combining two or three independent functions in a objective function with desired influence percentage combinations. Thus, it is possible to determine the influence rate of the system reliability in ED problem solution. The results obtained from the experimentations depict the fact that the system tends to utilize power units, which have lower values of FOR or units have higher reliability considering power supplied to system.

Today, the power plants outage and power interruption would cause considerable financial damages, which can be sometimes irrecoverable. Therefore, it is necessary to pay attention to the reliability issue in power systems planning. Since one of the initial efforts in supplying demanded power is system ED, this paper proposes the idea of applying reliability indices in economic dispatching to create more reliability in supplying power until the end of utilization and planning. The ED problem solution including the system reliability can be utilized at least in systems that some units of them have high outage rates because of natural disasters such as flood or earthquakes or due to internal difficulties to minimize system EENS amount by receiving less power from them.

REFERENCES

[1] N. Saman and C. Singh, "Reliability Assessment of Composite Power System Using Genetic Algorithms," Springer, *Computational Intelligence in Reliability Engineering*, vol. 39, pp. 237-286, 2007.

[2] A. K. Vermaa, A. Srividyaa and M. V. Bhatkarb, "A Reliability Based Analysis of Generation/Transmission System Incorporating TCSC Using Fuzzy Set Theory," *Proceedings of the IEEE International Conference on Industrial Technology*, Mumbai, India, pp. 2779-2784, 2006.

[3] D. Elmakias, "Reliability of Generation Systems," *Studies in Computational Intelligence,* vol. 111, pp. 171-238, 2008.

[4] R. Billinton and R. Allan "Reliability Evaluation of Power Systems," Second Edition, New York: Plenum Press, 1996.

[5] W. Qin, P. Wang, X. Han and X. Du, "Reactive Power Aspects in Reliability Assessment of Power Systems," *IEEE Transactions Power Systems*, vol. 26, no. 1, pp. 85-92, 2011.

[6] A. K. Barisal, "Dynamic search space squeezing strategy based intelligent algorithm solutions to economic dispatch with multiple fuels," *International Journal of Electrical Power and Energy Systems*, vol. 45, pp. 50-59, 2013.

[7] N. Ghorbani and E. Babaei, "Exchange Market Algorithm," *Applied Soft Computing*, vol. 19, pp:177-187, 2014.

[8] Y. Zhang, D. W. Gong, N. Geng and X. Sun, "Hybrid bare-bones PSO for dynamic economic dispatch with valve-point effects," *Applied Soft Computing*, vol. 18, pp. 248-260, 2014.

[9] T. Jayabrathi, P. Bahil, H. Ohri, A. Yazdani and V. Ramesh, "A hybrid BFA-PSO algorithm for economic dispatch with valve-point effects," *Front. Energy*, vol. 6, no. 2, pp. 155-163, 2012.

[10] V. Hosseinnezhad and E. Babaei, "Economic load dispatch using θ-PSO," *International Journal of Electrical Power and Energy System,* vol. 49, pp. 160-169, 2013.

[11] B. Mohammadi-Ivatloo, M. Moradi-Dalvand and A. Rabiee, "Combined heat and power economic dispatch problem solution using particle swarm optimization with time varying acceleration coefficients," *Electric power Systems Research,* vol. 95, pp. 9-18, 2013.

[12] N. Ghorbani, S. Vakili, E. Babaei and A. Sakhavati, "Particle Swarm Optimization with Smart Inertia Factor for Solving Nonconvex Economic Load Dispatch Problems," *International Transaction on Electrical Energy System,* vol. 24, pp. 1120-1133, 2014.

[13] J. Kennedy and R. C. Eberhart, "Particle swarm optimization", *Proceedings of the IEEE International Conference on Neural Networks*, pp. 1942-1948, 1995.

[14] J. B. Park, Y. W. Jeong, J. R. Shin and K. Y. Lee, "An Improved Particle Swarm Optimization for Nonconvex Economic Dispatch Problems", *IEEE Transactions on Power Systems*, vol. 25, no. 1, pp. 156-166, 2010.

[15] H. Khorramdel, B. Khorramdel, M. Tayebi Khorrami and H. Rastegar, "A Multi-Objective Economic Load Dispatch Considering Accessibility of Wind Power with Here-And-Now Approach", *Journal of Operation and Automation in Power Engineering*, vol. 2, no. 1, pp. 49-59, 2014.

[16] A. Hatefi and R. Kazemzadeh, "Intelligent Tuned Harmony Search for Solving Economic Dispatch Problem with Valve-point Effects and Prohibited Operating Zones", *Journal of Operation and Automation in Power Engineering*, vol. 1, no. 2, pp. 84-95, 2013.

[17] Y. Zhang, F. YAO and H. Ho-Ching-IU, "Sequential quadratic programming particle swarm optimization for wind power system operations considering emissions," *Journal of Modern Power Systems and Clean Energy*, vol. 1, no. 3, pp. 231-240, 2013.

[18] S. Niioka, A. Kozu. M. Ishimaru and R. Yokoyama, "Supply Reliability Evaluation Method for Deregulated Electric Power Market Considering Customers Uncertainty," *Proceedings of the International Conference on Power System Technology*, pp. 1782-1786, 2002.

[19] H. Shayeghi and M. Ghasemi, "FACTS Devices Allocation Using a Novel Dedicated Improved PSO for Optimal Operation of Power System," *Journal of Operation and Automation in Power Engineering*, vol. 1, no. 2, pp. 124-135, 2013.

[20] C. Wang and S.M. Shahidehpour, "Effects of ramp rate limits on unit commitment and economic dispatch," *IEEE Transactions on Power Systems*, vol. 8, pp. 1341-1350, 1993.

[21] C. Grigg, P. Wong, P. Albrecht, R. Allan, M. Bhavaraju and R. Billinton, "A report prepared by the reliability test system task force of the application of probability methods subcommittee," *IEEE Transactions on Power Systems*, vol. 14, pp. 1010-1020, 1999.

8

Robust Agent Based Distribution System Restoration with Uncertainty in Loads in Smart Grids

N. Zendehdel*, R. Asgarian Gannad Yazdi

Department of Electrical Engineering, Faculty of Engineering, Ferdowsi University of Mashhad, Mashhad, Iran

ABSTRACT

This paper presents a comprehensive robust distributed intelligent control for optimum self-healing activities in smart distribution systems considering the uncertainty in loads. The presented agent based framework obviates the requirements for a central control method and improves the reliability of the self-healing mechanism. Agents possess three characteristics including local views, decentralizations and autonomy. The message, exchanged among neighboring agents, is used to develop a global information discovery algorithm and updates the topology information of out-of-service areas, available supply capacity and routing information. Fuzzy description is employed to take into account the uncertainties of measurements in which are exchanged between agents. Moreover, to find the optimal restoration plan, incorporating the discovered data, a routing problem is developed as a fuzzy binary linear optimization problem. This problem is approached by a novel method using a specific ranking function. Finally, robustness and applicability of the proposed self-healing method is tested on two standard case studies. The obtained results emphasize that ignoring the uncertainties may lead to non-realistic solutions.

KEYWORDS: Agent based self-healing framework, Fuzzy binary linear optimization, Smart grid, Uncertain load.

1. INTRODUCTION

Smart distribution grids are distinguished from conventional distribution systems by their ability to automatically detect the fault location using the digital measurements and two-way communication-ns, to isolate the faulted areas using the remote digital switching technologies and quickly restore as many non-faulted outage loads as possible by minimum number of switching operations. These features improve the power system reliability and service quality.

Various approaches are available for implem-enting self-healing control method. This includes two main categories, namely, centralized and decentralized methodologies. Different centralized restoration techniques such as mathematical programming, knowledge-based systems and heu-ristic methods have been proposed to solve this problem. Mathematical programming formulates the restoration problem as a mixed integer programming problem. In [1], a two-stage algorithm has been proposed which decomposes the restoration problem into two sub-problems. In [2], an optimization technique, called differential evolution, has been used to solve distribution feeder reconfiguration and service restoration problem in a centralized way. In [3], an iterative centralized mechanism has been developed for system reconfiguration during normal operation and service restoration after single or multiple fault occurrences. In [4], optimal restoration problem is solved using a dynamic programming approach. In [5], a communication based algorithm was proposed for restoration problem to decrease the restoration time in the smart grid distribution management system.

On the other hand, heuristic techniques and

*Corresponding author:
N. Zendehdel (E-mail: n.zendehdel@gmail.com)

knowledge based systems search the solution space to solve the combinatorial restoration problem. In [6], an expert system algorithm has been utilized for restoration and loss reduction of distribution systems. In [7-9], the distribution restoration problem in presence of distributed generators is modeled as nonlinear mixed integer optimization problem and solved by heuristic methods. A heuristic approach is utilized to solve the proposed problem in [7], while a modified binary particle swarm optimization was applied to solve network reconfiguration problems in [8] and a multi-objective particle swarm optimization was used in [9]. A classifying system and co-evolutionary algorithm were used in [10] for solving the problem of power delivery recovery in case of the network failure. The elaborated method uses the theoretical background of genetic-based machine learning systems and fuzzy sets theory. Moreover, artificial neural network [11], ant immune system-ant colony optimization [12], genetic algorithm [13], among other heuristic methods were also used extensively for restoration problems. Although, centralized methods obtain optimal solutions, they require low-latency communication system and large amount of data transfer. Their computing center needs expensive computational capabilities, while the accuracy of obtained results may be influenced by the uncertain behavior of loads and distributed generations in smart grids. Therefore, the centralized methods may not be practical for large smart distribution systems. Meanwhile, multi agent systems (MASs) as decentralized approaches distribute the control and intelligence in every component level of the grid using agents, to fulfill self-healing duties of smart grids.

Decentralized approaches based on the MAS technology has been investigated in the literature [14-20]. In [14], MAS architecture has been utilized for only the service restoration without considering the load shedding and priorities and obtaining of extra available capacity through load transfer in the restoration procedure. In [15, 16], MAS framework has been developed for the service restoration problem, incorporating the load shedding concept but load variation and prioritization have not been considered in these two studies. In [15], the cooperation of agents is centrally regulated by a master agent, while in [16] each control agent individually solves a NP-hard complex combinatorial restoration problem to make restoration decisions. The grid information is entirely provided during initialization process and agents' negotiation. In [17, 18], the restoration problem has been investigated in the MAS architectures, considering load prioritization and shedding concepts. The focus of [17] is on restoring out-of-service areas using exchanged information among load agents and supply agents. Timely load variation is not considered in [17]. In [18], self-healing mechanism is studied based on Taipower distribution system rules using the knowledge based system and typical load patterns. A completely distributed algorithm has been proposed in [19] for the self-healing mechanism in distribution systems with distributed energy resources (DES). This approach takes load shedding and partial restoration into account. A decomposed agent-based self-healing control of an urban smart power grid has been proposed in [20] without considering the load prioritization. Investigating satisfaction of power system operating constraints imposes expensive centralized computation activities on the agent-based algorithm. The mentioned studies lack a control structure for self-healing mechanism in smart grids. In [21, 22], an agent based control framework has been presented for controlling the self-healing process considering the peak load in duration of the fault repair. A MAS architecture including agents with local views has been proposed in [23] to realize the self-healing mechanism.

Existing mentioned papers and similar researches on this subject in literature have not properly addressed how to design a comprehensive well-defined MAS framework which obtains an optimal control for the self-healing in smart grids incorporating the agents' decentralizations. The decision making policies in available studies are based on the learning methods or expert-based systems which often achieve near global objectives and require huge databases to restore statistical data. In spite of the available online sensor measurements in the smart grids, decision makers in the literature utilize historical data and load patterns to address the load variation that may lead to the limited restoration of some loads.

This paper proposes a new self-healing control framework considering the agents' decentralizations, local topology information and uncertainty in measurements. The developed control algorithm is based on the MAS architecture including agents in two classes: zone and switch. Their communication policy does not impose a low latency communication on the system and decreases the dependence of algorithm on the accuracy of the data transfer. In addition, in the proposed framework, online monitored information is used to set the parameters of decision making activities and perform the distributed calculations individually to investigate the satisfaction of the power system operating constraints. Hence, the suggested method is scalable, self-adaptable and self-updatable. At the same time, to distribute the computational activities an auxiliary grid, which refers to the out-of-service areas is introduced and a new mathematical model for restoration problem in the developed auxiliary grid is developed. The identified parameters in the proposed model can be automatically generated and updated based on the online measurements incorporating their uncertainties. To guarantee the optimality and robustness of obtained switching sequence during agents' decision makings in the designed MAS architecture, a hybrid policy is developed consisting of an expert-based technique and a mathematical programming method. To this end, the uncertainty in measurements is considered using fuzzy description and a new fuzzy binary linear programming (BLP) approach is proposed to solve the presented optimization problem and find the optimal and robust restoration plan. Considering the possibility of any variation in sensor measurements during the decision making process, guarantees the robustness of the algorithm. For the sake of comparison and implementation, two smart distribution systems are selected and the proposed method is tested on them. The results demonstrate the efficacy of the proposed method as well as the robustness of the obtained self-healing plan against the load variations.

The rest of this article is organized in six sections. Section 2 indicates the structure of the proposed distributed control framework. Section 3 explains a mathematical programming approach utilized to find a robust and optimal plan. Section 4 introduces the designed self-healing control rules to regulate the performance of agents in the proposed framework. Section 5 illustrates the results of testing the proposed control algorithm on two distribution test systems. Finally, the paper conclusions are drawn in Section 6.

2. DISTRIBUTED CONTROL FRAMEWORK

With respect to location of the Intelligent Electronic Devices (IEDs), some line segments, known as zones are formed in each distribution feeder such that some IEDs are placed on the boundaries of zones. Furthermore, considering the direction of current which passes from X side to Y side of an IED, two sides are defined for every IED. Fig. 1 shows this consideration.

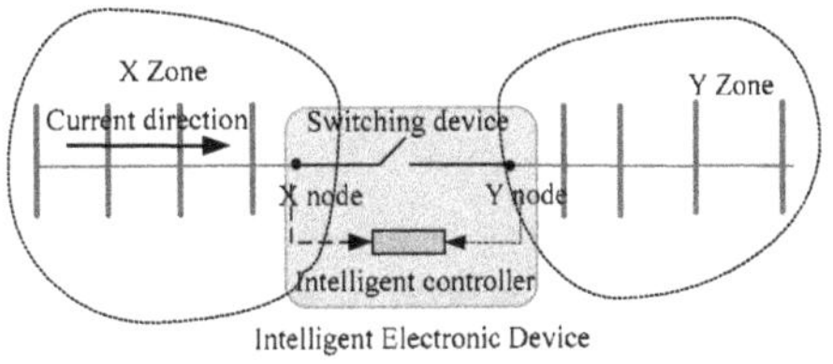

Fig. 1. X and Y sides of an intelligent electronic device

In order to propose a distributed control framework, each zone is appointed with a zone agent, which monitors local measurements and provides the fuzzy description of uncertainty. In addition, each agent-controlled switching device is assigned with a switch agent which utilizes the received measurements and cooperates with other agents to achieve self-healing mechanism. Moreover, the team concept is also defined in the MAS including the agent-controlled IEDs placed on the boundary side zone. Team concept provides the scalability of the proposed framework because if any intelligent electronic device is connected to the grid, then its IP address and operating situation can be identified for its teammates as a new agent which belongs to the team. The introduced MAS includes two types of switch agents, namely maneuvering agent which corresponds to normally opened switching device and sectionalizing agent which corresponds to normally closed switching device. These agents move among five operating modes defined based on local measurements, when they perform actions and change the operating condition of power system. The agents' movement with their actions is shown in Fig. 2. In this figure the numbered arrows indicate the agent actions as (1) Receiving out of range measurements; (2) Fault detection; (3) Isolation; (4) Restoration. Next, five operating modes are explained.

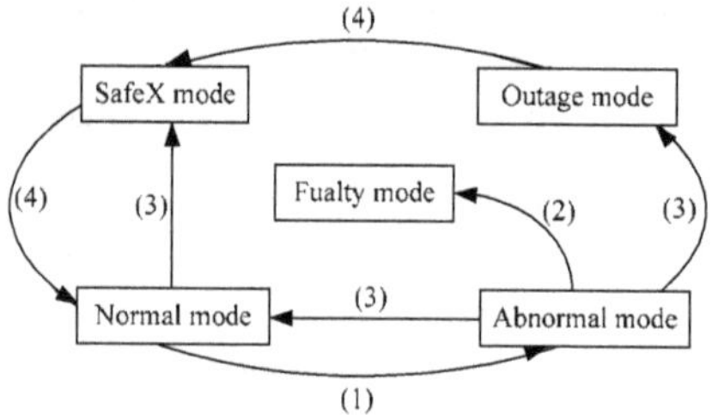

Fig. 2. The operating state transition

Abnormal mode: Large current passes from IED or an exceeding voltage drop is measured at neighbor zone.
Faulty mode: Agent has a faulted zone.
Normal mode: Neighbor zones operate normally.
Outage mode: There is a problem in both side zones.
SafeX mode: Exceeding voltage drop is measured. at Y zone

In the following, the proposed self-healing control framework including control structure, regulating rules and operating mechanism is presented regarding less of system rules or special case studies to provide the reusability and flexibility.

2.1. The agent architecture

In the MAS based control framework, the architecture of every agent as shown in Fig. 3 is composed of two layers, namely planning and operating layers.

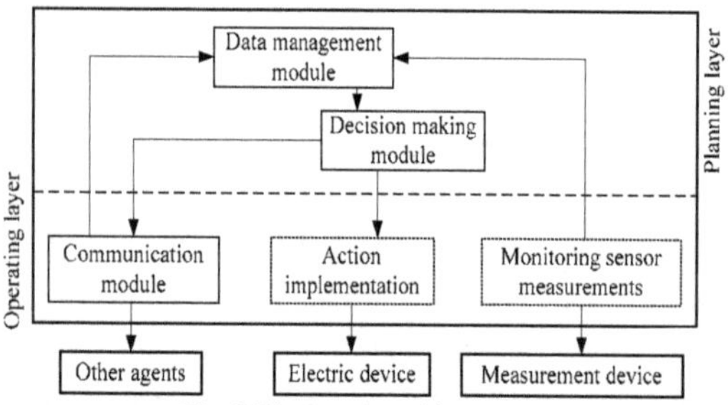

Fig. 3. The agent's architecture

The planning layer is the center of deliberation including data management, classification and decision making tasks, while the operating layer is related to the environmental perception and performance of agents containing communication based on Agent Communication Language (ACL) [24] and either monitoring measurements or action implementation. Considering agents' responsibility-es, the operating layer contains different modules. The major responsibility of zone agent is monitoring measurements at their correspondent zone and sending provided information to its neighbor zone. Therefore, their operating layer is equipped with "monitoring sensor measurements" and "comm-unication" modules. However, switch agents utilize the received information during their decision makings and carry out actions during their cooperation to heal the grid. Thus, their operating layer consists of "communication" and "action implementation" modules to give switch agents the ability of recognition and action. In addition, to this constructive difference, the functions are also defined for different modules in various levels.

2.2. Operating layer

Here, the performance of modules in operating layer is explained by the agents' type. On one hand, zone agent monitors the local measurements using the "monitoring sensor measurements" module and gives the provided uncertain description of local information to neighbor agents via "communi-cation" module. On the other hand, various types of switch agents receive the local information from their neighbor zones and communicate with other switch agents in different levels to be aware of next zones information. Sectionalizing agents (ScAgs) continuously communicate with their *X* and *Y* teammates, while Maneuvering Agents (MvAgs) communicate with other agents of the same type without considering their positions as well as their teammates. These coordination policies decrease the dependency of the algorithm to the accurate data transfer. Hence, the obtained results are more fault tolerance against any communication failure. During the negotiation process, every switch agent updates its information using received message as:

$$X_i=[ZInx_i, \Delta V_i, \Delta I_i]$$
$$X_{i+1}=[ZInx_{i+1}, \Delta V_{i+1}, \Delta I_{i+1}, \ldots, X_i] \quad (1)$$

Where, $ZInx_i$ denotes the ith agent's neighbor zone index, ΔV_i and ΔI_i represent the calculated voltage and current differences at both sides of ith zone. Moreover, switch agents affect on the power system when they carry out some actions such as closing their interfaces via "implementation action" module.

2.3. Planning layer

This layer contains two important modules which enable the agent to regulate its performance.

2.3.1. Data management module

In this module of zone agent architecture, online zone measurements including nodal voltage potential and branch current are managed and categorized separately. In switch agent architecture, the received data is classified and global information is recovered via a novel method to be used in optimal decision makings, considering the agents' local views. In this module, to discover global information about some parts of the grid some data packages are developed inspired by the link state packages which are introduced by routers in computer networks [25]. These packages consist of zones' information (i.e. zone index, operating measurements) exchanged among agents during their negotiation. Each data package starts with the neighbor zone index, and for each neighbor the voltage difference and current difference of both sides are given. In other words, this package includes data in three columns as zone index, voltage and current differences at both sides of the indexed zone. The data package rows are arranged in a particular order that points to the connectivity of zones next to agent which constructs the package. A sample of a data package constructed by $(i+1)^{th}$ agent with respect to Eq. (1) is shown in Table 1.

Table 1. Data package

The $(i+1)^{th}$ agent		
$ZInx_{i+1}$	ΔV_{i+1}	ΔI_{i+1}
$ZInx_i$	ΔV_i	ΔI_i

According to the order of rows in a data package and the zone indices presented in the first column, the global information including connectivity of zones placed next to the agent and their operating measurements are discovered.

2.3.2. Decision making module

Zone agent calculates the voltage and current differences at both sides of the relative zone and models the available uncertainties in measurements using proper methods. Different techniques have been proposed to deal with the uncertainties in power systems which can be generally classified in two categories: possibility theory [26-28] and probabilistic techniques. The probabilistic methods itself include two categories: simulation [29, 30] and analytical [31, 32]. In [26], for consideration of load uncertainty in load, loads in buses are assumed as triangular fuzzy numbers, while in [27] the uncertainty in load is modeled as trapezoidal fuzzy number. In [28] fuzzy variables of different distributions (triangular, trapezoidal or Gaussian distributions) are applied to reflect the power demand curves, which obviously change over 24-hour. In the first category of probabilistic methods (i.e. simulation methods) Monte Carlo simulation in conjunction with genetic algorithm has been used to solve the optimal Distributed Generation (DG) placement problem in presence of uncertainties [29]. In [30], the Monte Carlo simulation has been used to analyze the optimal power flow under the uncertainty of load. However in analytical approaches a Bayesian-based method has been proposed to forecast the uncertain power production of wind and photovoltaic generators and phase load demands [31]. Discrete Time Markov Chain (DTMC) process has been utilized to model the difference between photovoltaic generations and Load [32]. In this work, the uncertainties in loads and measurements are represented by fuzzy numbers with the same membership functions such that the defuzzifier, which has the highest membership value, is equal to online measurements at a zone.

In "decision making" module of the switch agent architecture, agents' decentralization is taken into account. This module enables such agents to perform their computational activities in a distributed manner. This policy avoids the extension of the efficacy of any agent's mistakes and provides MAS fault tolerance. Considering the switch agent types, the "decision making" module is equipped with different capabilities. For instance, ScAgs can make some comparison, simple decision makings and simple calculations using fuzzy arithmetic. These functions are regulated by expert-based rules related to following operational aspects of restoration problem. These expert-based rules utilize the operator's knowledge and experience to govern the control agents in order to achieve the self-healing objectives and guarantee the satisfaction of constraints.

- Radial topology of system should be maintained;
- The highest priority loads should be restored as soon as possible;
- The restoration is accomplished using minimum switching operations;
- Due to the radial topology of the system, the voltage drop is increased at the farthest node;
- In the radial faulted feeder, fault current flows from the substation to the lowest potential point;
- Due to the probabilistic nature of the DGs and hourly variations of the loads in the smart grids, online sensor measurements may vary during the restoration period. Hence, it is desirable to incorporate the uncertainty in the distribution system analysis.
- The online measured current difference at both sides of a zone (ΔI) points to an operational characteristic of that zone. Maximum possible voltage drop at a zone (ΔV) is calculated offline and indicates another operational characteristic of that zone.
- In order to satisfy the operational constraints during the transferring loads in restoration process, the available capacity of a backup feeder is calculated with voltage and line current limitations;
- The Allowable Additional Current (*AAC*) which passes feeder without any constraint violation is equal to minimum spare capacity of lines placed in the restoration path (i.e. path between the substation node and a connection node of a normally opened IED);

$$ZCur_i = I_l^{\max} - I_l$$
$$AAC = \min_i\{ZCur_i\} \qquad (2)$$

Where ZCuri indicates the additional current which is allowed to flow from the ith zone, I_l^{max} and I_l denote the lth line maximum current limitation and the current value flows from l^{th} line.

- The additional Allowable Voltage Drop (*AVD*) at a feeder indicates its voltage capacity;

$$AVC = 0.9 - v_i^{\min} \qquad (3)$$

Where v_i^{min} denotes the voltage magnitude at the lowest voltage point of the restoration path.

- Regarding to the last three mentioned points, an outage zone is transferred to a backup feeder to be restored if the feeder has sufficient capacity to deal with the required operational characteristics of an outage zone;
- If a feeder has sufficient capacity, it restores outage areas as a group; otherwise these areas are divided into some sections to be restored via some feeders. The remaining outage zones should be shed;

On the other hand, in MvAg's architecture, the decision making module is equipped with hybrid control policies to find the applicable and optimal solutions during healing process. These policies are designed using mathematical programming approaches combined with the knowledge-based systems. Although the knowledge-based systems obtain a good plan for scheduling switching operations during the healing process, they suffer from shortcomings in accomplishment. They require large and static data warehouse as well as, expensive offline computational activities to be extended. Moreover, they are not self-updatable and non-adaptable.

To propose a proper and easy applicable decision making method with the distributed structure of the proposed MAS, a mathematical programming approach is used in a more simplified manner. To this end, the information of out-of-service areas are discovered form the exchanged message and a new simple optimization model is constructed based on discovered data. During decision making process, limited information related to a part of the power system is used and some computational activities are distributed among various agents using the expert-based systems. Although, the introduced optimization model is formulated based on limited information of the power system, it gives an optimal and robust switching sequence for healing the main system.

3. MATHEMATICAL PROGRAMMING APPROACH

Mathematically, restoration problem has been modeled in the literature as a complex combinatorial optimization problem which requires the whole information of the grid to be solved in a centralized way. However, in this work a more simplified mathematical programming method is proposed as a decision making policy incorporating the agents' local views. To this end, the main grid is replaced by an auxiliary grid which is developed based on the classified information about the out-of-service areas. Then, the restoration problem in the auxiliary grid is mathematically modeled as a routing problem via a particular method. Routing problem is identified with the network topological characteristics and operational aspects of restoration. The obtained solution denotes an optimal restoration switching sequence.

3.1. Auxiliary grid construction

The auxiliary grid is a virtual grid which refers to un-faulted out-of-service areas. In the proposed distributed computing topology, the main power system is replaced by this auxiliary grid during the restoration decision making process. Constructing such grid needs some global information about out-of-service areas discovered using data packages. In this auxiliary grid, the outage zones, which their indices are presented in the first columns of the data packages, are connected together in a manner similar to package rows. Two mentioned characteristics of each zone, namely ΔV and ΔI, included in the second and third columns of data packages are considered as the representatives of zone's consumption information.

The auxiliary grid contains a capacity node connected to the grid at some zones which correspond to Y side zones of MvAgs in SafeX mode. The open/close situation of agents in the main grid identifies the connecting/disconnecting status of the correspondent outage zones and the capacity bus in the developed grid. The required information about the connected MvAgs is discovered during the MvAgs' negotiation.

3.2. Routing problem formulation

For the sake of simplicity simplification, the restoration problem in the main grid is replaced by a constrained routing problem in the auxiliary grid. The solution of routing problem is equivalent to a restoration plan which is applicable in the main grid. The proposed routing algorithm as a multi objective optimization problem seeks to find the lowest number of paths starting from the capacity bus to connect as many outage zones as possible in the grid, considering topological constraints and operational limitations related to restoration problem. Indeed, the results obtained by routing problem denote some paths which are considered as restoration paths in power grid. In other words, the outage zones placed on the obtained path are considered to be restored by a backup feeder which its relative MvAg is connected to the capacity us via the obtained path in the auxiliary grid.

In this routing problem, a binary variable, known as s_{ij}, is identified for the ith zone to indicate the connecting situation of this zone and the capacity bus via its jth output branch. This variable is introduced to see if there is a path in the auxiliary grid which starts from j^{th} output branch of capacity bus and contains the ith zone without forming any ring. This variable is "1" if in the determined path, i^{th} zone is connected to capacity bus from j^{th} output branch; otherwise it is considered as "0". The proposed method considers the radial topology of the grid during the definition of binary variables.

The objective of restoration problem is restoring as many outage loads as possible using minimum switching operations. From the perspective of the proposed routing problem, this objective equals to connecting as many zones as possible in the auxiliary grid via minimum number of paths starting from capacity bus. The objective function of routing problem is defined as

$$\max_{\min j} \sum_j \sum_i s_{ij} \tag{4}$$

The decision can be obtained by satisfying the

following constraints which are introduced based on the power system operational limitations and relative aspects of restoration problem. In the following formulations, the notation "~" indicates the uncertainty description of the correspondent parameters and "$\succeq$" represents the fuzzy ranking.

1- Voltage limits at the buses in the restored zone.

$$Z_j^{Rpath}\sum_i s_{ij}\Delta\tilde{I}_i+\sum_i s_{ij}\Delta\tilde{V}_i \preceq V_j^{Cn}-0.9^{p.u.} \tag{5}$$

where, V_j^{Cn} is the voltage at a junction node of jth backup feeder with the normally opened switching IED, Z_j^{Rpath} is the total impedance of restoration path from the substation to the junction node.

2- Voltage limits at buses placed in restoration path.

$$Z_j^{Rpath}\sum_i s_{ij}\Delta\tilde{I}_i - A\tilde{V}D_j \preceq 0 \tag{6}$$

where, $A\tilde{V}D_j$ represents the allowable voltage drop at the lowest voltage bus in the *jth* backup feeder.

3- Line current limits.

$$\sum_i s_{ij}\Delta\tilde{I}_i - A\tilde{A}C_j \preceq 0 \tag{7}$$

where, $A\tilde{A}C_j$ denotes the allowable additional current passes from branches of jth feeder.

4- Radial topology maintenance

An outage zone in auxiliary grid should be connected to capacity bus only via one path to not form a ring.

$$\sum_i s_{ij}-1\le 0,\forall j \tag{8}$$

5- Sequence of restored zones

From the perspective of routing problem, zones are connected to a capacity bus from the closest zone to the farthest zone, considering their connection order.

$$s_{i+1j}-s_{ij}\le 0 \tag{9}$$

6- Supply as many out-of-service zones as possible

If it is possible outage areas are supplied via lateral backup feeders.

$$\sum_i \Delta\tilde{I}_i - \sum_j\sum_i s_{ij}\Delta\tilde{I}_i \preceq 0 \tag{10}$$

The solution of this problem indicates each outage zone is re-energized by which feeder. According to the obtained solution, if some zones are still disconnected in auxiliary grid, their corresponding out-of-service areas should be shed.

3.3. Fuzzy binary linear programming

The proposed model is a fuzzy BLP problem with fuzzy constraints including fuzzy technological coefficients as well as the right hand side numbers. To the best of our knowledge, there is no work to approach such special type of fuzzy BLP problems. In this work a proper approach is presented to solve such problems to find the optimal restoration plan.

In the proposed model, the objective function is formulated as a weighted summation of the decision variables as well as the left hand side of the constraints. In this weighted summation, if any crisp binary decision variable is obtained as "1", its coefficient is kept in the summation; otherwise, the related coefficient is eliminated from the summation. Considering this fact, some constraints of the proposed model, satisfied by the obtained optimal solution, are formulated as the ranking operations on two fuzzy numbers. These two fuzzy numbers present at both sides of these constraints. As it clear, fuzzy arithmetic is utilized in the proposed restoration model. Therefore, some preliminaries are briefly reviewed in this section and then an approach is introduced for this special type of fuzzy BLP problem.

3.3.1. Basic definitions

Here, some preliminaries are discussed for the purpose of introducing a fuzzy BLP method.

Definition1. (Parametric form) [33]: Parametric form of a fuzzy number $\tilde{z}$ is a pair, $(L(r), R(r))$, of functions $L(r)$, $R(r)$, $0\le r\le 1$, satisfying the following requirements:

1. $L(r)$ is a bounded monotonically increasing left continuous function,
2. $R(r)$ is a bounded monotonically decreasing left continuous function,
3. $L(r)\le R(r)$, $0\le r\le 1$.

In decision makings, fuzzy number ranking is the most usable function that compares and orders fuzzy numbers. Various ranking fuzzy number methods have been reviewed in G. Bortolan and R. Degani [34]. Most of techniques select an alternative set and compare the alternatives instead of the fuzzy sets. In [35], the magnitude of fuzzy number $\tilde{z}$ with parametric form $\tilde{z}=(L(r),R(r))$ is defined as an alternative calculated using the following formulation,

$$\text{Mag}(\tilde{z}) = \frac{1}{2}\left(\int_0^1 (L(r) + R(r) + 2z_0)r\,dr\right) \quad (11)$$

As it is described in [22], the resulting scalar value of Eq. (11) is used to rank the fuzzy numbers. Indeed, the ranking of $\tilde{z}$ and $\tilde{v}$ is defined by magnitude function as:

1. $\text{Mag}(\tilde{z}) > \text{Mag}(\tilde{v})$ if and only if $\tilde{z} \succ \tilde{v}$.
2. $\text{Mag}(\tilde{z}) = \text{Mag}(\tilde{v})$ if and only if $\tilde{z} \approx \tilde{v}$.

3.3.2. Fuzzy BLP approach

In this work, the routing problem is modeled as a fuzzy BLP with a crisp and linear objective function as well as several linear equality and inequality constraints. Some of these constraints include the fuzzy technological coefficients as well as the right hand side numbers. They are taken into account as fuzzy number ranking operations satisfied by the optimal solution of the optimization problem. Such a fuzzy problem can be defuzzified by operating the raking function on the problem constraints, and then the surrogate crisp model can be solved using any crisp BLP approach. Consequently, the proposed fuzzy BLP approach uses the magnitude function to rank the fuzzy numbers on the both sides of any constraint and develops an auxiliary crisp BLP which can be solved by branch-and-bound technique [36].

4. DISTRIBUTED SELF-HEALING CONTROL FRAMEWORK

In the presented MAS, due to the fuzzy description of uncertainty in measurements, switch agents conduct some distributed computational activities to evaluate their situation using the arithmetic and ordering operations on fuzzy numbers. Furthermore, enough rules are developed to provide a feasible sequence of switching operations to guarantee the applicability of the given self-healing plan. The requirements of the decision making process such as power system operating constraints and restoration objectives are entirely considered with the expert-based knowledge.

4.1. Distributed calculation

Switch agents carry out some distributed calculations to determine the available capacity of backup feeders and discover the information of out-of-service areas using the following rules.

Rule1: When a ScAg in normal mode is asked about the available capacity of its related feeder, it forwards this message to its *X* teammates unless it belongs to the substation team.

Rule2: ScAg in substation team replies the capacity query with a message including the total transmission line impedance, the lowest bus voltage and the lowest value of spare current capacity of lines at its *X* zone.

Rule3: ScAg in normal mode determines some parameters and replies the capacity query after receiving the replication of its *X* teammates. ScAg calculates the Thevenin impedance related to path between substation and its *X* zone. It uses the impedance of its *X* zone lines and received impedance from its *X* teammates. It determines the lowest value among spar current capacity of lines at its *X* zone and the minimum current received from its *X* teammates related to the next zone. The lowest value among bus voltages at its *X* zone and the received minimum voltage is also calculated. This agent sends these parameters to its *Y* teammates.

Rule4: If MvAg in SafeX mode receives the replication of its *X* teammates, which are in normal mode, it repeats the distributed calculations mentioned in Rule3 to determine the allowable voltage drop and additional current of its correspondent backup feeder.

Rule5: If ScAg in Outage mode is asked about its *Y* zone requirements, it forwards the message to its *Y* teammates unless it belongs to the end or faulty team.

Rule6: ScAg in Outage mode replies the consumption query after attaching the information about its *Y* zone including the zone index, its $\Delta\tilde{V}$ and $\Delta\tilde{I}$ to the received message.

4.2. Fault location detection

The efficacy of a permanent fault is appeared only in the faulted feeder in a radial topology of a distribution system. Considering this fact, in the proposed MAS fault is localized by agents using following rules:

Rule1: If ScAg moves into the abnormal mode, it queries current measurements at its next zone by sending out "QUERY_IF" to its *Y* teammates.

Rule2: ScAg in abnormal mode replies the query with "INFORM True" if it measures exceeding current; otherwise it sends "INFORM False".

Rule3: If ScAg in abnormal mode only receives "False" it localizes the fault in its *Y* side zone.

4.3. Fault isolation

After detecting the fault location, the smallest possible area is isolated by agents using the following rules.

Rule1: ScAg in abnormal mode, who localizes the fault, opens its interface and moves into Faulty mode. It wants its *Y* teammates to perform similar actions.

Rule2: The interfaces of ScAgs in Faulty mode are allowed to be closed only if fault is repaired.

Rule3: Switch agents, who receive an extended loss of voltage from their neighbor zone agents open their interfaces and move into Outage mode.

The overall procedure of fault location detection and restoration is shown in Fig. 4.

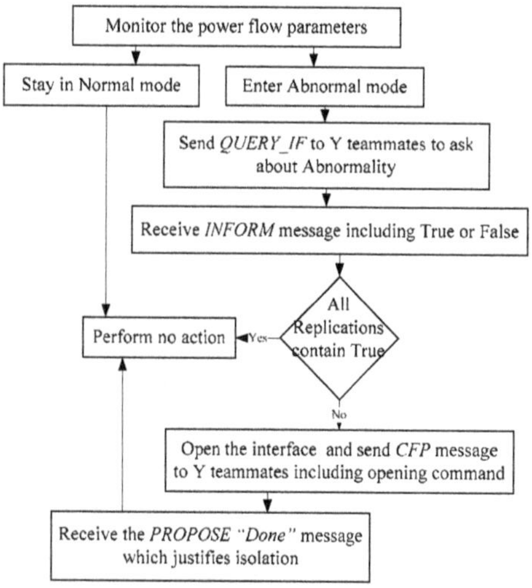

Fig. 4. Overall procedure of agents' operation to detect the fault location and isolate it

4.4. Priority load restoration

Regarding to the limited supply capacity, if the highest priority loads such as hospitals or big industrial centers place at a zone, that zone is considered as a highest priority zone for restoration.

Switch agents correspond to IEDs placed on the boundary of the priority zone are considered as high priority agents: High priority Sectionalizing Agents (HScAg) and High priority Maneuvering Agents (HMvAg). These high priority switch agents are responsible to find the best restoration path to reenergized high priority loads as soon as possible. They use the following policies.

Rule1: HScAg in Outage mode declares its request about re-energizing its neighbor zone by sending out "REQUEST" message containing the index of priority zone, its $\Delta\tilde{V}$ and $\Delta\tilde{I}$.

Rule2: If HScAg in Outage mode receives "CONFIRM close interface" from its *X* teammates, it closes its interface and moves into the normal mode.

Rule3: HMvAg in SafeX mode with maximum capacity has high priority to restore priority loads.

In addition, non-priority agents coordinate together using following rules:

Rule1: ScAgs in outage mode attaches the information of its *Y* zone to "REQUEST" message and forwards it to its X teammates.

Rule2: ScAg in outage mode closes its interface and forwards "CONFIRM" message.

Rule3: MvAg in SafeX mode considers its correspondent feeder. In addition, this agent utilizes data packages including information about outage zones placed in the priority path, to evaluate the possibility of restoration incorporating Eqs. (5) to (7). From the topological perspective, the path between MvAg and priority zone is known as priority-path. Rule4: MvAg in SafeX mode queries MnAgs' capacity by sending "QUERY_IF", if Eqs. (5) to (7) are satisfied; otherwise, it determines its capacity as zero.

Rule5: MvAg in SafeX mode replies the capacity queries with "INFORM" message including its non-zero capacity; otherwise it replies with "REFUSE".

Rule6: MvAg in SafeX mode with enough and maximum capacity closes its interface as a main maneuvering agent and sends out "CONFIRM" message to other ScAgs in Outage mode placed in priority-path to give them closing order.

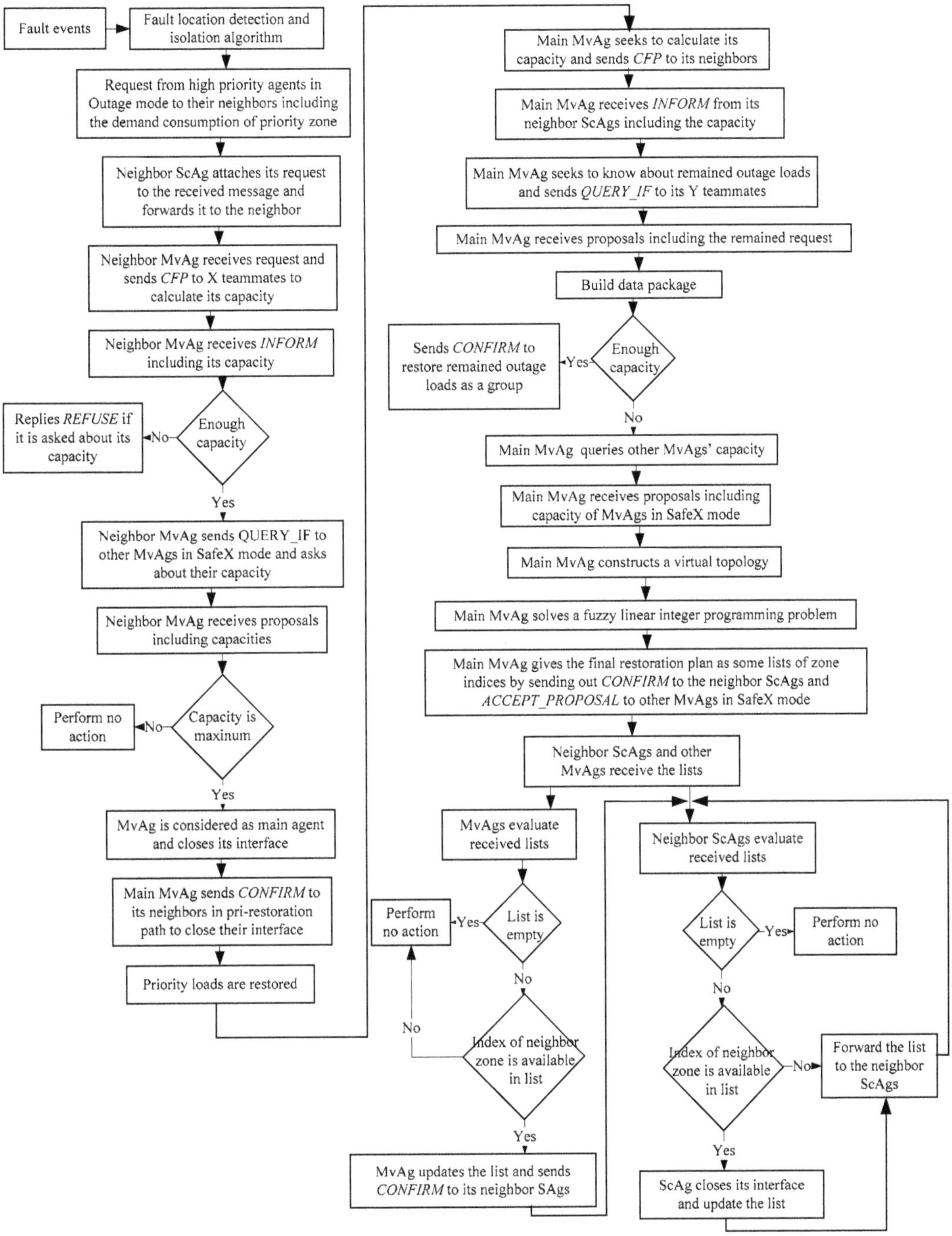

Fig. 5. Overall procedure of restoration phase using the proposed control agents

4.5. Non-priority load restoration

After restoring priority loads, the control switch agents cooperate together based on the following rules to restore as many remaining outage loads as possible.

Rule1: Main agent in normal mode updates its knowledge about its available capacity and the demand consumption of remaining outage zones.

Rule2: Main agent seeks to find the restoration plan with the performance of its decision making module.

Rule3: Main agent gives the founded plan as a list

of outage zone indices to MvAg in SafeX mode by sending "ACCEPT_PROPOSAL" message.

Rule4: Main agent sends "CONFIRM" message to its *Y* teammates and gives them a list of outage zone indices which should be restored by the main agent.

Rule5: ScAgs in Normal mode placed in priority-path forward the "CONFIRM" message.

Rule6: ScAgs in SafeX mode asks its *Y* teammates about the requirements of their *Y* side zones.

Rule7: ScAgs in SafeX mode forward the "CONFIRM" message to its *Y* teammates.

Rule8: ScAgs in Outage mode repeats rules 5 and 6 of subsection 4.1.

Rule9: ScAg in Outage mode closes its interface if it finds its *Y* zone index in the received list.

Rule10: MvAgs in SafeX mode close their interfaces if the received restoration plan is not empty.

Rule11: MvAgs in SafeX mode forwards the restoration plan to its *Y* teammates.

Rule10: MvAgs in SafeX mode close their interfaces if the received restoration plan is not empty.

Rule11: MvAgs in SafeX mode forwards the restoration plan to its *Y* teammates via a message.

Considering the mentioned rules, the overall procedure of agents' operating in the restoration phase of algorithm is shown in Fig. 5. As can be seen, the restoration phase is started by re-energizing the highest priority loads and then other outage loads are allowed to be restored. Furthermore, for clarification, the relationship between agents and their cooperation during the healing process is shown in Fig. 6.

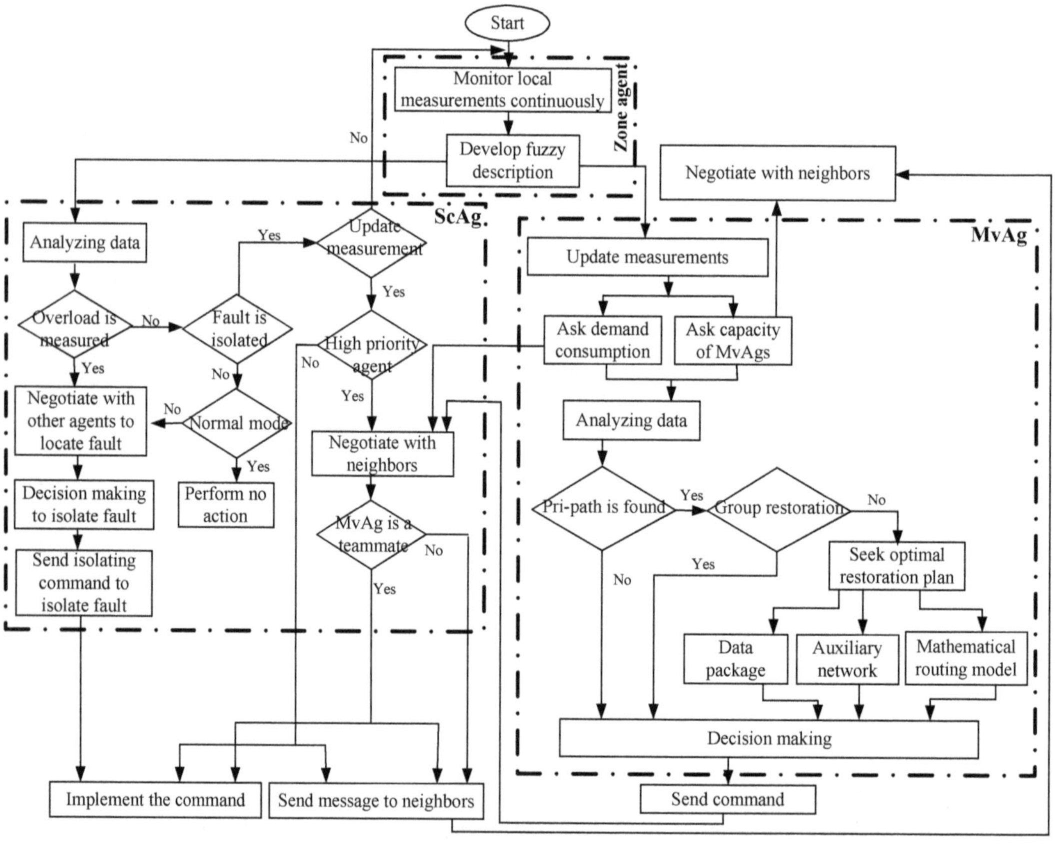

Fig. 6. The relationship among agents and their cooperation

As can be seen from Fig. 6, sectionalizing agents receive the measurements provided by neighbor zone agents and evaluate them to localize and isolate the fault. They cooperate with other switch agent types by sending data to them and implementing the received commands. On the other hand, maneuvering agents classify the received network information and develop a mathematical optimization for decision making problem using the managed data. According to the performance of agents and their cooperation, a whole pseudo-code for program of each agent is provided in appendix.

5. CASE STUDY

To validate the proposed robust self-healing MAS control algorithm, two distribution test systems including 70-nodes, 4-feeder [37] and IEEE 33-node [38] are selected as a case studies. It is assumed, the main distribution feeder has enough capacity to supply loads in normal operating condition. In this study, the distribution system is modeled using MATLAB simulator as a pilot system and the iterative simulation gives the online information as sensor measurements. Furthermore, the agents' coordination and communication services are provided in Java Agent Development Framework (JADE). Microsoft Excel is used as an interface to exchange information between Matlab and JADE platforms.

5.1. Case1. 70-node, 4-feeder distribution system

The proposed method is tested on a sample distribution system including two substations, 70 nodes and four feeders [37] as shown in Fig. 7 with zone numbers. In this system, feeders 1 and 2 are residential feeders, feeder 3 is an industrial feeder and feeder 4 is a commercial feeder. It is assumed that fault occurs in zone 1 at feeder 4 in duration of load growth and zone 10 contains large industrial loads which are considered as high priority loads for restoration. The uncertainty in loads is described as triangular fuzzy numbers shown in Fig. 8. Other shapes for fuzzy numbers based on operator insight or gathered information can also be used. Minimum load and maximum bus loads are 95% and 120% of the load with the highest membership value, respectively. With this assumption, the voltage and current differences at the both sides of a zone, branch current and also loading capacity of a backup feeder are considered as triangular fuzzy numbers with the same membership functions.

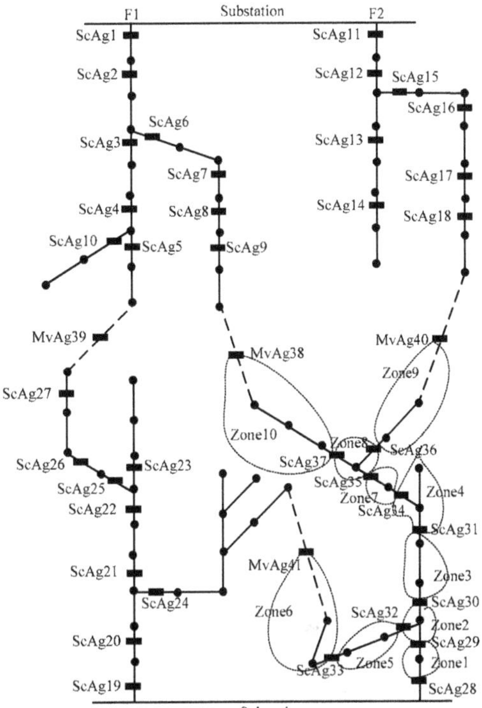

Fig. 7. The 70-node, 4-feeder test system

Zone agent creates these fuzzy descriptions by placing the sensor measurements at the center of fuzzy numbers with the highest membership degree. The measurements are monitored from the pilot system in the pre-fault condition. According to the possibility of load variation two separated periods are defined based on the load consumption, which are the lightest loading period (i.e. from 1 AM to 6 AM) and the highest loading period (i.e. from 5 PM to 10 PM). The proposed control method is tested on the mentioned distribution system in two scenarios considering these two periods and the fuzzy description of uncertainties in online measurements. The obtained switching sequences in each scenario are compared with the restoration plan presented in [21] to show the capability of the proposed method.

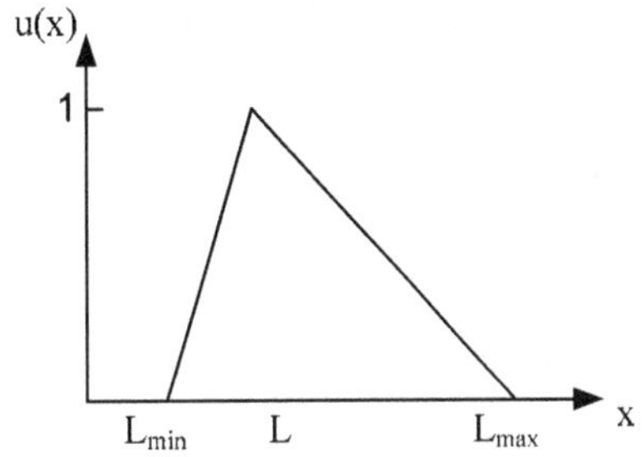

Fig. 8. Load as a triangular fuzzy number

5.1.1. Scenario 1

The fault current passes through switch 28 in the lightest loading period, and thus ScAg 28 concludes its situation is in the abnormal mode. It negotiates with its *Y* teammate, known as ScAg 29, to query about its operating situation. Fig. 9 illustrates the switch agents' negotiation during their cooperation to localize and isolate the fault. The fault is isolated by opening switches 28 and 29. After that, ScAgs 30 to 37 open their interfaces due to the extended loss of voltage and enter the outage mode. HScAg 37 in outage mode seeks a capable backup feeder to restore highest priority loads as soon as possible, so it sends the information of zone 10 to its teammates. MvAg 41 with the highest and enough supply capacity is selected as the main agent to close its interface and restore zone 10 considering the satisfaction of operating constraints modeled as Eqs. (5) to (7). This main agent sends a closing order to its Y teammate to be implemented and forwarded to other parts in restoration path. Now, MvAg 41 in normal mode starts negotiation to repeat the computation activities which obtain the new capacity of its correspondent feeder and the requirements of remaining outage zones to investigate the possibility of the group restoration.

In this case, due to satisfaction of Eqs. (5) to (7), MvAg concludes to implement the group restoration, and therefore it requests its Y teammates to close their interfaces. The "CONFIRM" message is forwarded and the ScAgs close their interfaces to restore outage areas. The obtained healing switching sequence is shown in Table 2. In this case study, the proposed method achieves the same healing plan in comparison with [21], while it presents fully distributed robust control framework including agents with local views. The proposed control algorithm uses the online sensor measurements and considers the possibility of their timely variations as uncertainty to provide enough robustness. In addition to using expert-based systems, it utilizes the mathematical programming to guarantee the optimality of obtained results and reduce the requirements to the expensive pre-computations, historical data and huge data bases which are necessary in [21].

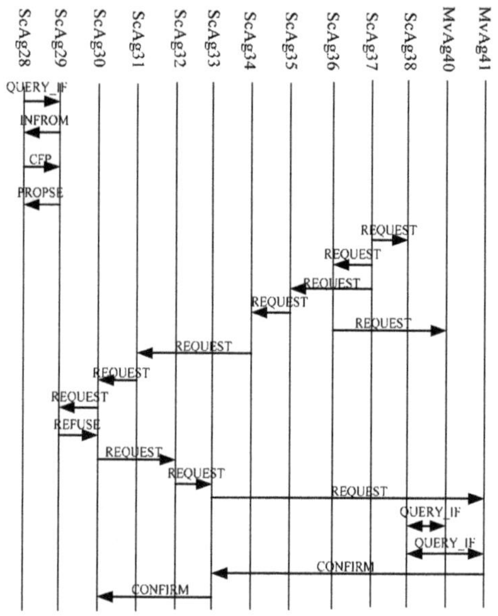

Fig. 9. The shortened list of exchanged message

5.1.2. Scenario 2

Here, it is assumed that fault occurs in highest loading period. Agents cooperate with each other in the same way as described in scenario 1 to locate and isolate the fault. The downstream agents also open their interfaces and move into the outage mode. HScAg 37 in outage mode seeks to find a backup feeder and sends its request to its teammates. MvAgs 38, 40 and 41 negotiate to its X teammates to calculate their available capacity after receiving the request of HScAg 37. Furthermore, these three MvAgs negotiate together and exchange information about their capacities. Consequently, MvAg 38 closes its interface as a main agent and restores zone 10. It repeats its distributed calculations explained in section 4.1 to investigate how remaining outage areas can be restored. Due to the lake of capacity for restoring whole of the outage loads via feeder F1, the main agent asks other MvAgs about their capacities. Moreover, it creates data packages to discover the global information about the outage areas.

MvAg 38 constructs an auxiliary grid shown in Fig. 10 and constructs the mathematical routing model for restoration problem in auxiliary grid as explained in section 4.

The solution of the mathematical programming problem gives the optimal paths in the auxiliary grid such that each path connects some zones in the network to the capacity node.

Table 2. The healing plan using switching operations

Step	ScAg28	ScAg29	ScAg30	ScAg31	ScAg32	ScAg33	ScAg34	ScAg35	ScAg36	ScAg37	MvAg38	MvAg40	MvAg41
I	0	0	1	1	1	1	1	1	1	1	0	0	0
II	0	0	0	0	0	0	0	0	0	0	0	0	0
III	0	0	1	1	1	1	1	1	0	1	0	0	1
IV	0	0	1	1	1	1	1	1	1	1	0	0	1

I. Isolation, II. Extended loss of voltage at downstream nodes, III. Priority load restoration, IV. Non-priority load restoration.

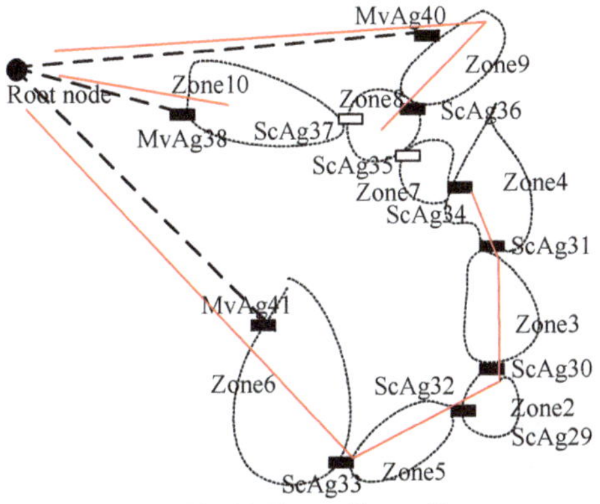

Fig. 10. The auxiliary grid

Considering the auxiliary grid and the obtained paths, shown in Fig. 10 by red lines, the optimal healing plan divides outage loads into separated groups such that each one is restored via a backup feeder. The switching sequences during the self-healing mechanism are illustrated in Fig. 11. The healing mechanism is divided into some sub-process such as I. Fault detection, II. Fault isolation, III. Extended loss of voltage, IV. Highest priority load restoration, V. Restoring non-priority outage loads via the main agent and VI. Complete the restoration process.

Comparing with [21], a similar switching sequence is obtained by the proposed control framework such that it utilizes three backup feeders to restore out-of-service areas in three separated groups using similar numbers of switching operations. This fact emphasizes the capability of the proposed method.

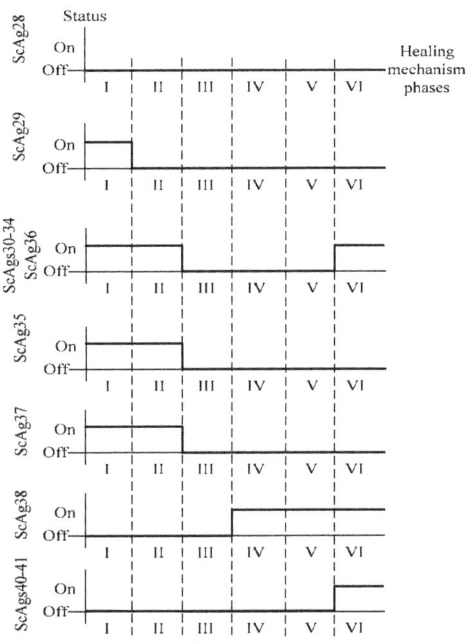

Fig. 11. The switching operations

5.2. Case 2. IEEE 33-node standard test system

The IEEE 33-node radial distribution test system includes 33 buses and 32 branches [38]. The 12.66 kV sub-station is located at node 1. All calculations are carried out in the p.u. system with 12.66 kV and 100 MVA as base voltage and power quantities, respectively. Considering the location of intelligent switching devices, 18 agent-controlled switching devices, including five MvAgs and 13 ScAgs, are added. Fig. 12 presents system with its zones and teams. According to the position of the intelligent switching devices, 13 teams are formed and 13 zone agents, ZAgs, are defined to monitor the measurements and present the fuzzy description of the available uncertainties. In the test system, the team 6 load is critical while others are not. A permanent balanced fault is considered at node 7, zone 3.

A comparison between two various scenarios is presented to demonstrate the robustness of the proposed self-healing plan in presence of the load uncertainties.

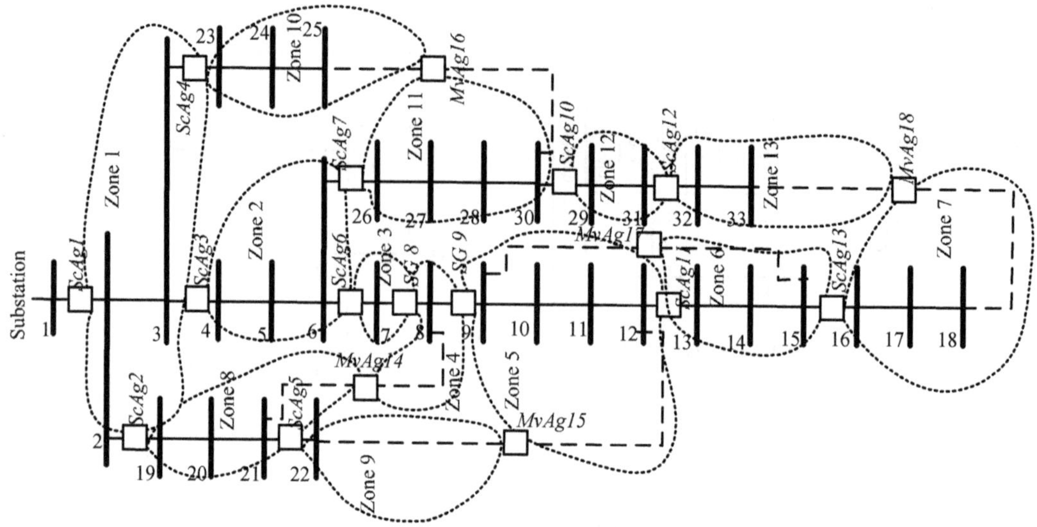

Fig 12. IEEE 33-node test system.

In the first scenario, the proposed method is implemented on the distribution system with the uncertain loads. The obtained results illustrate the capability of the proposed MAS architecture to control the self-healing process. The consideration of the uncertainties in the loads increases the robustness, reliability and applicability of the obtained self-healing plan. To demonstrate this fact, the viewpoint of the proposed MAS is deliberately changed in the second scenario and the responsibility of ZAgs is modified.

ZAgs monitor the pre-fault measurements and ignore the possibility of the load variations during the repair process of the faulted area. The agents in the modified MAS structure cooperate together with respect to measurements monitored by ZAgs, and obtain a new self-healing plan. The comparison between these two obtained self-healing plans demonstrates the robustness of the proposed method and shows the importance of the consideration of the load variations in decision makings. The results emphasize that disregarding the uncertainties may lead to a non-realistic self-healing policy.

5.2.1. Scenario 1. Consideration of uncertainty in loads

In this scenario, the time variation of loads in a smart distribution system and the uncertainty in the measurements are considered to increase the flexibility of the self-healing process and guarantee the applicability of the resulting switching plan. In this case the description of uncertainty is the same as case 1. With this assumption, the voltage difference and current difference on the both sides of each zone, branch current and also loading capacity of backup feeder are considered as triangular fuzzy numbers with the same membership functions. ZAgs are responsible to create these fuzzy descriptions, while, other agents carry out their tasks using this provided data during the various phases of the proposed method. The proposed approach controls the self-healing method using the switching device operations. The obtained sequence of the switching operations during the healing process is illustrated in Table 3.

As can be seen in Table 3, the given restoration policy divides the outage zones into two groups and utilizes two switching operations. Although this policy is more costly than a single-group restoration plan, the effectiveness of the obtained plan is guaranteed against the load variations during the faulted area repair time.

In this case, a permanent fault occurs at node 7 and the simulation ends while the MvAg 18 closes its interface and re-energizes the disconnected zone 7. To show the effectiveness of the proposed control framework, Table 4 illustrates the variations of the voltage magnitude of bus 14 as a critical node at zone 6.

Table 3. Switching sequences in scenario 1

Step	ScAg1	ScAg2	ScAg3	ScAg4	ScAg5	ScAg6	ScAg7	ScAg8	ScAg9	ScAg10	ScAg11	ScAg12	ScAg13	MvAg14	MvAg15	MvAg16	MvAg17	MvAg18
Pre-Fault																		
	1	1	1	1	1	1	1	1	1	1	1	1	1	0	0	0	0	0
Post-Fault																		
I	1	1	1	1	1	0	1	1	1	1	1	1	1	0	0	0	0	0
	1	1	1	1	1	0	1	0	1	1	1	1	0	0	0	0	0	0
II	1	1	1	1	1	0	1	0	0	1	0	1	0	0	0	0	0	0
III	1	1	1	1	1	0	1	0	0	1	0	1	0	1	0	0	0	0
	1	1	1	1	1	0	1	0	1	1	0	1	0	1	0	0	0	0
	1	1	1	1	1	0	1	0	1	1	1	1	0	1	0	0	0	0
IV	1	1	1	1	1	0	1	0	1	1	1	1	0	1	0	0	0	1

I. Isolation, II. Extended loss of voltage at downstream nodes, III. Priority load restoration, IV. Non-priority load restoration

As can be seen that, the voltage magnitude returns to its normal range and stays within this range after restoring the loads placed in pre-restoration path.

Table 4. Voltage magnitude of node14

Voltage (p.u.)	0.9205	0.0001	0.9644	0.9643
Self-healing phase	Pre-fault	Isolation	Priority load restoration	Complete restoration

5.2.2. Disregarding of uncertainty

In this scenario, consideration of the load uncertainties by the zone agents is ignored. In other words, ZAgs monitor the on-line measurements and provide them to the agents placed on the boundary of the relative zone. Indeed, the ZAgs ignore the possibility of the load variation during the fault repair time. According to this modification, switching device agents cooperate together to obtain a self-healing plan.

In this scenario, the performance of the agents is similar to the first scenario, except for when the non-priority out-of-service loads are allowed to be restored. In this part of the MAS algorithm, MvAg14 as a main manoeuvring agent updates its knowledge about the requirements of remaining disconnected loads and the available capacity by negotiating with its teammates, and investigates the possibility of the group restoration. Considering the satisfaction of Eqs. (5) to (7), the MvAg 14 concludes to restore the remaining disconnected outage loads as a single group and sends a closing command to its teammates asks other agents in restoration path to close their interfaces to complete the restoration process.

The sequence of the switching operations during the self-healing process is shown in Table. 4. As can be seen that, the modified MAS algorithm obtains a self-healing plan which restores loads as a single group. However, the accuracy and applicability of this plan in presence of the load variations, as an unavoidable concept of power systems, should be examined.

5.2.3. Robustness of the Proposed Method

In the real world, the probabilistic nature of the DGs and time varying loads lead to the uncertain behavior of the smart grids. Therefore, disregarding the uncertainty during the self-healing decision making may cause a non-realistic plan which is not implementable in real smart grids. To illustrate this fact, in this section the simulation conducted in the first scenario is repeated using the self-healing plan obtained in the second scenario to restore the out-of-service areas in the smart distribution system with the uncertain and time varying loads. Table 5 includes the on-line branch currents in the backup feeder after applying the policy.

As can be seen that in Table 5, group restoration causes backup feeder overloading in the distribution system with uncertain loads. The current magnitudes of the "line1" and "line18" violate their limitations when the switching sequence developed in the second scenario, is applied to the system with the uncertain in loads. Indeed, controlling the self-healing mechanism without the consideration of the uncertainty is not robust enough and may violate the operational constraints. It emphasizes that, considering the uncertainty in the measurements increases the reliability and guarantees the applicability of switching policy in healing process.

Table 5. Current magnitude of some lines

Line Num.	Zone	Cur. Mag. (Amp)	Max. Allowable Cur. (Amp)
Line1	Zone1	**512.14**	500
Line2		**330.43**	500
Line18		168.23	142
Line19	Zone8	155.55	200
Line20		143.11	200
Line21		12.37	200
Line8	Zone4	90.41	400
Line9	Zone5	82.50	200
Line10		74.66	200
Line11		68.24	200
Line12		59.88	200
Line13	Zone6	51.58	200
Line14		34.97	200
Line15		27.50	200
Line16	Zone7	19.71	200
Line17		11.95	200

6. CONCLUSION

A distributed comprehensive framework is introduced for implementing optimum and robust self-healing control activities in smart distribution systems with uncertain in loads. The contributions of this control algorithm are (1) Building a fully distributed architecture which considers agents with local views, decentralizations and limited communication; (2) Designing a control method which is self-updatable automatically using continues online measurements; (3) Designing a global information discovery algorithm using a proper data classification; (4) Proposing a routing problem as a simplified mathematical optimization method for restoration problem to guarantee the optimality of the obtained restoration plan; (5) Considering the possibility of load variation during the restoration period using fuzzy description of uncertain measurements to guarantee the robustness. The obtained results of testing the proposed algorithm on two standard test systems verify the applicability of the proposed method.

APPENDIX

In this section, pseudo-code for programming each type of agents is provided with its special responsibilities. Fig. 13 illustrates pseudo-code for programming zone agents. Given various capabilities of each class of switch agents, the relative pseudo-code is developed incorporating agents' operating situations. The pseudo-codes for programming sectionalizing and maneuvering agents are shown in Figs. 14 and 15 respectively.

```
Initialize voltage to zero;
Initialize line current to zero;
Initialize maximum limit of current;
Initialize line impedance to zero;
Initialize delta-v to zero;
Initialize delta-i to zero;
Repeat
 Initialize counter to zero;
 Input the voltage, line current, impedance;
 Compute the minimum spar capacity of lines;
 Compute minimum voltage;
 Set the delta-v to the calculated voltage difference at
both side of zone;
 Set the delta-i to the calculated current difference at
both side of zone;
 Generate the fuzzy description of measured data;
 Compute the total impedance of lines at zone;
 Send the determined parameters to the neighbors via a
message;
 Repeat
   Increment counter;
 Until counter reaches to its limit;
Until the self-healing mechanism is completed;
```

Fig 13. Pseudo-code for programming zone agents

```
Initialize the AgCnt to zero;
Initialize the MgCnt to zero;
Receive the uncertainty description of measurements;
Determine the operating situation;
If operating mode is Abnormal
  Set step to zero;
  Switch(step){
   Case 0:
     Send QUERY_IF" to its Y teammates to ask their
     situation;
     Repeat
       Increment AgCnt;
     Until the message sent to all the teammates;
```

```
    Increment step;
   Break;
  Case 1:
   Repeat
    Receive a replication;
    Investigate the content of message;
    Increment MgCnt;
   Until MgCnt is equal to AgCnt;
   Increment step;
  Break;
  Case 2:
   If all messages include "False"
    Fault is localized at Y side zone;
    Opening the interface;
    Entering the Faulton mode;
    Sending opening command to its Y teammates;
   Elseif receive opening command
    Opening the interface;
    Entering the Faulton mode;
    Forward confirm message to ensure the accurate
     procedure;
   Endif
   Increment step;
  Break;}
 If voltage drop is extended
  Open the interface;
  Enter the outage mode;
 Endif;
Elseif operating mode is Normal
 If agent is asked about the available capacity;
  If belongs to substation team;
   Compute the spar current capacity at X side zone;
   Compute allowable voltage drop at X side zone;
   Send calculated parameters to the Y teammates;
  Else
   Forward the message to its X teammates;
  Endif;
Endif;
If agent receives a replication from its X teammates
 Compute the spar current capacity at X side zone;
 Compute the allowable voltage drop at X side zone;
 Compute minimum value between received and
 calculated current capacity;
 Compute minimum value between received and
 calculated allowable voltage drop;
 Send the calculated values to Y teammates;
Endif;
If agent receives ordering command
 Implement the command or forward it;
Endif;
Elseif operating mode is Outage
 If agent is a high priority agent
  It sends the request message including the
  requirements of the high priority zone to its
  teammates;
 Endif;
 If agent is asked about the demand consumption of
 outage zone
  If agent is in Faulton mode
    It replies with REFUSE message and does not
    perform any action;
   Elseif
    If agent belongs to the end zone or receives
    REFUSE replication
     It replies by sending the information about the
     neighbor zone;
    Else
    It forwards the message;
    If it receives the replication of their teammates
     It attaches the information of the neighbor zone to
       the received message;
     It replies by sending the provided attachments;
    Endif;
   Endif;
  Endif;
Endif;
If agent receives the request of high priority agent
 Repeat
  It sums its neighbor zone requirements with the
  content of received message;
  It forward provided information to its teammates;
 Until the message received by an agent in Faulton
 mode;
Endif;
If agent receives a closing order
 It closes its interface;
 It forwards the received ordering message;
Endif
If agent receives a list of zone indices
 Processing the received list;
 If the index of the outage neighbor zone is available in
 the list
  Closing the interface;
  Eliminating the zone index from the list;
  Forwarding the list to the teammates;
 Else
  Performing nothing;
 Endif;
Endif;
Esleif operating mode is SafeX
 If agent is asked about the remaining outage zones'
 demand
  It sends a CFP to Y teammates and asks them about
 their requirements;
  It waits to receive the answer to calculate the sum of
 voltage differences at both sides of outage zones;
  It forwards the calculated information back;
 Endif;
Elseif operating mode is Faulton
 Agent does not perform any action;
Endif;
Endif;
```

Fig 14. Pseudo-code for sectionalizing agents

```
Initialize MvAg-Cnt to zero;
Receive the uncertainty description of measurements;
Determine the operating situation;
If operating situation is not SafeX mode
 Agent does not perform any action;
Else
 If agent receives the request of priority zone
   It sends a CFP message to its X teammates and asks
   them about the capacity;
   If it receives the replication from its X teammates
     It investigates the available capacity to see if it is
     sufficient for restoring the highest priority loads
     and the outage zones placed in the path (using (5)
     to (7));
   If the capacity is enough
     It asks other maneuvering agents about their
     capacity;
     It waits to receive replications;
     Repeat
       Receiving maneuvering agent replication;
       Increment MvAg-Cnt;
     Until MvAg-Cnt equals to the numbers of
     maneuvering agent minus one
     Processing the received capacities to find out
     maximum value;
     If agent has the highest capacity
       It closes its interface and enters to the Normal
       mode;
       It is considered as main maneuvering agent
       It sends the closing command to its Y teammates
       which gives the request message
     Else
       It does not perform any action
     Endif
   Else
     Set its capacity to zero
   Endif
 Endif
Endif
If agent is asked about its capacity
 It sends a CFP message to its X teammates and asks
 them about the capacity
 If it receives the replication from its X teammates
   It replies with sending its available capacity
 Endif
Endif
If agent receives a list of zone indices from the main
agent
 If the list is not empty
   Closing the interface
   Sending the message to its Y teammates including
   the list
 Else
   Performing no action
 Endif
End if
If agent is assigned as main agent
 It sends a message to its Y teammates to ask them
 about the requirements of remaining outage zones
 It sends a message to its X teammates to update its
 knowledge about the capacity of correspondent
 backup feeder
 It waits to receive replications
 If it receives replication
   Generating data packages
   Generating auxiliary grid
   Generating the rooting optimization problem
   Computing the solution of the rooting optimization
   problem using fuzzy BLP approach
 Endif
 Generating lists of zone indices by dividing given
 solution into some categories such that each one
 related to a maneuvering agent
 Sending the relative list of zone indices to another
 maneuvering agent
 If a list assign to it, is not empty
   Sending CONFIRM message including the list to its
   Y teammates
 Else
   Performing no action
 Endif
Endif
Endif
```

Fig 15. Pseudo-code for maneuvering agents

REFERENCES

[1] T. Nagata, S. Hatakeyama, M. Yasouka and H. Sasaki, "An efficient method for power distribution system restoration based on mathematical programming and operation strategy," *Proceedings of the International Conference of Power System Technology*, Perth, WA, pp. 1545-1550,2000.

[2] D. Pal, S. Kumar, B. Tudu, K.K. Mandal and N. Chakraborty, "Efficient and automatic reconfiguration and service restoration in radial distribution system using differential evolution," *Proceedings of the International Conference on Frontiers of Intelligent Computing: Theory and Applications Advances in Intelligent Systems and Computing*, vol. 199, pp. 365-372, 2013.

[3] T. Ananthapadmanabha, R. Prakash, M. Kumar Pujar, A. Gangadhara and M. Gangadhara, "System reconfiguration and service restoration of primary distribution systems augmented by capacitors using a new level-wise approach," *Journal of Electrical and Electronics Engineering Research*, vol. 3, no. 3, pp. 42-51, 2011.

[4] R. Perez-Guerrero, G. Heydt, N. Jack, B. Keel, and A. Castelhano, "Optimal restoration of distribution systems using dynamic programming," *IEEE Transactions on Power Delivery*, vol. 23, no. 3, pp. 1589-1596, 2008.

[5] A. Hussain, M.-S. Choi and S.-J. Lee "A novel algorithm for reducing restoration time in smart distribution systems utilizing reclosing dead time," *Journal of Electrical Engineering & Technology*, vol. 9, pp. 742-748, 2014.

[6] C.-C. Liu, S. Lee, and S. Venkata, "An expert system operational aid for restoration and loss reduction of distribution systems," *IEEE Transactions on Power Systems*, vol. 3, no. 2, pp. 619-626, 1988.

[7] W.P. Mathias-Neto "Distribution system restoration in a DG environment using a heuristic constructive multi-start algorithm," *Proceedings of the* IEEE/PES *Transmission and Distribution Conference and Exposition*, Sao Paulo, pp. 86-91, 2010.

[8] Y. Shi, S. Yao and Y. Wang, "Research on distribution system restoration considering distributed generation," *Proceedings of the China International Conference on Electricity Distribution*, pp. 1-5, 2010.

[9] A. Arya, Y. Kumar and M. Dubey, "Reconfiguration of Electric Distribution Network Using Modified Particle Swarm Optimization," *International Journal of Computer Applications*, vol. 34. no. 6, pp. 54-62, 2011.

[10] J. Stepien and S. Filipiak, "Evolutionary approach to restoration service of electric power distribution networks," *American Journal of Engineering Research*, vol. 2, no. 6, pp. 94-102, 2013.

[11] Y. Hsu and H. Huang, "Distribution system service restoration using the artificial neural network approach and pattern recognition method," *IEE Proceedings on Generation, Transmission, Distribution*, vol. 142, no. 3, pp. 251-256, 1995.

[12] A. Ahuja, S. Das and A. Pahwa "An ais-aco hybrid approach for multi-objective distribution system reconfiguration," *IEEE Transactions on Power Systems*, vol. 22, no. 3, pp. 1101-1111, 2007.

[13] Y. Kumar, B. Das and J. Sharma, "Multiobjective, multiconstraint service restoration of electric power distribution system with priority customers," *IEEE Transactions on Power Delivery*, vol. 23, no. 1, pp. 261-270, 2008.

[14] M. Tsai and Y. Pan, "Application of BDI-based intelligent multi-agent systems for distribution system service restoration planning," *European Transactions on Electrical Power*, vol. 21, pp. 1783-1801, 2011.

[15] T. Nagata and H. Sasaki, "A multi-agent approach to power system restoration," *IEEE Transactions on Power Systems*, vol. 17, no 2, pp. 457-462, 2002.

[16] X. Yinliang and L. Wenxin, "Novel multiagent based load restoration algorithm for microgrids," *IEEE Transactions on Smart Grid*, vol. 2, no. 1, pp. 152-161, 2011.

[17] J. M. Solanki, S. Khushalani and N. Schulz, "A multi-agent solution to distribution systems restoration," *IEEE Transactions on Power Systems*, vol. 22, no. 3, pp. 1026-1034, 2007.

[18] C.H. Lin, C.S. Chen, T.T. Ku, C.T. Tsai and C. Y. Ho, "A multiagent-based distribution automation system for service restoration of fault contingencies," *European Transactions on Electrical Power*, vol. 21, pp. 239-253, 2011.

[19] C.P. Nguyen and A.J. Flueck, "Agent Based Restoration With Distributed Energy Storage Support in Smart Grids," *IEEE Transactions on Smart Grid*, vol. 3, no. 2, pp. 1029-1038, 2012.

[20] H. Liu, X. Chen, K. Yu and Y. Hou, "The control and analysis of self-healing urban power grid," *IEEE Transactions on Smart Grid*, vol. 3, no. 3, pp. 1119-1129, 2012.

[21] E. Zidan and F. El-Saadany, "A cooperative multiagent framework for self-Healing mechanisms in distribution systems," *IEEE Transactions on Smart Grid*, vol. 3, no. 3, pp. 1525–1539, 2012.

[22] S.A. Arefifar, Y. A-R. Mohamed and T.H.M. EL-Fouly, "Comprehensive operational planning framework for self-healing control actions in smart distribution grids," *IEEE Transactions on Power Systems.*, vol. 28, no. 4, pp. 4192-4200, 2013.

[23] W. Ling and D. Liu, "A distributed fault localization, isolation and supply restoration algorithm based on local topology," *International Transactions on Electrical Energy Systems*, 2014.

[24] M. Wooldridge, "*Developing Multi-Agent Systems with JADE*", John Wiley & Sons Ltd, NY, USA 2007.

[25] A.S. Tanenbaum, "*Computer Networks*", Pearson Education LTD, 4th edn, 2003.

[26] M. R. Haghifam and O. P. Malik, "Genetic algorithm-based approach for fixed and switchable capacitors placement in distribution systems with uncertainty and time varying loads," *IET Proceedings on Generation, Transmission & Distribution*, vol. 1, no. 2, pp. 244-252, 2007.

[27] Y.Y. Hong and P.H. Chen, "Genetic-Based under frequency load shedding in a stand-alone power system considering fuzzy loads," *IEEE Transactions on Power Delivery*, vol. 27, no. 1, pp. 87-95, 2012.

[28] B. Wang, Y. Li and J. Watada, "Supply reliability and generation cost analysis due to load forecast

uncertainty in unit commitment problems," *IEEE Transactions on Power Systems*, vol.28, no.3, 2014.

[29] Z. Liu, F. Wen and G. Ledwich "Optimal siting and sizing of distributed generators in distribution systems considering uncertainties," *IEEE Transactions on Power Delivery*, vol. 26, no. 4, pp. 2541-2551, 2011.

[30] A. Bracale, G. Carpinelli, D. Proto, A. Russo and P. Varilone, "New approaches for very short-term steady-state analysis of an electrical distribution system with wind farms," *Energies*, vol. 3, no. 4, pp. 650-670, 2010.

[31] A. Bracale, P. Caramia, G. Carpinelli, A.R.D. Fazio and P. Varilone, "A bayesian-based approach for a short-term steady-state forecast of a fmart grid," *IEEE Transaction on Smart Grid*, vol.4, no. 4, pp. 1760-1771, 2013.

[32] M.N. Kabir, Y. Mishra, G. Ledwich, Z.Y. Dong and K.P. Wong, "Coordinated control of grid connected photovoltaic reactive power and battery energy storage systems to improve the voltage profile of a residential distribution feeder," *IEEE Transactions on Industrial Informatics*, vol.10, no.2, 2014.

[33] M. Ma, M. Friedman and A. Kandel, "A new fuzzy arithmetic," *Fuzzy Sets and Systems*, vol. 108, pp. 83-90, 1999.

[34] G. Bortolan and R. Degani, "A Review of Some Methods for Ranking Fuzzy Subsets," *Fuzzy Sets and Systems*, vol. 15, pp. 1-19, 1985.

[35] S. Abbasbandy and T. Hajjari "A new approach for ranking of trapezoidal fuzzy numbers," *Computers and Mathematics with Applications*, vol. 57, pp. 413-419, 2009.

[36] L. A. Wolsey, "*Integer Programming*", John Wiley & Sons Ltd , NY, USA, 1998.

[37] D. Das, "Reconfiguration of distribution system using fuzzy multi-objective approach," *Electrical Power and Energy Systems*, vol. 28, pp. 331-338, 2006.

[38] M. M. Hamada, M.A.A. Wahab and N.G.A. Hemdan, "Simple and efficient method for steady-state voltage stability assessment of radial distribution systems," *Electric Power Systems Research*, vol. 80, pp. 152-160, 2010.

Probabilistic Multi Objective Optimal Reactive Power Dispatch Considering Load Uncertainties Using Monte Carlo Simulations

S.M. Mohseni-Bonab[1], A. Rabiee[1], S. Jalilzadeh[1,*], B. Mohammadi-Ivatloo[2], S. Nojavan[2]

[1]Departemant of Electrical Engineering, University of Zanjan, Zanjan, Iran
[2]Faculty of Electrical and Computer Engineering, University of Tabriz, Tabriz, Iran

ABSTRACT

Optimal Reactive Power Dispatch (ORPD) is a multi-variable problem with nonlinear constraints and continuous/discrete decision variables. Due to the stochastic behavior of loads, the ORPD requires a probabilistic mathematical model. In this paper, Monte Carlo Simulation (MCS) is used for modeling of load uncertainties in the ORPD problem. The problem is formulated as a nonlinear constrained multi objective (MO) optimization problem considering two objectives, i.e., minimization of active power losses and voltage deviations from the corresponding desired values, subject to full AC load flow constraints and operational limits. The control variables utilized in the proposed MO-ORPD problem are generator bus voltages, transformers' tap ratios and shunt reactive power compensation at the weak buses. The proposed probabilistic MO-ORPD problem is implemented on the IEEE 30-bus and IEEE 118-bus tests systems. The obtained numerical results substantiate the effectiveness and applicability of the proposed probabilistic MO-ORPD problem.

KEYWORDS: Monte Carlo simulation, Multi objective optimal reactive power dispatch, Real power loss, Voltage deviation.

NOMENCLATURE

k	k-th network branch that connects bus i to bus j	J_{pu}	Normalized objective function
i / j	Bus number where $i, j = 1, 2, \ldots, N_B$	$J_r^{\max} / J_r^{\min}$	Maximum /minimum value for r-th objective function
g_k	Conductance of the line i-j	w_1	Weight of objective 1 (real power loss)
V_i	Voltage magnitude of bus i	w_2	Weight of objective 2 (voltage deviation)
θ_i	Voltage angle at bus i	PL	Real power loss
x	Vector of dependent variables	VD	Voltage deviation
u	Vector of control variables	N_D	Set of load bus
J	Total objective function	N_B	Number of buses
J_1	First objective function (PL=Real power loss)	ψ_k	Set of buses adjacent branch k
J_2	Second objective function (VD=Voltage deviation)	P_G	Active power in bus i

*Corresponding author:
S. Jalilzadeh (E-mail: jalilzadeh@znu.ac.ir)

Q_{Gi}	Reactive power generation in bus *i*
P_{D_i}	Real power of the *i*-th bus
Q_{D_i}	Reactive power of the *i*-th bus
$Y_{ij} = G_{ij} + jB_{ij}$	ij-th element of system Y_{bus} matrix
S_ℓ	power flow of ℓ -th transmission lin
S_ℓ^{max}	Maximum value of power flow of ℓ -th transmission line
V_i^{min} / V_i^{max}	Minimum/ Maximum value for voltage magnitude of the i-th bus
$Q_{Gi}^{min} / Q_{Gi}^{max}$	Minimum/ Maximum value for reactive power of the i-th bus

1. INTRODUCTION

Optimal Power Flow (OPF) affects both security and economy of power systems, and hence, it has to be considered as an integral part of power system operation and planning studies. The OPF can be divided into two sub-problems, Optimal Reactive Power Dispatch (ORPD) and optimal real power dispatch [1], [2].

1.1. Literature review

The ORPD problem is a complex problem in power systems and has attracted great attention in recent years, because it is strongly related to both economy and security of the system [3]. In most cases, the aim of ORPD is to optimize the following objective functions:

- Minimization of the network real power losses (as an economical objective).
- Optimization of voltage profile of the network, by minimizing voltage deviations from their nominal values in the load buses.

The aforementioned objectives are attained by regulating generator bus voltages, VAr compensators switching on/off, and optimization of transformer tap settings, with respect to various operational constraints such as load flow equations [4].

The ORPD problem is extensively studied in the literature. For instance, management and rescheduling of reactive power support via an ORPD model is presented in [3]. The objective function in [3] is to maximize voltage stability margin, at the same time as taking care of the economic dispatch of active power, by rescheduling the reactive power injection of synchronous generators and synchronous condensers. An objective function which depends on a voltage stability index is offered in [4], for solving ORPD problem. A model for ORPD is proposed in [5] for minimization of total costs, including energy loss of transmission network and costs of adjusting the control devices. A solution for the ORPD problem by Particle Swarm Optimization (PSO) based on multi-agent systems is proposed in [6]. A Seeker Optimization Algorithm (SOA) is suggested for ORPD taking into consideration static voltage stability [7]. In [8], a harmony search algorithm is proposed for partially solution of ORPD problem. A steady-state voltage stability constrained ORPD model is studied in [9]. In [10], an evolutionary-based approximation is presented for ORPD solution. This approach uses a differential evolution algorithm in order to determination of optimal settings of ORPD control variables. A particle swarm optimization, combined with a feasible solution search used for dealing with the ORPD problem in the presence of Wind Farms (WF) is presented in [11]. The proposed approach optimizes the reactive power dispatch, considering the reactive power requirement at the WF point of connection. A hybrid approach based on the evolutionary planning and particle swarm optimiz-ation is proposed in [12] to solve the ORPD problem. In [13], the behavior of different constraint controlling methods such as superiority of feasible solutions, self-adaptive penalty, ε-constraint, stochastic ranking, and the ensemble of constraint handling techniques on ORPD are investigated. A heuristic algorithm is introduced in [14] by combining modified teaching learning algorithm and double differential evolution algorithm until to handle the ORPD problem. Furthermore, in [15], a reliable and effective algorithm based on hybrid modified imperialist competitive algorithm and invasive weed optimization is proposed for solving the ORPD problem. Furthermore, a hybrid algorithm combining firefly algorithm and Nelder mead simplex method is represented in [16] for solution of ORPD problem.

A number of literatures study Multi Objective ORPD problem, considering the uncertainties. For example, a strength Pareto evolutionary algorithm is proposed in [17] to handle the MO-ORPD. A hierarchical clustering algorithm was suggested to provide a representative and manageable Pareto-optimal set. In [18], a reformed version of NSGA-II was applied by incorporating controlled elitism and dynamic crowding distance strategies in NSGA-II. The approach is utilized to solve the MO-ORPD problem by minimizing real power loss and maximizing the system voltage stability. A hybrid fuzzy multi objective evolutionary algorithm for solving complicated MO-ORPD problem is reported in [19], which considers voltage stability. A well-organized genetic algorithm method for solution of MO-ORPD problem is represented in [20], which considers fuzzy goal programming in uncertain environment. In [21], an advanced teaching learning based optimization algorithm is presented to solve MO-ORPD problem by minimizing real power loss, voltage deviation and voltage stability index. Chaotic improved PSO based multi-objective optimization and improved PSO-based multiobjective optimization approaches are prop-osed in [22], for solving MO-ORPD problem. The objective functions considered are power losses and L index. In [23], a multi objective chaotic parallel vector evaluated interactive honey bee mating optimization is presented to find the optimal solution of MO-ORPD problem considering operational restrictions of the generators.

It should be noted that few references have considered the possible uncertainties in the MO-ORPD problem. For example, in [24], a chance-constrained programming formulation is proposed to solve the MO-ORPD problem that considers uncertain nodal power injections and random branch outages.

1.2. Contributions

It is observed from the above literature survey that the MO-ORPD problem has been solved so far with lots of intelligent algorithms, but the uncertainty of load demand which is key factor in MO-ORPD problem was not investigated so far.

Load forecasting is usually performed based on the past and future information of the system such as weather condition, temperature and demand requirement. But, because of the random nature of load, the nonlinear relationship between the load and climate, and lack of precision in the prediction of climate, always the forecasted real and reactive demands are inaccurate and a certain degree of prediction errors exist. Therefore, it is necessary to consider the uncertainty of loads in the MO-ORPD problem.

Since this paper focuses on the uncertainties associated with the load, it is assumed that the statistical model of loads are estimated or measured. Due to the composite load modeling in the ORPD problem, the load is modeled by normal Probability Distribution Function (PDF) with a known mean and standard deviation, which are obtained from historical data and load forecasting programs.

The following well suited objective functions are considered in this paper for MO-ORPD:

- Minimization of real power losses
- Minimization of voltage deviation from the corresponding nominal value.

The main contributions of this study are summarized as follows:

1- The effect of uncertain nature of loads is studied in the MO-ORPD problem. The normal PDF is used for this aim.

2- Monte Carlo simulation (MCS) is used to solve the probabilistic MO-ORPD problem.

The numerical results substantiate the superiority of the proposed probabilistic MO-ORPD model in comparison with the existing heuristic algorithms.

1.3. Paper organization

The rest of this paper is organized as follows: Sections 2 and 3 describe the ORPD and MO-ORPD problem formulations, respectively. Implementation of deterministic MO-ORPD, MCS-based MO-ORPD problems and numerical results are presented in Sec. 4. Finally, the findings and conclusions of this paper are summarized in Sec. 5.

2. RPD PROBLEM FORMULATION

A system operator usually has various objectives such as minimization of sum of system transmission loss, and voltage deviation of load buses from their desired values etc. These objective functions may conflict with each other. Hence, at the first, the confliction between them is investigated.

2.1. ORPD objective functions

In this paper, the objective functions are minimization of real power losses and voltage deviations from the corresponding nominal values, in load buses.

2.1.1. Minimization of total real power losses

With the increasing rate of energy consumption, the amount of power losses are increased too, making the reduction of power losses as an important aim for system operators [25]. The active power losses can be expressed as follows [26].

$$J_1 = PL(x,u) = \sum_{\substack{k=1 \\ i,j\in\Psi_k}}^{N_L} g_k\left[V_i^2 + V_j^2 - 2V_iV_j\cos\left(\theta_i - \theta_j\right)\right] \quad (1)$$

2.1.2 Minimization of voltage deviations at load bus

The second aim of ORPD problem is to maintain a proper voltage level at load buses. Any electrical equipment is designed for optimum operation at a nominal voltage. Any deviation from this specified voltage decreases its efficiency, damages it, and reduces its useful lifetime. Thus, the voltage profile of the system should be optimized. This is accomplished by minimization of sum of voltage deviations from the corresponding rated values at load buses. This objective function is stated as follows [27]:

$$J_2 = VD(x,u) = \sum_{i=1}^{N_D}\left|V_i - V_i^{spc}\right| \quad (2)$$

2.2. Constraints

2.2.1. Equality constraints

The AC active/reactive power flows equations are expressed as follows.

$$P_{G_i} - P_{D_i} = V_i\sum_{j=1}^{NB}V_j\left[G_{ij}\cos\left(\theta_i - \theta_j\right) + B_{ij}\sin\left(\theta_i - \theta_j\right)\right] \quad (3)$$

$$Q_{G_i} - Q_{D_i} = V_i\sum_{j=1}^{NB}V_j\left[G_{ij}\sin\left(\theta_i - \theta_j\right) - B_{ij}\cos\left(\theta_i - \theta_j\right)\right]$$

2.2.2. Operational limits

The generators reactive power output and bus voltages should be hold in a pre-specified interval, as follows:

$$Q_{Gi}^{\min} \le Q_{Gi} \le Q_{Gi}^{\max} \quad (4)$$

$$V_i^{\min} \le V_i \le V_i^{\max} \quad (5)$$

Also, the line flow limits are as follows.

$$\left|S_\ell\right| \le S_\ell^{\max} \qquad \forall \ell \in NL \quad (6)$$

Besides, transformers' tap settings must be restricted by their lower and upper limits as follows:

$$T_i^{\min} \le T_i \le T_i^{\max} \quad (7)$$

3. MO-ORPID

Various methods are available to solve multi-objective optimization problems such as weighted sum approach [28], ε-constraint method [29] and evolutionary algorithms [30]. In this paper, the proposed multi-objective model of the MO-ORPD is solved using the weighted sum method. In this method, different weights are used for the conflicting objective functions to generate different Pareto optimal solutions. Hence, the overall objective function (which should be minimized) is the weighted sum of individual objective functions as follows:

$$\min\left[J(x,u)\right] = w_1J_{1,pu}\left(x,u\right) + w_2J_{2,pu}\left(x,u\right) \quad (8)$$

where,

$$w_1 + w_2 = 1 \quad (9)$$

The aforementioned MO-ORPD problem is mathematically a nonlinear constrained optimization problem. The decision variables including the control variables (i.e. u) and state variables (i.e. x) are as follows:

$$u^T = \left[\left[V_G\right]^T, \left[Q_C\right]^T, \left[T\right]^T\right]$$
$$x^T = \left[\left[V_L\right]^T, \left[Q_G\right]^T, \left[S_L\right]^T\right] \quad (10)$$

Since the objective functions Eqs. (1) and (2), do not have the same dimensions, in this paper, fuzzy satisfying method [31] is utilized to calculate the normalized (or per unit) form of both individual objective functions in Eq. (8). In the fuzzy satisfying method, a fuzzy membership number is defined for each objective function, which maps it to the interval [0, 1]. More generally, the *ith* objective function, J_i is normalized as follows.

$$J_{i,pu} = \begin{cases} 1 & J_i \le J_i^{\min} \\ \dfrac{J_i - J_i^{\max}}{J_i^{\min} - J_i^{\max}} & J_i^{\min} \le J_i \le J_i^{\max} \\ 0 & J_i \ge J_i^{\max} \end{cases} \quad (11)$$

In this paper for objective functions Eqs. (1) and (2), the normalized values are expressed as:

$$PL_{pu} = J_{1,pu} = \frac{PL - PL^{\max}}{PL^{\min} - PL^{\max}} \quad (12)$$

$$VD_{pu} = J_{2,pu} = \frac{VD - VD^{\max}}{VD^{\min} - VD^{\max}} \quad (13)$$

After running the MO-ORPD for different values of weighting factors, to select the best compromising solution, fuzzy satisfying method based on logistic membership function is used. After normalization the objective functions best solution is selected by using min-max operator.

4. CASE STUDY

All coding is implemented in General Algebraic Modeling System (GAMS) environment and solved by SBB solver. Simulations are carried out on the IEEE 30-bus and IEEE 118-bus systems. The IEEE 30-bus system consists of 30 buses, which its 6 buses are generator bus. The network has 41 branches, 4 transformers and 9 capacitor banks [32]. Hence, according to Eq. (9), total number of control variables is 25.

The IEEE 118-bus system consists of 118 buses, with 54 generator buses. Bus 69 is the slack bus. The network has 186 branches, 9 transformers and 14 capacitor banks [32]. The total number of control variables is 78. The initial operating point of the systems are given in [33]. In order to clearly illustrate the effectiveness of proposed method, a comparison is made between the results of two different cases:

(A) Deterministic optimization (ignoring the uncertainty in load).

(B) Uncertainty characterization using Monte Carlo simulation.

The simulation results are described as follows.

4.1. Case I – IEEE 30-bus test system

4.1.1 Deterministic Optimization

In deterministic case, the actual value of load is considered in the multi objective optimal reactive power dispatch problem. Real power loss and voltage deviation are considered as conflicting objective functions through Eq. (8). In order to solve the MO-ORPD problem by weighted sum method, maximum and minimum values of the expected real power loss (i.e. J_1) and voltage deviation (i.e. J_2) are calculated, which are 1.6012MW, 1.2577MW, 0.034pu and 0.0011pu, respectively.

These border values are achieved by maximize-ing and minimizing J_1 and J_2 individually as the objective function of MO-ORPD. Table 1 shows the values of both objective functions for all 21 Pareto optimal solutions. As explained in Sec. 3, in order to select the best solution from the obtained Pareto optimal set, a fuzzy satisfying method is utilized here. It is evident from the last column of Table 1 that the best solution is *Solution#2*, with the maximum weakest membership number of 0.8291. The corresponding *PL* and *VD* are equal to 1.316 MW and 0.0056 pu, respectively. For the above Pareto optimal set, the Pareto optimal front is depicted in Fig. 1. In this figure, the optimal compromise solution (i.e. *Solution#2)* is also specified.

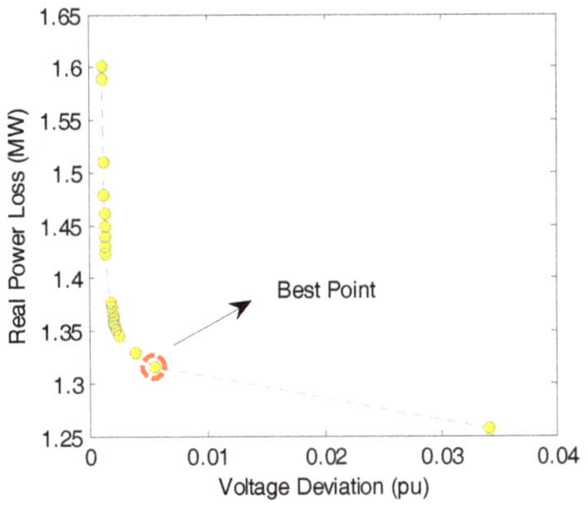

Fig. 1. Pareto optimal front for the IEEE 30-bus test system (Deterministic case)

Table1. Pareto optimal solutions for the IEEE 30-bus test system (Deterministic case)

#	W_1	W_2	*PL* or (J_1) (MW)	*VD (or J_2)* (pu)	$J_{1,pu}$	$J_{2,pu}$	Min ($J_{1,pu}$, $J_{2,pu}$)
1	1	0	1.2577	0.034	1	0	0
2	0.95	0.05	1.3164	0.0056	0.8291	0.8632	0.8291
3	0.9	0.1	1.329	0.0039	0.7924	0.9149	0.7924
4	0.85	0.15	1.3463	0.0025	0.7421	0.9574	0.7421
5	0.8	0.2	1.3511	0.0023	0.7281	0.9635	0.7281
6	0.75	0.25	1.3551	0.0022	0.7164	0.9666	0.7164
7	0.7	0.3	1.3587	0.0021	0.706	0.9696	0.706
8	0.65	0.35	1.362	0.002	0.6964	0.9726	0.6964
9	0.6	0.4	1.3655	0.002	0.6862	0.9726	0.6862
10	0.55	0.45	1.3692	0.0019	0.6754	0.9757	0.6754
11	0.5	0.5	1.3733	0.0019	0.6635	0.9757	0.6635
12	0.45	0.55	1.3778	0.0018	0.6504	0.9787	0.6504
13	0.4	0.6	1.423	0.0014	0.5188	0.9909	0.5188
14	0.35	0.65	1.4306	0.0014	0.4967	0.9909	0.4967
15	0.3	0.7	1.4393	0.0013	0.4713	0.9939	0.4713
16	0.25	0.75	1.4495	0.0013	0.4416	0.9939	0.4416
17	0.2	0.8	1.4622	0.0013	0.4047	0.9939	0.4047
18	0.15	0.85	1.4799	0.0012	0.3531	0.997	0.3531
19	0.1	0.9	1.5101	0.0012	0.2652	0.997	0.2652
20	0.05	0.95	1.5901	0.0011	0.0323	1	0.0323
21	0	1	1.6012	0.0011	0	1	0

4.1.2. Uncertainty modeling using MCS

In this section, a MCS-based procedure is considered to deal with the aforementioned load uncertainty [34]. The MCS is a numerical simulation procedure applied to the problems involving random variables with known or assumed probability distributions. It consists of repeating a deterministic simulation process, where in each simulation, a particular set of values for the random variables are generated according to their corresponding probability distributions. The results obtained in each iteration of MCS are similar to a deterministic simulation case. By collecting the results of many such MCS runs, it is possible to analyze the obtained results by statistical indices, such as mean (or average) value, standard deviation etc.

In the MCS the mean value (μ_{MCS}) and standard deviation (σ_{MCS}) for a given variable (or parameter) X are calculated as follows.

$$\mu_{MCS} = \frac{1}{N}\sum_{i=1}^{N} X_i \qquad (14)$$

$$\sigma_{MCS} = \sqrt{\frac{1}{N}\sum_{i=1}^{N}(X_i - \mu_{MCS})^2}$$

For load buses, the random variable to be considered in the MCS is load demand, due to its stochastic behavior. It is assumed that loads are normally distributed with a known mean value (corresponding to the forecasted value) and a known standard deviation in each bus. It is worth to note that, the mean value considered for each load, is its forecasted value which may be the peak or non-peak load. The appropriate values for each random variable are generally achieved from its probability distribution function or cumulative distribution function. In particular, the MATLAB function *randn* provides normally distributed random numbers directly. In this case, 10,000 random samples are picked up for considering the stochastic behavior of loads.

Here, for the sake of brevity just some statistical parameters such as mean, standard deviation and variance of the Pareto optimal solutions are reported. Table 2 gives the mean value of both objective functions for all 21 Pareto optimal solutions. Again, by using fuzzy satisfying method, *Solution#2* is the best. The Pareto optimal front of the objective functions is depicted in Fig. 2. The numerical values on the figure are mean, Standard Deviation (SD) and variance (Var) of the compromise optimal solution (i.e. *Solution#2*).

4.1.3. The obtained control variables

For *Solution#2* (i.e. the best compromise solution) the optimal values of control variables for both deterministic and probabilistic methods are given in Table 3. In the probabilistic case (i.e. by

MCS), the mean values for the control variables are given in this Table. One of the main advantages of MSC approach is that it gives the probability distribution of all uncertain variables. Histogram diagram is a proper tool for illustration of probability distribution. In this case, 4 randomly selected control variables and their corresponding probability distribution histograms are shown. The probability distributions for voltage of bus 11 (V_{g11}), real power output of generator located at bus 11 (P_{g11}), reactive power compensation in bus 24 and transformers tap changer between bus 28-27 are shown in Figs. 3-6, respectively.

Table 2. Pareto optimal solution for IEEE 30-bus test system (probabilistic case with MSC)

#	W_1	W_2	*PL (or J_1)* MW)	*VD (or J_2)* (pu)	$J_{1,pu}$	$J_{2,pu}$	Min ($J_{1,pu}$, $J_{2,pu}$)
1	1	0	1.262454	0.033635	1	0	0
2	**0.95**	**0.05**	**1.321376**	**0.005165**	**0.8291**	**0.8728**	**0.8291**
3	0.9	0.1	1.334024	0.003597	0.7924	0.9208	0.7924
4	0.85	0.15	1.351389	0.002306	0.7421	0.9604	0.7421
5	0.8	0.2	1.356207	0.002121	0.7281	0.9661	0.7281
6	0.75	0.25	1.360222	0.002029	0.7164	0.9689	0.7164
7	0.7	0.3	1.363836	0.001937	0.706	0.9717	0.706
8	0.65	0.35	1.367148	0.001845	0.6964	0.9746	0.6964
9	0.6	0.4	1.370662	0.001845	0.6862	0.9746	0.6862
10	0.55	0.45	1.374376	0.001752	0.6754	0.9774	0.6754
11	0.5	0.5	1.378491	0.001752	0.6635	0.9774	0.6635
12	0.45	0.55	1.383008	0.00166	0.6504	0.9802	0.6504
13	0.4	0.6	1.428379	0.001291	0.5188	0.9915	0.5188
14	0.35	0.65	1.436008	0.001291	0.4967	0.9915	0.4967
15	0.3	0.7	1.444741	0.001199	0.4713	0.9944	0.4713
16	0.25	0.75	1.454979	0.001199	0.4416	0.9944	0.4416
17	0.2	0.8	1.467727	0.001199	0.4047	0.9944	0.4047
18	0.15	0.85	1.485494	0.001107	0.3531	0.9972	0.3531
19	0.1	0.9	1.515808	0.001107	0.2652	0.9972	0.2652
20	0.05	0.95	1.596111	0.001015	0.0323	1	0.0323
21	0	1	1.607253	0.001015	0	1	0

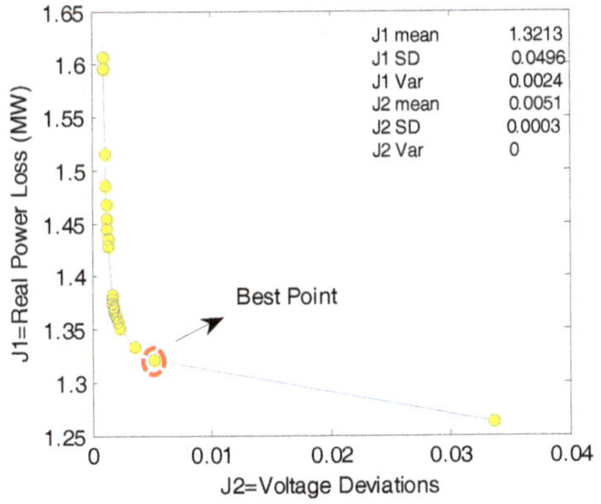

Fig. 2. Pareto optimal front for IEEE 30-bus test system (probabilistic case with MSC)

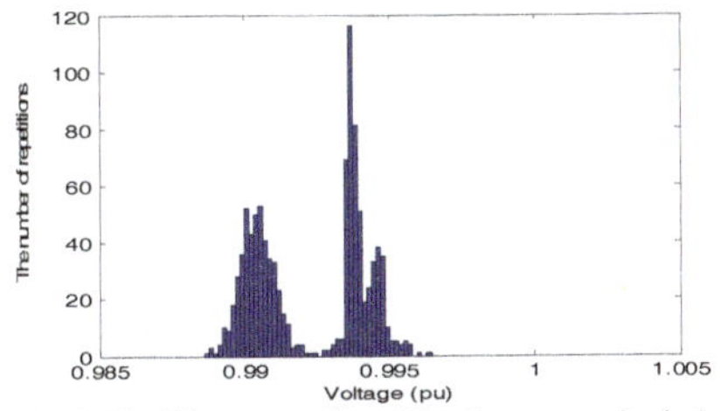

Fig. 3. The histogram of bus 11 voltage magnitude (pu)

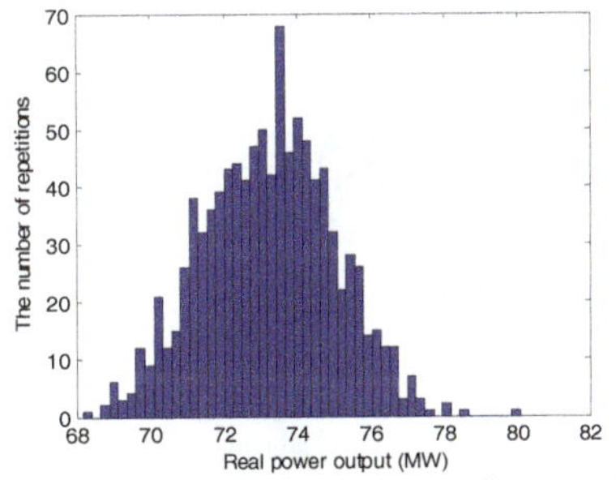

Fig. 4. The histogram of real power output of generator located at bus 11

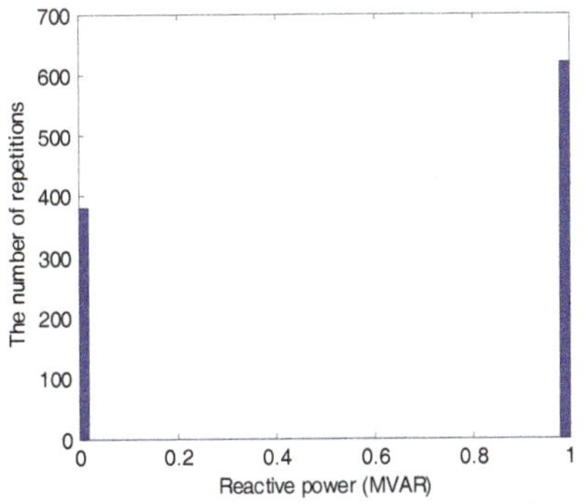

Fig. 5. The histogram of reactive power compensation at bus 24

Table 3. The obtained control variables for best compromise deterministic and probabilistic solutions (IEEE 30-bus system)

Control Variable	Deterministic	probabilistic
V_{g1}(pu)	0.998	1.0002
V_{g2}(pu)	0.998	1.0003
V_{g5}(pu)	0.9969	0.9989
V_{g8}(pu)	0.9986	1.0009
V_{g11}(pu)	0.9904	0.9923
V_{g13}(pu)	1.0098	1.01
P_{g1}(MW)	3.507	3.5177
P_{g2}(MW)	29.2979	29.8066
P_{g5}(MW)	100	99.5501
P_{g8}(MW)	45.3168	45.6626
P_{g11}(MW)	73.4394	73.2205
P_{g13}(MW)	33.1679	33.1967
Q_{c10}(MVar)	0	0
Q_{c12}(MVar)	0	0
Q_{c15}(MVar)	0	0
Q_{c17}(MVar)	10	8.05
Q_{c20}(MVar)	0	0.42
Q_{c21}(MVar)	10	10
Q_{c23}(MVar)	0	0
Q_{c24}(MVar)	10	6.2
Q_{c29}(MVar)	0	0
$T_{6\text{-}9}$	0.9657	0.9672
$T_{6\text{-}10}$	1.1	1.0895
$T_{4\text{-}12}$	0.9906	0.9898
$T_{28\text{-}27}$	0.9936	0.9926

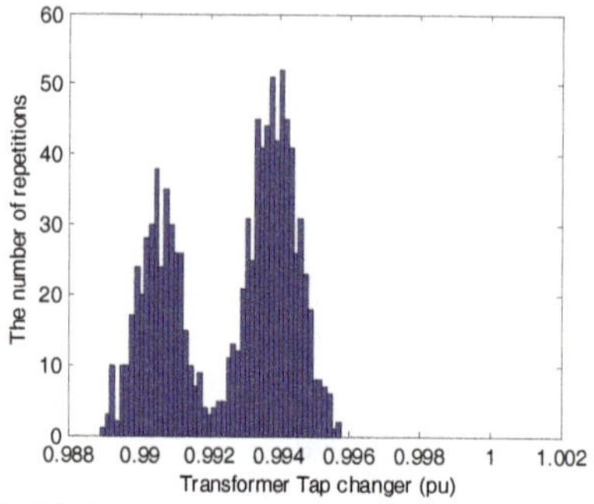

Fig. 6. . The histogram of tap ratio for the transformer between buses 27 and 28

4.2. Case II – IEEE118 bus test system

For the sake of brevity, in this case only the results obtained in probabilistic case are presented and for deterministic case only a summary of the obtained results are presented for the aim of comparison with MCS solutions.

4.2.1. Uncertainty modeling using MCS

In this case 10,000 different samples with normal PDF are selected. In this case, 11 different Pareto optimal solutions are derived. Table 4 summarizes the obtained results using MCS for case II. The corresponding Pareto front is depicted in Fig. 7.

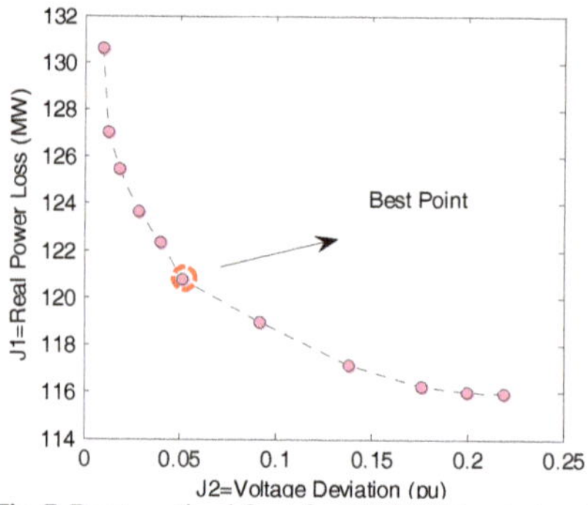

Fig. 7. Pareto optimal front for IEEE 118-bus test system

4.2.2. The obtained control variables

Table 5 summarizes the obtained control variables for the best compromise solutions in the deterministic and probabilistic approaches. It is noted worthy that in the probabilistic case the mean values for the control variables are given in this Table.

Table 4. Pareto optimal solution for IEEE 118-bus test system (probabilistic case with MSC)

#	W_1	W_2	*PL (or J_1)* (MW)	*VD (or J_2)* (pu)	$J_{1,pu}$	$J_{2,pu}$	Min ($J_{1,pu}$, $J_{2,pu}$)
1	1.0	0.0	115.9607	0.2191	1	0	0
2	0.9	0.1	116.02456	0.1992	0.9956	0.0951	0.0951
3	0.8	0.2	116.2456	0.1761	0.9805	0.2054	0.2054
4	0.7	0.3	117.1425	0.1378	0.9192	0.3884	0.3884
5	0.6	0.4	118.9547	0.0914	0.7952	0.6101	0.6101
6	**0.5**	**0.5**	**120.7648**	**0.0512**	**0.6714**	**0.8022**	**0.6714**
7	0.4	0.6	122.3565	0.0395	0.5625	0.8581	0.5625
8	0.3	0.7	123.6466	0.0286	0.4742	0.9102	0.4742
9	0.2	0.8	125.4579	0.0184	0.3503	0.9589	0.3503
10	0.1	0.9	127.0545	0.0122	0.2411	0.9885	0.2411
11	0	1.0	130.5794	0.0098	0	1	0

Table 5. The obtained control variables for best compromise deterministic and probabilistic solutions (IEEE 118-bus system)

Control Variable	Deterministic	probabilistic
Vg1	0.9902	1.028
Vg4	1.0237	0.9985
Vg6	1.0137	1.0145
Vg8	1.0325	1.041
Vg10	1.0459	1.0482
Vg12	1.0109	1.0141
Vg15	1.0072	1.0079
Vg18	1.0091	1.0095
Vg19	1.0058	1.0062
Vg24	1.0202	1.0225
Vg25	1.0473	1.0512
Vg26	1.0547	1.0552
Vg27	1.0116	1.0222
Vg31	1.0037	1.0045
Vg32	1.0084	1.0096
Vg34	1.0166	1.0215
Vg36	1.0138	1.0211
Vg40	0.9989	1.0882
Vg42	1.0032	1.0085
Vg46	1.0145	1.0215
Vg49	1.0316	1.0385
Vg54	1.0063	1.0021
Vg55	1.0048	1.0085
Vg56	1.0051	1.0251
Vg59	1.0253	1.0524
Vg61	1.0256	1.0595
Vg62	1.0225	1.0925
Vg65	1.0344	1.0352
Vg66	1.0443	1.0487
Vg69	1.0503	1.0821
Vg70	1.0158	1.0165
Vg72	1.0111	1.0194
Vg73	1.0111	1.0185
Vg74	0.9949	0.9954
Vg76	0.9802	0.9882
Vg77	1.0212	1.0338
Vg80	1.0325	1.0327
Vg85	1.0315	1.0419
Vg87	1.0105	1.0228
Vg89	1.0574	1.0612
Vg90	1.0293	1.0298
Vg91	1.0282	1.0325
Vg92	1.0352	1.0654
Vg99	1.0215	1.0502
Vg100	1.0298	1.0315
Vg103	1.0191	1.0223
Vg104	1.0089	1.0199
Vg105	1.0063	1.0092
Vg107	0.9944	0.9904
Vg110	1.0021	1.0121
Vg111	1.0094	1.0099
Vg112	0.9879	0.9808
Vg113	1.0166	1.0187
Vg116	1.0262	1.0254
Pg69(MW)	500.7689	550.8577
Qc5(MVar)	0	0
Qc34(MVar)	2.8801	
Qc37(MVar)	0	0
Qc44(MVar)	10	
Qc45(MVar)	10	
Qc46(MVar)	10	
Qc48(MVar)	4.1331	
Qc74(MVar)	12	
Qc79(MVar)	20	
Qc82(MVar)	20	
Qc83(MVar)	10	
Qc105(MVar)	18.5654	
Qc107(MVar)	0	0
Qc110(MVar)	10	
$T_{8\text{-}5}$	0.9982	0.9987
$T_{26\text{-}25}$	1.056	1.058
$T_{30\text{-}17}$	1.0025	1.0029
$T_{38\text{-}37}$	0.9994	0.9995
$T_{63\text{-}59}$	1.0022	1.0028
$T_{64\text{-}61}$	1.0023	1.0151
$T_{65\text{-}66}$	1.0023	1.0098
$T_{68\text{-}69}$	1.0121	1.0138
$T_{81\text{-}80}$	1.0102	1.0115

4.3. Discussion on the obtained results

Since the probabilistic MO-ORPD considering the load uncertainty is not reported in the previous literature, investigation of the performance of the proposed method is only possible by comparison of the obtained results in the deterministic case with the previously reported results in the literature.

Table 6 compares the obtained deterministic results for Cases I and II with the previously published works. In this table the results for minimization of both objective functions (J_1 and J_2) are compared with the heuristic methods.

In Tables 1, 2 and 4 the *solution*#1 is the case in which the only real power loss is minimized. *Solution* #11 is the case where the voltage deviations are minimized. It is observed from Table 6 that in both cases the probabilistic MCS-based approach results better solutions than the previously published methods.

Table 6. Comparison of obtained results for deterministic cases with previously published methods

J_1	Real Power Loss (MW)			
Method	Proposed	DE [13]	CPVEIHBMO [23]	QOTLBO [21]
Case I	1.3164	4.8623	5.3243	5.2594
Case II	119.7686	129.579	124.0983	134.4059
J_2	Voltage Deviation (pu)			
Method	Proposed	DE [13]	CPVEIHBMO [23]	QOTLBO [21]
Case I	0.0056	0.0911	0.87664	0.121
Case II	0.0498	---	0.7397	0.24

5. CONCLUSIONS

This paper proposes a probabilistic approach for MO-ORPD problem. In this model, the technical constraints as well as the load uncertainty are taken into consideration. The stochastic nature of load is modeled using Monte Carlo simulations. Mixed integer nonlinear programming is used to solve the proposed probabilistic MO-ORPD problem. In order to evaluate the effectiveness of the proposed model, it is implemented on the IEEE 30-bus and IEEE-118 bus test systems. The numerical results demonstrate the effectiveness of the proposed methodology. The main advantages of this study are summarized as follow:

- Using MCS for load uncertainty modeling is a help system operator to have realistic decisions.
- The solutions obtained in the deterministic case, are better than the results attained by heuristic algorithms.
- The MCS approach gives the probability distribution of all output variables such as bus voltages, line flows etc. This is an important result, since the probability distribution of any uncertain parameter shows the statistical behavior of it. Hence, system operator can use the proposed MCS-based MO-ORPD problem for determination of optimal probability distribution of important variables such as power losses, voltage levels etc. “”

REFERENCES

[1] L. Shi, C. Wang, L. Yao, Y. Ni and M. Bazargan, “Optimal power flow solution incorporating wind power, ” *IEEE Systems Journal,* vol. 6, no. 2, pp. 233-241, 2012.

[2] T. Niknam, M. Narimani, J. Aghaei, S. Tabatabaei, and M. Nayeripour, “Modified Honey Bee Mating Optimisation to solve dynamic optimal power flow considering generator constraints, ” *IET Proce. on Generation, Transmission & Distribution,* vol. 5, no. 10, pp. 989-1002, 2011.

[3] A. Rabiee and M. Parniani, “Optimal reactive power dispatch using the concept of dynamic VAR source value,” *Proceedings of the IEEE Power & Energy Society General Meeting,* pp. 1-5, 2009.

[4] A. Rabiee, M. Vanouni and M. Parniani, “Optimal reactive power dispatch for improving voltage stability margin using a local voltage stability index, ”*Energy Conversion and Management,* vol. 59, pp. 66-73, 2012.

[5] Y.j. Zhang and Z. Ren, “Optimal reactive power dispatch considering costs of adjusting the control devices, ” *IEEE Transactions on Power Systems,* vol. 20, no. 3, pp. 1349-1356, 2005.

[6] B. Zhao, C. Guo and Y. Cao, “A multiagent-based particle swarm optimization approach for optimal reactive power dispatch,” *IEEE Transactions on Power Systems,* vol. 20, no. 2, pp. 1070-1078, 2005.

[7] C. Dai, W. Chen, Y. Zhu and X. Zhang, “Reactive power dispatch considering voltage stability with seeker optimization algorithm, ” *Electric Power Systems Research,* vol. 79, no. 10, pp. 1462-1471, 2009.

[8] A. Khazali and M. Kalantar, “Optimal reactive power dispatch based on harmony search algorithm,” *International Journal of Electrical Power & Energy Systems,* vol. 33, no. 3, pp. 684-692, 2011.

[9] J. M. Ramirez, J. M. Gonzalez and T. O. Ruben, “An investigation about the impact of the optimal reactive power dispatch solved by DE,” *International Journal of Electrical Power & Energy Systems,* vol. 33, no. 2, pp. 236-244, 2011.

[10] A. Ela, M. Abido and S. Spea, “Differential evolution algorithm for optimal reactive power dispatch, ” *Electric Power Systems Research,* vol. 81, no. 2, pp. 458-464, 2011.

[11] M. Martinez-Rojas, A. Sumper, O. Gomis-Bellmunt and A. Sudrià-Andreu, “Reactive power dispatch in wind farms using particle swarm optimization technique and feasible solutions search,” *Applied energy,* vol. 88, no. 12, pp. 4678-4686, 2011.

[12] C.M. Huang and Y.C. Huang, “Combined differential evolution algorithm and ant system for optimal reactive power dispatch,” *Energy Procedia,* vol. 14, pp. 1238-1243, 2012.

[13] R. Mallipeddi, S. Jeyadevi, P. N. Suganthan and S. Baskar, “Efficient constraint handling for optimal reactive power dispatch problems,” *Swarm and Evolutionary Computation,* vol. 5, pp. 28-36, 2012.

[14] M. Ghasemi, M.M. Ghanbarian, S. Ghavidel, S. Rahmani and E. Mahboubi Moghaddam, “Modified teaching learning algorithm and double differential evolution algorithm for optimal reactive power dispatch problem: A comparative study,” *Information Sciences,* vol. 278, pp. 231-249, 2014.

[15] M. Ghasemi, S. Ghavidel, M.M. Ghanbarian and A. Habibi, “A new hybrid algorithm for optimal reactive power dispatch problem with discrete and continuous control variables,” *Applied Soft Computing,* vol. 22, pp. 126-140, 2014.

[16] A. Rajan and T. Malakar, “Optimal reactive power dispatch using hybrid Nelder–Mead simplex based firefly algorithm,” *International Journal of*

Electrical Power & Energy Systems, vol. 66, pp. 9-24, 2015.

[17] M. Abido and J. Bakhashwain, "Optimal VAR dispatch using a multiobjective evolutionary algorithm," *International Journal of Electrical Power & Energy Systems,* vol. 27, no. 1, pp. 13-20, 2005.

[18] S. Jeyadevi, S. Baskar, C. Babulal and M.W. Iruthayarajan, "Solving multiobjective optimal reactive power dispatch using modified NSGA-II," *International Journal of Electrical Power & Energy Systems,* vol. 33, no. 2, pp. 219-228, 2011.

[19] A. Saraswat and A. Saini, "Multi-objective optimal reactive power dispatch considering voltage stability in power systems using HFMOEA," *Engineering Applications of Artificial Intelligence,* vol. 26, no. 1, pp. 390-404, 2013.

[20] B.B. Pal, P. Biswas and A. Mukhopadhyay, "GA based FGP approach for optimal reactive power dispatch," *Procedia Technology,* vol. 10, pp. 464-473, 2013.

[21] B. Mandal and P.K. Roy, "Optimal reactive power dispatch using quasi-oppositional teaching learning based optimization," *International Journal of Electrical Power & Energy Systems,* vol. 53, pp. 123-134, 2013.

[22] G. Chen, L. Liu, P. Song and Y. Du, "Chaotic improved PSO-based multi-objective optimization for minimization of power losses and L index in power systems," *Energy Conversion and Management,* vol. 86, pp. 548-560, 2014.

[23] A. Ghasemi, K. Valipour and A. Tohidi, "Multi objective optimal reactive power dispatch using a new multi objective strategy," *International Journal of Electrical Power & Energy Systems,* vol. 57, pp. 318-334, 2014.

[24] S. Fang, H. Cheng, Y. Song, P. Zeng, L. Yao and M. Bazargan, "Stochastic optimal reactive power dispatch method based on point estimation considering load margin," *Proceedings of the IEEE PES General Meeting| Conference & Exposition,* pp. 1-5, 2014.

[25] C. Wang, Y. Liu, Y. Zhao and Y. Chen, "A hybrid topology scale-free Gaussian-dynamic particle swarm optimization algorithm applied to real power loss minimization," *Engineering Applications of Artificial Intelligence,* vol. 32, pp. 63-75, 2014.

[26] G. Kannan, D. P. Subramanian and R. U. Shankar, "Reactive Power Optimization Using Firefly Algorithm," *Power Electronics and Renewable Energy Systems,* pp. 83-90, 2015.

[27] F. Namdari, L. Hatamvand, N. Shojaei and H. Beiranvand, "simultaneous RPD and SVC placement in power systems for voltage stability improvement using a fuzzy weighted seeker optimization algorithm," *Journal of Operation and Automation in Power Engineering,* vol. 2, no. 2, pp. 129-140, 2014.

[28] R. Salgado and E. Rangel Jr, "Optimal power flow solutions through multi-objective programming," *Energy,* vol. 42, no. 1, pp. 35-45, 2012.

[29] A. Rabiee, A. Soroudi, B. Mohammadi-ivatloo, and M. Parniani, "Corrective voltage control scheme considering demand response and stochastic wind power," *IEEE Transactions on Power Systems,* vol. 29, no. 6, pp. 2965 - 2973, 2014.

[30] K. Deb, *Multi-objective optimization using evolutionary algorithms,* John Wiley & Sons, 2001.

[31] A. Soroudi, B. Mohammadi-Ivatloo and A. Rabiee, "Energy hub management with intermittent wind power," *Large Scale Renewable Power Generation,* pp. 413-438, 2014.

[32] R.D. Christie, "Power systems test case archive," *Electrical Engineering department, University of Washington,* 2000.

[33] R.D. Zimmerman, C.E. Murillo-Sanchez and D. Gan," Matlab power system simulation package," Version 4.1, *Available at http://www.pserc.cornell.-edu/ matpower/* 2011.

[34] M. Allahnoori, S. Kazemi, H. Abdi and R. Keyhani, "Reliability assessment of distribution systems in presence of microgrids considering uncertainty in generation and load demand," *Journal of Operation and Automation in Power Engineering,* vol. 2, no. 2, pp. 113-120, 2014.

10

An Intelligent Approach Based on Meta-Heuristic Algorithm for Non-Convex Economic Dispatch

F. Namdari, R. Sedaghati*

Department of Electrical Engineering, Faulty of Engineering, Lorestan University, Khorram abad, Iran

ABSTRACT

One of the significant strategies of the power systems is Economic Dispatch (ED) problem, which is defined as the optimal generation of power units to produce energy at the lowest cost by fulfilling the demand within several limits. The undeniable impacts of ramp rate limits, valve loading, prohibited operating zone, spinning reserve and multi-fuel option on the economic dispatch of practical power systems are scrutinized in this paper. Thus, the proposed nonlinear non-convex formulation is solved by a new modified version of bio-inspired bat algorithm. Due to the complexities associated with the large-scale optimization problem of economic dispatch, adaptive modifications are added to the original bat algorithm. The modification methods are applied at two separate stages and pledge augmentation in convergence rate of the algorithm as well as extricating the algorithm from local optima. Veracity of the proposed methodology are corroborated by performing simulations on three IEEE test systems.

KEYWORDS: Non-convex economic dispatch, Modification mechanism, Meta-heuristic algorithm, Nonlinear constrained optimization.

1. INTRODUCTION

Scarcity of energy resources, increasing generation cost of the power systems, and ever-growing load demand for electric energy necessitate optimal Economic Dispatch (ED) in the current electric power systems. As power demand increases and considering that the fuel cost of the power generation is exorbitant, reducing the operation costs of power systems has turned to a significant topic [1]. Economic dispatch is an essential and significant optimization problems task for the operation of power systems. The main objective of ED in power systems is to economically distribute the total required generation between the generation units, while satisfying the load demand and system equality and inequality constraints [2]. Improvements in the scheduling of the unit power outputs can contribute to significant cost savings. Furthermore, it offers information in forming market-clearing prices.

Previously, different algorithms to find rate of optimum product for each power generation unit are proposed in the literature. Conventional algorithms such as gradient method, Lambda Iteration Method (LIM), Linear Programming (LP), Quadratic Programming (QP), Lagrangian multiplier method, and classical technique based on coordination equations can solve the ED problems.

There are complex and nonlinear characteristics with equality and inequality constraints associated with the practical ED, which may be imposed to the problem to ensure the system operator of system reliability during disturbances and a secure operation. Many generating units are supplied with multiple fuel sources and should be scheduled by the most economic fuel to burn [3]. Taking everything

*Corresponding author:
R. Sedaghati (E-mail: reza_sedaghati@yahoo.com)

into account, the system operator must consider ramp rate limits, Prohibited Operating Zones (POZs), system spinning reserve, valve loading effects, and multiple fuel source options to solve a realistic ED problem, which makes hard the finding of the optimum solution difficult [4].

Recently, as an alternative for conventional mathematical approaches, the modern stochastic optimization algorithms based on operational research and artificial intelligence concepts, such as Genetic Algorithms (GA), Tabu search, simulated annealing (SA) , differential evolution and Particle Swarm Optimization (PSO) are considered as realistic and powerful solution schemes for obtaining the global optimums in power system optimization problems and due to their ability to find an almost global optimal solution for ED problems with operating constraints. While each of the above studies has considered different parts of the practical ED problem, none of them have considered all the practical constraints simultaneously [5]. In addition, the utilized algorithms are not robust at all and have found different results for different runs. In response to these deficiencies, a practical formulation was devised for ED and then a new method is proposed as its solution in this paper. Thus, a modified version of bat algorithm (BA) as an evolutionary meta-heuristic algorithm was employed to solve the proposed realistic ED problem. BA tries to formulate and simulate the journey of bats in search of nutritious resources or chasing preys. The algorithm is simple in concept; thus, it is easy to implement, since many adjusting parameters are not included in the formulation. However, the original algorithm suffers low convergence rate and it is destined to get trapped in local optima due to the lack of diversity in the population. Thus, two modification stages are emplaced in the original algorithm to help increase the convergence rate of the algorithm and diversify the population. Interspersing the population to the entire search space improved the odds of finding the global optima. The robustness and capability of the proposed methodology is demonstrated by applying the procedure to two various IEEE standard test systems. In sum, the main contributions of this study can be summarized as follows:

- Proposing a comprehensive model for ED problem to consider practical constraints in real systems.
- Modifying the original BA to enable it of seeking the search space faster and more precisely.

2. PRACTICAL ED MATHEMATICAL DISCRIPTION

The ED problem is a nonlinear optimization problem, the objective of which is to determine the optimal combination of power generations, which minimizes the cost function while satisfying an equality constraint and an inequality constraint. The mathematical representation of the classical ED problem and the proposed practical ED are described in this section.

2.1. Classic ED

ED in its classical formulation aims to minimize the summing costs of thermal generating units which are generally considered as the second order polynomial function of the generation [2], as follows:

$$f(X) = Cost(X) = \sum_{i=1}^{n} F_i(P_i) = \sum_{i=1}^{n} (a_i + b_i P_i + c_i P_i^2) \qquad (1)$$

in which P_i denotes output power of the ith unit and n stands for the number of generators in the network. The polynomial coefficients of cost for the ith unit are represented by a_i, b_i, and c_i as well. The conventional ED optimization problem is subjected to the following constraints forcing generators to produce power within specific limits so that their total generation equals to total power demand in the network (D).

$$\sum_{i=1}^{n} P_i = D \qquad (2)$$

$$P_i^{\min} \le P_i \le P_i^{\max} \qquad (3)$$

In the above formulation, the lower and upper bounds of power generation for the ith unit are denoted by P_i^{min} and P_i^{max}, respectively.

2.2. Proposed practical ED formulation

- **Effects of valve-point loadings**

However, it is more practical to consider the effect of valve point loading for thermal power plants [6]. These effects, which occur as each steam admission valve in a turbine, create a rippling influence on the unit's cost curve. Typically, as each steam valve starts to open, the valve point results in the ripples like in Fig. 1. Considering the valve-point effects, the fuel cost function of the ith thermal generating unit is expressed as the sum of a quadratic and a sinusoidal function in the following form:

$$F_i(P_i) = a_i + b_i P_i + c_i P_i^2 + | e_i \sin(f_i(P_i^{\min} - P_i)) | \quad (4)$$

Costs of valve loading effect are represented by coefficients e_i and f_i in the sinusoidal term.

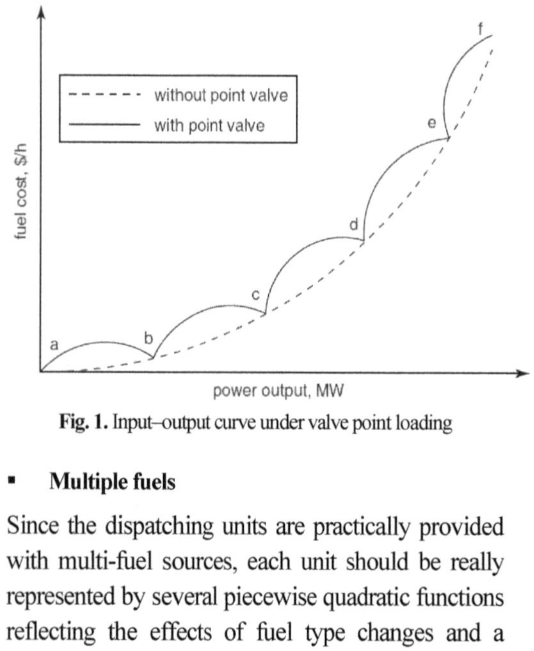

Fig. 1. Input–output curve under valve point loading

- **Multiple fuels**

Since the dispatching units are practically provided with multi-fuel sources, each unit should be really represented by several piecewise quadratic functions reflecting the effects of fuel type changes and a generator must identify the economic fuel to burn. Thus, since different fuels possess various costs, the final generation cost of the units will depend on their choice of fuel, leading to divided cost function for generators as follows:

$$F_i(P_i) = \begin{cases} a_{i1} + b_{i1}P_i + c_{i1}P_i^2 & ; \text{Fuel 1: } P_i^{min} \le P_i \le P_{i1}^{max} \\ a_{i2} + b_{i2}P_i + c_{i2}P_i^2 & ; \text{Fuel 2: } P_{i2}^{min} \le P_i \le P_{i2}^{max} \\ \ldots & \\ a_{ij} + b_{ij}P_i + c_{ij}P_i^2 & ; \text{Fuel j: } P_{ij}^{min} \le P_i \le P_i^{max} \end{cases} \quad (5)$$

By adding the term related to valve-loading effect, the above formulation is turns to the following form:

$$\begin{aligned} & F_i(P_i) = a_{ij} + b_{ij}P_i + c_{ij}P_i^2 + | e_{ij} \sin(f_{ij}(P_{ij}^{\min} - P_i)) | \quad ; \\ & \text{Fuel j: } P_{ij}^{min} \le P_i \le P_i^{max} \\ & j = 1,2,..,n_f \end{aligned} \quad (6)$$

The number of different fuel types that are provided is denoted by n_f.

Apart from the two previously mentioned conventional constraints, the secure operation of network mandates respecting the following constraints:

- **Ramp rate limits**

$$\begin{cases} P_i - P_{0i} \le UR_i; & \text{If generation increases} \\ P_{0i} - P_i \le DR_i; & \text{If generation decreases} \end{cases} \quad (7)$$

Ramp-up and ramp-down rate limits of the ith unit are denoted by UR_i and DR_i, respectively. Also, P_{0i} is the active power output of the ith unit in the previous hour. It is an significant matter that, due to the consideration of ramp rates, the output power of each unit is now bounded by a new limit as follows:

$$\begin{aligned} & \max(P_i^{\min}, P_{0i} - DR_i) \le P_i \le \min(P_i^{\max}, P_{0i} + UR_i) \\ & i = 1,2,..,n \end{aligned} \quad (8)$$

- **Prohibited Operating Zones (POZs)**

Each generator has its generation capacity limitation, which cannot be exceeded. The prohibited operating zones in the input-output performance curve, due to steam valve operating in shaft bearing, were considered in this paper in order to determine the optimum ED problem. In practice, when adjusting the generation output of a unit, operation in the prohibited zones must be avoided. Thus, the shape of the input-output curve in the neighborhood of the prohibited zones is difficult to be determined and the best economical approach is achieved by avoiding the operation in these areas, as shown in Fig. 2. The POZ restrictions might be represented for the ith unit as follows:

$$\begin{cases} P_i^{\min} \le P_i \le P_{i,1}^{LB} \\ P_{i,j-1}^{UB} \le P_i \le P_{i,j}^{LB}; \; j = 2,3,..,NP_i \\ P_{i,j}^{UB} \le P_i \le P_i^{\max} \end{cases} \quad (9)$$

$$For \quad i = 1,2,..,N_{GP}$$

N_{GP} is the number of generators incorporating POZ and NP_i is the number of POZs of the ith unit. Besides, P_{ij}^{LB} and P_{ij}^{UB} refer to the lower and upper boundaries of the jth POZ of the ith generator respectively.

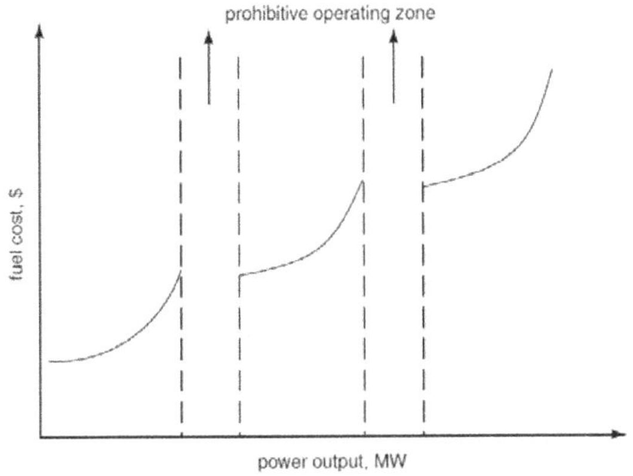

Fig. 2. The prohobited operating zones and generation limits for a generator

- **Spinning reserve**

Due to the inclusion of POZ in the formulation of ED problem, spinning reserve of the system should be written in the following form:

$$S_i = \begin{cases} \min\{(P_i^{\max} - P_i), S_i^{\max}\} & \forall i \in (\Omega - \Theta) \\ 0 & \forall i \in \Theta \end{cases} \quad (10)$$

$$S_r = \sum_{i=1}^{n} S_i \quad (11)$$

$$S_r \geq S_R \quad (12)$$

in which, S_i and S_i^{max} refer to the spinning reserve of the ith unit and its maximum value, respectively. The set of operating units is denoted by Ω and the set of operating units with POZ is represented by Θ. S_R refers to the total spinning reserve required by the system.

3. OPTIMIZATION TECHNIQUE

3.1. Original bat algorithm

Bats locate the position of food or prey by echolocation. Echolocation is the process of figuring out the position of objects from their response to some sort of subsonic signals. Bats spreads a signal in the perimeter and waits to receive the reverberated signal from food or prey. This idea has been simulated to form a meta-heuristic evolutionary optimization algorithm called the (BA) [7]. The ruling ideas behind the BA are listed as bellow:

1) Reverberated signals from various objects are different and thus bats are capable of distinguishing between food and prey;
2) The generated signal by a bat stationed in position X_i or flying at the velocity of V_i is specified by f_i and A_i as signal frequency and amplitude, respectively;
3) Amplitude of the signal denoted by A_i is reduced gradually;
4) Frequency of the signal denoted by f_i and its rate denoted by r_i are changed automatically.

The algorithm is initiated by generating a preliminary population of bats which are randomly interspersed in the search space. The process of evolution is followed by updating the position of bats based on two separate steps. The first adjustment in the bat position is according to:

$$\begin{aligned} &\mathbf{V}_i^{new} = \mathbf{V}_i^{old} + f_i(X_G - X_i); i = 1,\ldots,N_{Bat} \\ &X_i^{new} = X_i^{old} + \mathbf{V}_i^{new} \quad ; i = 1,\ldots,N_{Bat} \\ &f_i = f_i^{\min} + \varphi_1(f_i^{\max} - f_i^{\min}) \quad ; i = 1,\ldots,N_{Bat} \end{aligned} \quad (13)$$

Where, X_G indicates the best global solution. The upper and lower frequency limits of the ith bat are represented by f_i^{max} and f_i^{min}, respectively. The population size is equal to the total number of bats denoted by N_{Bat} and φ_1 is a randomly generated number between 0 and 1.

The second movement in the bat position is simulated as follows:

$$X_i^{new} = X_i^{old} + \varepsilon A_{mean}^{old}; i = 1,\ldots,N_{Bat} \quad (14)$$

where, ε is a random number in the range of [-1,1] and A_{mean}^{old} is the mean value of amplitude of all the bats. Once the position of bats is improved by the above adjustments, a new random individual X_i^{new} is generated in case the rate of its signal r_i is greater than a random value β. This new solution will be inserted to the population in case the following constraint is respected:

$$[\beta < A_i] \& [f(X_i) < f(Gbest)] \quad (15)$$

As mentioned formerly, the value of signal amplitudes generated by bats has a gradual decrease formulated by:

$$\begin{aligned} &A_i^{new} = \alpha A_i^{old} \\ &r_i^{Iter+1} = r_i^0[1 - exp(-\gamma \times t)] \end{aligned} \quad (16)$$

Where t represents iteration number and α and γ are constant parameters.

3.2. Modified BA

The original BA suffers some drawbacks such as the possibility of getting trapped in local optima and low rate of convergence to the optimal solution. Two modifications are devised and added to the algorithm in order to improve its convergence rate and diversity as follows:

- **Modification method 1**

In the first modification step, it is attempted to diversify the bat population using Lévy flight, defined as a random walk with regular and dispersed step lengths according to heavy-tailed probability distribution [8]. The mathematical representation of the Lévy flight is formulated as:

$$Le'vy(\omega) \sim \tau = t^{-\omega} \quad ; \quad (1<\omega \leq 3) \tag{17}$$

This idea is borrowed to generate a new individual in each iteration as follows:

$$X_i^{new} = X_i^{old} + \varphi_1 \oplus Le'vy(\omega) \tag{18}$$

This new solution might replace the ith bat in the population in case it excels the objective function.

- **Modification method 2**

The second modification step is devised to intersperse randomly generate solutions in the population based on conventional GA operators of crossover and mutation. To do so, three bats X_{b1}, X_{b2} and X_{b3} are chosen randomly such that $b_1 \neq b_2 \neq b_3 \neq i$ are related for i^{th} bat in the population and two test solutions are generated as follows:

$$X_{Test,1} = X_{b_1} + \varphi_1 \times (X_{b_2} - X_{b_3}) \tag{19}$$

$$X_{Test,2} = \varphi_2 \times X_G + \varphi_3 \times (X_G - X_i) \tag{20}$$

The above individuals are compared to the ith bat and the one which enhances the objective function replaces X_i.

4. SOLUTION PROCEDURE

In order to apply MBA to solve the ED problem, the following steps should be implemented:

Step 1: Defining the input data. Here, all data including the network data, algorithm data (such as number of bats, initial positions, constant coefficients, etc.), objective function parameters, constraints parameters, and etc. are defined completely.

Step 2: Formation of the fitness function. It is noted that the fitness function includes the objective function and the penalty values related to the problem constraints.

Step 3: Generation of the initial population based on the information given in the previous section.

Step 4: Evaluation of the objective functions for each bat separately and finding the best solution.

Step 5: Movement of the bat population to the new improved positions.

Step 6: Application of the proposed modification methods according to Eqs. (17)-(20).

Step 7: Updating the value of the best individual.

Step 8: Checking the termination criterion. If the termination criterion is satisfied, then the algorithm is finished and the results are printed; else, step 5 and the rest of the steps of repeated.

5. SIMULATION RESULTS

Effectiveness of the proposed approach in solving the proposed realistic ED problem is illustrated by applying the method to two test systems. It is worth noting that the adjusting parameters of MBA are all determined experimentally by several running of the algorithm. But, the significant point is that the algorithm is not much dependent on the values of the setting parameters and the output results is robust. The first example included ten generating units considering fuels valve loading effects of multiple fuels simultaneously [13]. The optimization problem is solved 100 times to generate the results according to Table 1. For comparison, the obtained results from several recently published ED solution methods (with similar trial runs) are also represented in this Table. It should be noticed that the results of the other reported methods are directly quoted from their respective references. According to the results, the best, average, and worst solutions of the proposed MBA are better than the best, average, and worst results of all other methods in the first ED test systems (IEEE 10-unit system), respectively. In addition, the optimal operating points of the units are given in Table 2.

The impact of the modifications on the convergence rate of the BA is illustrated in Fig. 3.

Table 1. Cost function optimization using the proposed method on the IEEE 10-unit test system

Method	Best	Average	Worst
CGA-MU [13]	624.7193	627.6093	633.8652
IGA-MU [13]	624.5178	625.8692	630.8705
DE [14]	624.5146	624.5246	624.5458
RGA [14]	624.5081	624.5079	624.5088
PSO [14]	624.5074	624.5074	624.5074
PSO-LRS [15]	624.2297	624.7887	628.3214
NPSO[15]	624.1624	625.2180	627.4237
NPSO-LRS [15]	624.1273	625.9985	626.9981
BA	624.1763	625.8636	626.8660
Proposed MBA	623.8963	623. 9883	623.9964

Table 2. Optimal operating point for the generators on the IEEE 10-unit test system

Unit	Optimal Generation
1	216.0512
2	211.9071
3	280.6641
4	240.7613
5	279.6584
6	239.7952
7	292.2220
8	240.4925
9	424.2443
10	274.2036

It is obvious that the proposed MBA was effectively successful in approaching the global optimal solution in less than 100 iterations.

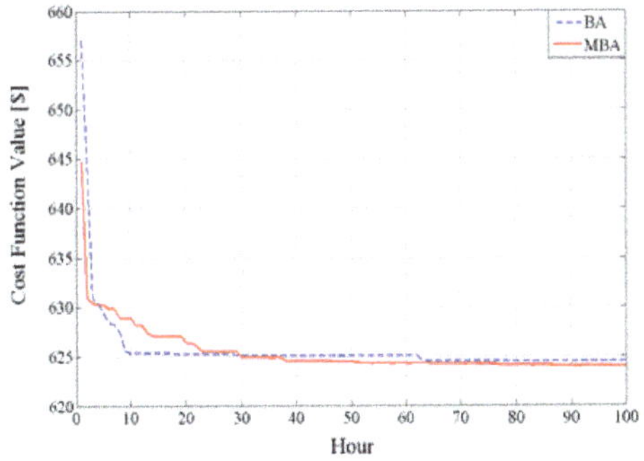

Fig. 3. Convergence speed of the proposed MBA on the IEEE 10-unit test system

For the sake of better demonstrating the robustness of the proposed methodology, the procedure is applied to the 40-unit IEEE power system and the second test case was fully introduced in [15]. Similar to the first case, 100 trials are done to the IEEE-40 unit system and the optimization results and the unit allotted outputs are illustrated in Table 3 and Table 4, respectively.

Investigation of the results revealed the dominance of the proposed MBA in terms of finding the optimal solution compared to other methods.

Table 3. Cost function optimization using the proposed method on the IEEE 40-unit test system

Method	Best	Average	Worst
SPSO [16]	124,350.40	126,074.40	NA
PSO [17]	123,930.45	124,154.49	NA
CEP [17]	123,488.29	124,793.48	126,902.89
HGAPSO [16]	122,780.00	124,575.70	NA
FEP [16]	122,679.71	124,119.37	127,245.59
IFEP [16]	122,624.3500	123,382.0000	125,740.6300
MPSO [18]	122,252.2650	NA	NA
ESO [19]	122,122.1600	122,558.4565	123,143.0700
PSO-LRS [18]	122,035.7946	122,558.4565	123,461.794
Improved GA [20]	121,915.9300	122,811.4100	123,334.0000
HPSOWM[21]	121,915.3000	122,844.4	NA
IGAMU[22]	121,819.2521	NA	NA
NPSO [18]	121,704.7391	122,221.3697	122,995.0976
HDE[23]	121,698.5100	122,304.3000	NA
NPSO-LRS [18]	121,664.4308	122,209.3185	122,981.5913
KH [24]	121,694.5938	121,721.0043	121,756.9473
BA	121,678.7742	122,218.8773	122,867.8832
Proposed MBA	121,578.4832	121,583.3029	121,601.0001

The improvements were obvious in all average and the worst and best solutions. The convergence behavior of the proposed MBA is depicted in Fig. 4 and it can be seen that the modifications enhanced convergence rate of the algorithm.

Table 4. Optimal operating point of the generators on the IEEE 40-unit test system

Unit	Optimal Generation	Unit	Optimal Generation
1	112.2460	21	523.2798
2	112.3141	22	523.2793
3	97.8202	23	523.2804
4	179.7331	24	523.2796
5	91.7458	25	523.2796
6	140.0000	26	523.2798
7	259.6055	27	10.0000
8	284.6495	28	10.0001
9	284.6061	29	10.0002
10	130.0000	30	92.3796
11	243.5996	31	190.0000
12	168.7997	32	190.0000
13	125.0000	33	190.0000
14	304.5195	34	200.0000
15	394.2796	35	192.1066
16	304.5196	36	200.0000
17	489.2798	37	109.9999
18	489.2794	38	109.9996
19	511.2794	39	109.9999
20	511.2792	40	511.2794

Table 6. Optimal operating point of the generators on the IEEE 10-unit with 24 periods and considering power losses

Unit	Load	P1(MW)	P2(MW)	P3(MW)	P4(MW)	P5(MW)	P6(MW)	P7(MW)	P8(MW)	P9(MW)	P10(MW)	P_{loss}(MW)
1	1036	150.0000	135.0000	206.2299	60.0000	122.8666	122.4500	129.5905	47.0000	20.0000	55	12.1370
2	1110	150.0000	135.0000	282.7915	60.0000	122.8666	122.4503	129.5904	47.0000	20.0000	55	14.6988
3	1258	226.6238	135.0000	307.4811	60.0000	172.7504	122.4596	129.5905	47.0000	20.0000	55	17.9054
4	1406	303.2465	215.0000	304.3687	60.0000	172.7337	122.4605	129.5906	47.0000	20.0000	55	23.4000
5	1480	379.8698	222.2664	297.3988	60.0000	172.7329	122.4495	129.5902	47.0000	20.0000	55	26.3076
6	1628	456.5075	226.3025	308.4565	72.4387	222.6419	122.8118	129.5904	47.0000	20.0000	55	32.7493
7	1702	456.5444	306.3025	307.1318	121.3848	172.7468	122.3960	129.5905	47.0000	20.0358	55	36.1326
8	1776	456.4974	309.5562	298.4469	171.3705	172.7428	122.4498	129.5906	77.0000	20.0000	55	36.6542
9	1924	456.4967	389.5562	296.8421	192.1576	222.6014	122.4640	129.5907	85.2939	20.0000	55	46.0026
10	2072	456.5033	396.7990	323.0990	242.1576	222.6067	160.0000	129.5905	115.2939	20.0000	55	49.0500
11	2146	457.1994	396.7997	340.0000	292.1576	225.9023	160.0000	129.5907	120.0000	20.2852	55	50.9349
12	2220	456.4968	460.0000	324.8436	300.0000	222.5994	160.0000	129.5909	120.0000	50.2501	55	58.7808
13	2072	456.4970	396.7995	306.6229	291.2521	222.6005	122.4516	129.5904	120.0000	20.2504	55	49.0644
14	1924	456.4976	316.8009	310.8607	241.2521	222.6085	122.4469	129.5904	90.0000	20.0000	55	41.0571
15	1776	379.8727	309.2594	296.1432	191.2536	222.5823	122.4639	129.5905	85.3111	20.0000	55	35.4767
16	1554	303.2464	229.2594	281.1420	179.7180	172.7089	122.4453	129.5902	85.3106	20.0000	55	24.4208
17	1480	226.6385	309.2594	305.1282	129.7337	122.8699	122.5016	129.5951	85.3116	20.0000	55	26.0380
18	1628	303.2405	309.5331	296.9850	163.0457	172.7322	122.4489	129.5903	85.3094	20.0000	55	29.8851
19	1776	379.8730	316.0412	299.8847	181.0377	222.6179	122.4680	129.5906	85.3138	20.0000	55	35.8269
20	2072	456.4940	396.0411	336.7905	231.0377	222.5993	160.0000	129.5945	85.3105	50.0000	55	50.8676
21	1924	456.4971	389.5926	307.6317	181.0377	222.6025	122.5375	129.5914	85.3119	20.0000	55	45.8024
22	1628	379.8759	309.5926	284.8721	131.0377	172.7304	122.4478	129.5904	55.3122	20.0000	55	32.4591
23	1332	303.2483	229.5926	204.8721	118.6365	122.8462	122.4423	129.5900	47.0000	20.0000	55	21.2280
24	1184	226.6173	222.2665	184.6396	120.3965	73.0000	122.4494	129.5903	47.0000	20.0000	55	16.9596

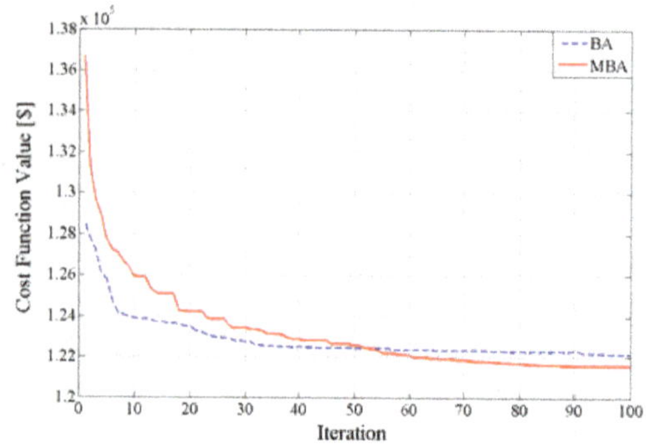

Fig. 4. Convergence speed of the proposed MBA on the IEEE 40-unit test system

Table 5. Optimal operating point of the generators on the IEEE 10-unit test system optimizing both cost and emission functions

Unit	Optimal Generation	Best	Average	Worst
Cost Function	EP [27]	1 054 685	1 057 323	NA
	EP-SQP [27]	1 052 668	1 053 771	NA
	IPSO [28]	1 046 275	1 048 154	NA
	AIS [29]	1 045 715	1 047 050	1 048 431
	TVAC-IPSO [30]	1 041 066	1 042 118	1 043 625
	EBSO [31]	1 038 915	1 039 188	1 039 272
	BA	1 040 203	1 041 091	1 043 467
	Proposed MBA	1037571	1037862	1038031
Emission Function	BA	289 309	289 367	289 413
	Proposed MBA	289208	289229	289246

According to the following results, the proposed MBA could achieve better results from both cost and emission targets.

Finally, Table 6 shows the optimal generation values of the power units associated with the best result found by MBA. It is worth noting that these results belonged to the cost function optimization and the emission results were not shown to avoid repetition.

6. CONCLUSION

This paper addressed the impacts and constrains imposed to the ED formulation of practical power systems in order to optimally solve it. In a practical system the existence of limitations such as multiple fuel option, valve loading effects, power generation limits, spinning reserve, ramp rate limits, and POZs made solving the ED problem complicated. The system operator faced a complicated nonlinear non-convex optimization problem mandating the use of power full optimization techniques.

As a result, the MBA was proposed to add diversity to the algorithm and ensure fast convergence to the global optimal solution. Contributions of the proposed methodology were corroborated through the numerical analysis of two various test systems. In order to see the effect of power units on the emission function, this target was also considered in the paper. Simulation results showed high capability of the algorithm for solving this target as well.

REFERENCES

[1] K. Afshar and A. Shokri Gazafroudi, "Application of stochastic programming to determine operating reserve with considering wind and load uncertainties," *Journal of Operation and Automation in Power Engineering* , vol. 1, no. 2, pp. 96-109, 2013.

[2] M.A. Ortega-Vazquez and D.S. Kirschen, "Estimating the spinning reserve requirements in systems with significant wind power generation penetration," *IEEE Transactions on Power Systems*, vol. 24, no.1, pp. 114-124, 2009.

[3] H. Khorramdel, B. Khorramdel, M. Tayebi Khorrami and H. Rastegar, "A multi-objective economic load dispatch considering accessibility of wind power with here-and-now approach," *Journal of Operation and Automation in Power Engineering*, vol. 2, no. 1, pp. 49-59, 2014.

[4] A. Hatefi and R. Kazemzadeh, "Intelligent tuned harmony search for solving economic dispatch problem with valve-point effects and prohibited operating zones," *Journal of Operation and Automation in Power Engineering*, vol. 1, no. 2, pp: 84-95, 2013.

[5] J. B. Park, K.S. Lee, J.R. Shin and K.Y. Lee, "A particle swarm optimization for economic dispatch with non-smooth cost functions," *IEEE Transactions on Power Systems*, vol. 20, no. 1, pp. 34-42, 2005.

[6] T. Niknam, R. Azizipanah-Abarghooee, R. Sedaghati and A. Kavousi-Fard, " An Enhanced hybrid particle swarm optimization and simulated annealing for practical economic dispatch," *Energy Education Science and Technology Part A: Energy Science and Research*, vol. 30, no. 1, pp. 553-564, 2012.

[7] G. Komarasamy, A. Wahi, "An Optimized k-means clustering technique using bat algorithm," *European Journal of Scientific Research*, vol. 84, no. 2, pp. 263-273, 2012.

[8] C. Brown, L.S. Liebovitch and R. Glendon, " Lévy flights in Dobe Juhoansi foraging patterns," *Human Ecology*, vol. 35, no. 1, pp. 129-138, 2007.

[9] A. Kavousi-Fard, T. Niknam, H. Taherpoor, A. Abbasi, "Multi-objective probabilistic reconfigure-tion considering uncertainty and multi-level load model," *IET Science, Measurment and Technology*, vol. 10, no. 1, pp. 1-8, 2014.

[10] A. Kavousi-Fard and T. Niknam, "Optimal stochastic capacitor placement problem from the reliability and cost views using firefly algorithm," *IET Science, Measurment and Technology*, vol. 8, no. 5, pp. 260-269, 2014.

[11] A. Kavousi-Fard and H. Samet, "Multi-objective performance management of the capacitor allocation Problem in distributed system based on modified HBMO evolutionary algorithm," *Electric Power and Component systems*, vol. 41, no. 13, pp. 1223-1247, 2013.

[12] A. Kavousi-Fard, T. Niknam, "Optimal distribution feeder reconfiguration for reliability improvement considering uncertainty," *IEEE Transactions on Power Delivery*, vol. 29, no. 3, pp. 1344-1353, 2014.

[13] C. L. Chiang, "Improved genetic algorithm for power economic dispatch of units with valve-point effects and multiple fuels," *IEEE Transactions on Power Systems*, vol. 20, no. 4, pp. 1690-1699, 2005.

[14] P. S. Manoharan, P.S. Kannan, S. Baskar, and M.W. Iruthayarajan, "Penalty parameter-less constraint handling scheme based evolutionary algorithm solutions to economic dispatch," *IET* Proceedings *on Generation Transmission and Distribution*, vol. 2, no. 4, pp. 478-490, 2008.

[15] A. I. Selvakumar and K. Thanushkodi, " A new particle swarm optimization solution to nonconvex economic dispatch problems," *IEEE Transactions on Power Systems*., vol. 22, no. 1, pp. 42-51, 2007.

[16] N. Sinha, R. Chakrabati and P.K. Chattopadhyay, " Evolutionary programming techniques for economic load dispatch," *IEEE Transactions on Evolutionary Computation*, vol. 7, no. 1, pp. 83-94, 2003.

[17] T.A.A. Victoire, A.E. Jeyakumar, "Hybrid PSO-SQP for economic dispatch with valve-point effect," *Electric Power Systems Research*, vol. 71, no. 1, pp. 51-59, 2004.

[18] J.-B. Park, K.-S. Lee, J.-R. Shin and K. Y. Lee, "A particle swarm optimization for economic dispatch with non-smooth cost functions," *IEEE Transactions on Power Systems*, vol. 20, no. 1, pp. 34-42, 2008.

[19] A. Pereira-Neto, C. Unsihuay and O. R. Saavedra, "Efficient evolutionary strategy optimization procedure to solve the non-convex economic dispatch problem with generator constraints," *IEE Proceedings on Generation, Transmission and Distribution*, vol. 152, no. 5, pp. 653-660, 2005.

[20] S.H. Ling and F.H.F. Leung, "An improved genetic algorithm with average-bound crossover and wavelet mutation operations," *Soft Computing*, vol. 11, no. 1, pp. 7-31, 2007.

[21] S.H. Ling, H.H.C. Iu, K.Y. Chan, H.K. Lam, B.C.

W. Yeung and F.H. Leung, "Hybrid particle swarm optimization with wavelet mutation and its industrial applications," *IEEE Transactions on Systems, Man, and Cybernetics, Part B: Cybernetics,* vol. 38, no. 3, pp. 743-763, 2008.

[22] C.-L. Chiang, "Genetic-based algorithm for power economic load dispatch," *IET Proceeding on Generation, Transmission and Distribution,* vol. 1, no. 2, pp. 261-269, 2007.

[23] S.-K. Wang, J.-P. Chiou, and C.-W. Liu, "Non-smooth/non-convex economic dispatch by a novel hybrid differential evolution algorithm," *IET Proceedings on Generation, Transmission and Distribution,* vol. 1, no. 5, pp. 793-803, 2007.

[24] A.H. Gandomi and A. H. Alavi, Krill herd: A new bio-inspired optimization algorithm, *Communications in Nonlinear Science and Numerical Simulation*, vol. 17, no. 12, pp. 4831-4845, 2012.

[25] C.K. Panigrahi, P.K. Chattopadhyay, R.N. Chakrabarti, and M. Basu, "Simulated annealing technique for dynamic economic dispatch," *Electric Power Components and Systems*, vol. 34, no. 5, pp. 577-586, 2006.

[26] T.A.A. Victoire and A.E. Jeyakumar, "A modified hybrid EP-SQP approach for dynamic dispatch with valve-point effect," *Electric Power & Energy Systems,* vol. 27, no. 8, pp. 594-601, 2005.

[27] P. Attaviriyanupap, H. Kita, E. Tanaka, and J. Hasegawa, "A hybrid EP and SQP for dynamic economic dispatch with nonsmooth fuel cost function," *IEEE Transactions on Power Systems,* vol. 17, no. 2, pp. 411-416, 2007.

[28] X. Yuan, A. Su, Y. Yuan, H. Nie, and L. Wang, "An improved PSO for dynamic load dispatch of generators with valve-point effects," *Energy,* vol. 34, no. 1, pp. 67-74, 2009.

[29] S. Hemamalini and S.P. Simon, "Dynamic economic dispatch using artificial immune system for units with valve-point effect," *Electric Power Energy Systems*, vol. 33, no. 4, pp. 868-874, 2011.

[30] B. Mohammadi-ivatloo, A. Rabiee and M. Ehsan, "Time-varying acceleration coefficients IPSO for solving dynamic economic dispatch with non-smooth cost function," *Energy Conversion and Management,* vol. 56, pp. 175-183, 2012.

[31] T. Niknam and F. Golestaneh, "Enhanced bee swarm optimization algorithm for dynamic economic dispatch," *IEEE Systems Journal,* vol. 23, pp. 219-228, 2012.

Simultaneous RPD and SVC Placement in Power Systems for Voltage Stability Improvement Using a Fuzzy Weighted Seeker Optimization Algorithm

F. Namdari*, L. Hatamvand, N. Shojaei, H. Beiranvand

Faculty of Engineering, Lorestan University, Khorram abad, Iran

ABSTRACT

Voltage stability issues are growing challenges in many modern power systems. This paper proposes optimizing the size and location of Static VAR Compensator (SVC) devices using a Fuzzy Weighted Seeker Optimization Algorithm (FWSOA), as an effective solution to overcome such issues. Although the primary purpose of SVC is bus voltage regulation, it can also be useful for voltage stability enhancement and even real power losses reduction in the network. To this aim, a multi-objective function is presented which includes voltage profile improvement, Voltage Stability Margin (VSM) enhancement and minimization of active power losses. Voltage stability is very close to Reactive Power Dispatch (RPD) in the network. Therefore, in addition to voltage regulation with locating SVCs, considering all of the other control variables including excitation settings of generators, tap positions of tap changing transformers and reactive power output of fixed capacitors in the network, simultaneous RPD and SVC placement will be achieved. Simulation results on IEEE 14 and 57-bus test systems, applying Genetic Algorithm (GA), Particle Swarm Optimization (PSO), Seeker Optimization Algorithm (SOA) and FWSOA verify the efficiency of FWSOA for the above claims.

KEYWORDS: Reactive power dispatch, Voltage stability margin enhancement, Voltage deviation reduction, Real power losses minimization.

1. INTRODUCTION

Voltage stability is one of attractive stability aspects in power systems and is considerably affected by Reactive Power Dispatch (RPD) in the network. Flexible AC Transmission System (FACTS) equipments are new fast compensator devices, which increase the power system capacity and make it more capable for controlling the power flow by enhancing the capacity of existing transmission system [1]. These power electronic converters control various electrical parameters in the network, both steady state power flow and dynamic stability. FACTS devices play an important role to overcome power flow and voltage stability problems like Thyristor- Controlled Series Compensator (TCSC), SVC, Unified Power Flow Controller (UPFC), Static Compensator (STATCOM), etc [2-4]. One of the most important of these devices is Static VAR Compensator (SVC). SVC is used widely because of its cheaper and proper operation. SVC is a shunt compensator that can be in an inductive reactor mode that consumes the reactive power or be a capacitive element, which generates reactive power for the system [5].

Many studies have focused on SVC placement in power networks by different analytical techniques [6-9] and many others, employed Evolutionary Algorithms (EAs) like Harmony Search Algorithm (HSA), Genetic Algorithm (GA), Particle Swarm Optimization (PSO), etc [5, 10-12] to achieve various goals.

Reference [5] has proposed HSA to optimize the size and location of shunt VAR compensation devices such as SVC so that improve voltage

*Corresponding author:
F. Namdari (E-mail: namdari.f@lu.ac.ir)

deviation and its stability along with active power losses and cost reduction of mentioned compensators. GA has been used to SVC placement considering all or some of the above objectives in [10-11]. Reference [12] also employed PSO in SVC allocation for voltage regulation along with transient rotor angle stability improvement.

While the primary duty for SVC is bus voltage regulation, this is possible that voltage stability enhancement and real power losses reduction also obtained by optimal placement of SVC [10].

On the other hand, if reactive power well dispatches all over the network, then voltage stability will be guaranteed. This objective will be obtained when all settings related to all control variables of compensator devices are adjusted optimally. Consequently, voltage deviation, VSM and active power losses take proper values in the system. In addition, there are two sets of variables, including continuous and discrete variables in such optimization problems. Considering these reasons, RPD is addressed as a nonlinear optimization problem [13].

Previous classic optimization approaches are based on gradient or mathematical methods. Recently, the EAs like Differential Evolution (DE) [14-16], Hybrid Differential Evolution (HDE) [17], HSA [18], GA [19], PSO [20-23] and Seeker Optimization Algorithm (SOA) [24] for RPD are more attractive because of more efficiency in handling the inequality constraints and discrete values. EAs do not rely on gradient information, so they rarely suffer from being trapped in local minima [25]. However, in the case of some usages, may be slower with respect to gradient-based methods. However, many studies have focused on RPD problems with EAs.

Reference [19] has improved the voltage stability using an improved GA approach. Optimization variables are generator voltages, capacitor bank sizes and tap of transformers. In [20-23] PSO algorithm has been considered for RPD problem. In [20] PSO is used to achieve the optimal reactive power flow in the system. To eliminate premature convergence in PSO, a new learning strategy is presented. Consequently, active power losses reduce, voltage profile improves and VSM increases. In [21] a novel PSO technique based on multi-agent systems (MAPSO) has suggested to RPD problem. Reference [22] also aims to reduce active power losses and voltage deviation by a multi-objective PSO algorithm. Reference [24] performs RPD using SOA that is a global optimization approach. Objectives are active power losses and voltage deviation reduction and increase the voltage stability margin.

There are two deficiencies in GA operation. GA converges precociously and is not enough capable in local search [26]. On the other hand, PSO also suffers from precocious convergence because of dependability on its parameters [27]. However, reference [24] claims that SOA performance is better than GA and PSO in RPD problem.

In this paper, a Fuzzy Weighted Seeker Optimization Algorithm (FWSOA) is proposed to optimize the size and location of SVC as well as the control variables of other compensator devices in the network. Therefore, a simultaneous RPD and optimal SVC placement is obtained.

The objective function includes minimization of active power losses, voltage profile improvement and VSM enhancement. To this aim, all control variables of the network are excitation settings of generators, tap position of tap changing transformers, reactive power output of fixed capacitors and voltage with location of SVCs in the network. RPD is performed by using GA, PSO, SOA and FWSOA. Therefore, in the present study, the effectiveness of GA, PSO, SOA and FWSOA for simultaneous RPD and SVC placement is compared. Simulation results verify that FWSOA is the best solution to solve RPD and therefore voltage stability improvement problem. Furthermore, simulation results show that FWSOA outperforms the GA, PSO and SOA in solving simultaneous RPD and SVC placement problem.

2. PROBLEM FORMULATION

2.1. SVC ideal model

Figure 1 illustrates the general circuit structure of a Static VAR Compensator (SVC). This structure is composed of a fixed capacitor (with susceptance B_C) parallel with a thyristor-controlled reactor (with susceptance B_L). The equivalent susceptance $B_{eq,SVC}$

is determined by the firing angle α of the thyristors. The equivalent susceptance is expressed as follows:

$$B_{eq,SVC} = B_L(\alpha) + B_C \quad (1)$$

where,

$$B_L(\alpha) = -\frac{1}{\omega L}\left(1 - \frac{2\alpha}{\pi}\right), \quad B_C = \omega * C \quad (2)$$

The reactive power amount that SVC consumes or generates is:

$$Q_{SVC} = -V_K^2 * B_{eq,SVC} \quad (3)$$

Subject to

$$B_{eq,SVC}^{\min} \le B_{eq,SVC} \le B_{eq,SVC}^{\max} \quad (4)$$

2.2. Objective function

In this section, the overall objective function for simultaneous RPD and SVC placement problem is presented.

2.2.1. Real power losses

First sub-objective function is real power losses minimization, which is defined as follows [24]:

$$\min P_{loss} = f(\vec{x_1}, \vec{x_2}) = \sum_{k \in N_E} g_K (V_i^2 + V_j^2 - 2 V_i V_j \cos\theta_{ij}) \quad (5)$$

Subject to

$$\begin{cases} P_{Gi} - P_{Di} = V_i \sum_{j \in N_i} V_j (G_{ij} \cos\theta_{ij} + B_{ij} \sin\theta_{ij}) & i \in N_0 \\ Q_{Gi} - Q_{Di} = V_i \sum_{j \in N_i} V_j (G_{ij} \sin\theta_{ij} - B_{ij} \cos\theta_{ij}) & i \in N_{PQ} \\ V_i^{\min} \le V_i \le V_i^{\max} & i \in N_B \\ T_k^{\min} \le T_k \le T_k^{\max} & k \in N_T \\ Q_{Gi}^{\min} \le Q_{Gi} \le Q_{Gi}^{\max} & i \in N_G \\ Q_{Ci}^{\min} \le Q_{Ci} \le Q_{Ci}^{\max} & i \in N_C \\ V_{SVC_i}^{\min} \le V_{SVC_i} \le V_{SVC_i}^{\max} & i \in N_{SVC} \\ L_{SVC_i} & i \in N_{PQ} \\ Q_{SVC_i}^{\min} \le Q_{SVC_i} \le Q_{SVC_i}^{\max} & i \in N_{SVC} \\ S_l \le S_l^{\max} & l \in N_l \end{cases} \quad (6)$$

where x_1 and x_2 are control and dependent variable vectors, respectively. Control vector includes V_G, K_T, Q_C, V_{SVC} and L_{SVC}. The dependent vector includes V_L and Q_G and Q_{SVC}. V_G and V_{SVC} denote generator and SVC voltages respectively, and are continuous variables. While, K_T, Q_C, L_{SVC} are discrete variables and represent transformer tap position, capacitor size and location of *SVCs*, respectively. g_k is the conductance of branch k, θ_{ij} is difference between voltage angles of bus i and bus j. P_G and P_D are active power generation and power demand respectively. Q_G and Q_D are reactive generation and demand. Q_{svc} also shows the reactive power amount that each SVC absorbs or injects to the network. G is conductance of the transfer branch and B is the susceptance. S_l also is power flow in transmission line l. N_E is the number of all network branches, $N0$ represents each bus except slack bus, N_{PQ}: load buses, N_B: all buses, N_T the number of tap changer transformers, N_G the number of generator buses, N_C the number of possible capacitor installation buses and N_{SVC} means the number of possible *SVCs* installation buses.

2.2.2. Voltage deviation

Second sub-objective function is voltage deviation minimization, which is defined as follows [24]:

$$\min \Delta V_L = \sum_{i=1}^{N_L} \left| \frac{V_i - V_i^*}{N_L} \right| \quad (7)$$

ΔV_L is the voltage deviation, N_L represents the number of all load buses, V_i is actual voltage magnitude and V_i^* is the expected corresponding value.

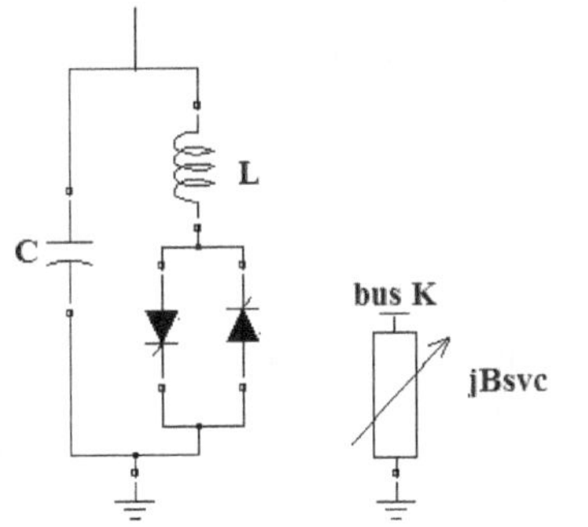

Fig. 1. General circuit structure for SVC [11]

2.2.3. Voltage stability margin

Tertiary sub-objective function is voltage stability margin, which is defined as follows [24]:

$$\max VSM = \max(\min|eig(Jacobi)|) \quad (8)$$

VSM is abbreviated of voltage stability margin and *Jacobi* is the power flow Jacobian matrix and *eig(Jacobi)* means all eigenvalues of Jacobian matrix. Equation (8) expresses that if can maximize the minimum eigenvalue of the power flow Jacobian matrix, in reality VSM increases.

2.2.4. Multi-objective conversion

In this section, f_1, f_2 and f_3 are normalized so as kept within [0, 1]. This fuzzy decision for sub-objective functions is because of each sub-objective function has different range of function values. Note that f_3 function is a maximization optimization problem [24].

$$f1=\begin{cases}0 & if\ Ploss<Ploss_{\min}\\ \dfrac{Ploss-Ploss_{\min}}{Ploss_{\max}-Ploss_{\min}} & if\ Ploss_{\min}\le Ploss\le Ploss_{\max}\\ 1 & if\ Ploss>Ploss_{\max}\end{cases} \quad (9)$$

$$f2=\begin{cases}0 & if\ \Delta VL<\Delta VL_{\min}\\ \dfrac{\Delta VL-\Delta VL_{\min}}{\Delta VL_{\max}-\Delta VL_{\min}} & if\ \Delta VL_{\min}\le\Delta VL\le\Delta VL_{\max}\\ 1 & if\ \Delta VL>\Delta VL_{\max}\end{cases} \quad (10)$$

$$f3=\begin{cases}0 & if\ VSM>VSM\max\\ \dfrac{VSM\max-VSM}{VSM\max-VSM\min} & else\end{cases} \quad (11)$$

where subscripts *min* and *max* denote corresponding expectant minimum and possible maximum values, respectively. Finally, the overall objective function for RPD problem is presented as follows:

$$\min\ f=w1\,f1+w2\,f2+w3\,f3+\lambda_V\sum_{N_V^{\lim}}\Delta V_L^2+\lambda_Q\sum_{N_Q^{\lim}}\Delta Q_G^2 \quad (12)$$

Dependent variables are constrained using penalty factors while the control variables are self-constrained.

w_i (i= 1, 2, 3) is the user-defined constant which represents the weight of contribution for different sub-objective functions. λ_V and λ_Q are penalty factors. N_V^{lim} is the number of load buses, which violate from the permitted voltage range and N_Q^{lim} is the number of generator buses that violate from permitted reactive power range. ΔV_L, ΔQ_G and ΔQ_{SVC} also are defined as follows:

$$\Delta V_L=\begin{cases}V_{L\min}-V_L & if\ V_L<V_L^{\min}\\ V_L-V_{L\max} & if\ V_L>V_L^{\max}\end{cases} \quad (13)$$

$$\Delta Q_G=\begin{cases}Q_{G\min}-Q_G & if\ Q_G<Q_G^{\min}\\ Q_G-Q_{G\max} & if\ Q_G>Q_G^{\max}\end{cases} \quad (14)$$

$$\Delta Q_{SVC}=\begin{cases}Q_{SVC\min}-Q_{SVC} & if\ Q_{SVC}<Q_{SVC}^{\min}\\ Q_{SVC}-Q_{SVC\max} & if\ Q_{SVC}>Q_{SVC}^{\max}\end{cases} \quad (15)$$

3. OPTIMIZATION ALGORITHMS

3.1. PSO and genetic algorithms

Particle swarm optimization algorithm was presented by Kennedy and Eberhart [28]. This algorithm is based on the social behavior of animals like birds or fish, which search the best locations for their food and after finding it, all of them, attack to the food. This seeking behavior is corresponding to the optimization search for solutions that are capable in solving the non-linear problems in a real-valued search space [29].

The main version of GA was proposed by Holland [30]. GA is a search algorithm based on the mechanics of natural genetics and natural selection. The GA is a population search method. A population of strings is kept in each generation. The simulation of the natural process of reproduction, gen crossover and mutation produces the next generation.

3.2. Seeker optimization algorithm

SOA operates on a set of solutions called search population. The individuals of this population called seeker or searcher. The total population is equally categorized into K subpopulations according to the indexes of the seekers and K=3 is selected in this study [24]. All the seekers in the same subpopulation constitute a neighborhood, which represents the social component for the social sharing of information. Search direction d_{ij} and step length α_{ij} are computed for each seeker i (1≤i≤s, s is the population size), on each dimension j by time step t where $\alpha_{ij} \geq 0$ and d_{ij} belongs to $\{-1, 0, 1\}$. When d_{ij}=1 this means seeker i goes toward positive direction of the coordinate axis on the dimension j, d_{ij}=0 i.e. seeker has no motions and d_{ij}= -1 means negative direction. At each time step or iteration t, the position of each seeker is updated by (16):

$$x_{ij}(t+1) = x_{ij}(t) + \alpha_{ij}(t).\, d_{ij} \tag{16}$$

On the other hand, to prevent of entrapment in local minima, at each iteration, the current positions of the worst two individuals of each subpopulation are combined with the best ones in each of the other two subpopulations, using binomial crossover operator as (17):

$$x_{knj,worst} = \begin{cases} x_{lj,best} & \text{if } R_j \le 0.5 \\ x_{knj,worst} & \text{else} \end{cases} \tag{17}$$

where R_j is a uniformly random real number within [0,1], $x_{knj,worst}$ is j-th dimension of the n-th worst position in the kth subpopulation, $x_{lj,best}$ is the j-th dimension of the best position in the l-th subpopulation, n, k, l=1,2,..., K−1 and $k \neq l$. In result, diversity of the population will increase [24].

3.2.1. Search direction

In SOA, search space can be considered as a gradient field where in the search empirical gradient is determined based on the position change and seeker follows the empirical gradient to guide his search. SOA is not dependent on empirical gradient magnitude and therefore, search direction can be determined by signum function of minus between best and worst positions.

Seeker search directions are determined based on evaluating their current or historical (previous) positions or their neighbors and three kinds of behaviors.

In the first kind i.e. egotistic behavior, each seeker likes to go toward his historical best position $\vec{p}_{i,best}(t)$ and therefore, egotistic direction $\vec{d}_{i,ego}(t)$ is determined by (18):

$$\vec{d}_{i,ego}(t) = sign(\vec{p}_{i,best}(t) - \vec{x}_i(t)) \tag{18}$$

The second case is altruistic behavior where in all seekers in the same neighborhood, are coordinated with each other to achieve desired goal. In this state, two kinds of search direction are defined based on neighbor's historical best position and the other is neighbor's current best position i.e. $\vec{d}_{i,alt_1}(t)$ and $\vec{d}_{i,alt2}(t)$.

$$\vec{d}_{i,alt_1}(t) = sign(\vec{g}_{best}(t) - \vec{x}_i(t)) \tag{19}$$

$$\vec{d}_{i,alt2}(t) = sign(\vec{l}_{best}(t) - \vec{x}_i(t)) \tag{20}$$

In the third kind i.e. pro-activeness behavior, seekers rarely influenced by their environment and they often focus to achieve their desired goal. In addition, next behavior of seeker can be determined by his previous behavior and therefore each seeker is pro-activeness in changing of his search direction. This pro-activeness $\vec{d}_{i,pro}(t)$ is as follows:

$$\vec{d}_{i,pro}(t) = sign(\vec{x}_i(t_1) - \vec{x}_i(t_2)) \tag{21}$$

where t_1, $t_2 \in$ {t, t−1, t−2}, and $x_i(t_1)$ is better than $x_i(t_2)$.

Finally, the actual search direction $\vec{d}_i(t)$ for each seeker i is determined by a compromise among four mentioned behaviors. Equation (22) shows this parameter:

$$d_{ij} = \begin{cases} 0 & \text{if } r_j \le p_j^0 \\ +1 & \text{if } p_j^0 < r_j \le p_j^0 + p_j^{+1} \\ -1 & \text{if } p_j^0 + p_j^{+1} < r_j \le 1 \end{cases} \tag{22}$$

where r_j is a uniform random number in [0,1], $pj(m)$ (m∈{0,+1,−1}) is the percentage of the number of "m" from the set{$d_{ij,ego}$, $d_{ij,alt1}$,$d_{ij,alt2}$, $d_{ij,pro}$}on each dimension j of all the four empirical directions, i.e. $p_j^{(m)}$= the number of $m/4$.

3.2.2. Step length

In the optimization problem with continuous search space, usually there is a neighborhood region close to an extremum point. Fitness values of input variables are proportional with their distances from this extremum point. So, near optimal solutions may be found in a neighborhood region with little width (narrow) and lower fitness values or in a spread neighborhood region (broad) containing higher fitness values. Since fuzzy logic is a capable solution in solving of if-then problems, this logic is employed to model the conditional section (if {fitness value is small}) and action part (Then {step length is short}) of the problem. All seeker fitness values are descendingly sorted and converted to consecutive numbers so as fuzzy system be applicable to wide range of optimization problem. A linear membership function is used for conditional part as follow:

$$\mu_i = \mu_{max} - \frac{s - I_i}{s - 1}(\mu_{max} - \mu_{min}) \quad (23)$$

where I_i is the sequence number of $x_i(t)$ after sorting the fitness values, μ_{max} is the maximum membership degree value which is equal to or a little less than 1.0.

For action part the Bell membership function $\mu(x)=e^{-x^2/2\delta^2}$ in Fig. 2 is used. One dimension is considered, membership degree values of the input variables are between [-3δ, 3δ] and other values are neglected. Parameter δ in Bell function is presented as:

$$\vec{\delta} = \omega * abs(\vec{x}_{best} - \vec{x}_{rand}) \quad (24)$$

where ω is weight parameter, $\vec{x}_{best}$ and $\vec{x}_{rand}$ are best seeker and a randomly selected seeker of the same subpopulation, respectively.

The parameter μ_i is changed to vector $\vec{\mu}_i$ by (25) to produce of randomicity on each dimension and improve local search capability. Finally action part α_{ij} will be presented by (26).

$$\mu_{ij} = RAND(\mu_i, 1) \quad (25)$$

$$\alpha_{ij} = \delta_j \sqrt{-\ln(\mu_{ij})} \quad (26)$$

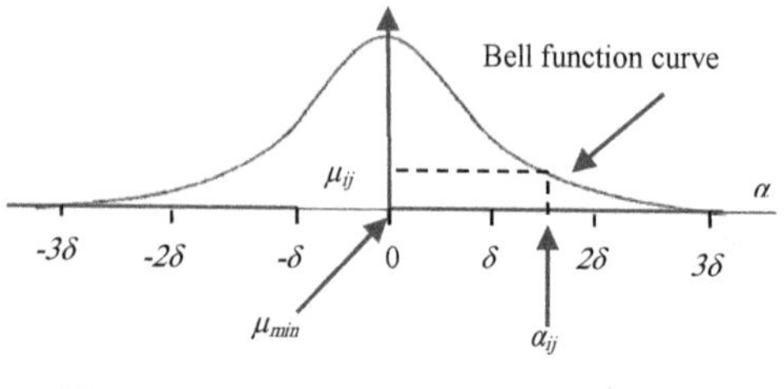

Fig. 2. The action part of the Fuzzy reasoning

3.3. Fuzzy weighted SOA

In the number of evolutionary algorithms like PSO and SOA a weight parameter (ω in (24)) is defined which decreases linearly by iteration increment. The parameter ω is used to decrease the step length with time step increasing so as to gradually improve the search precision [24]. So, the probability to find a better solution in adjacent of the recent optimal point increases. If the algorithm cannot find a proper solution at the last iterations, this update role may cause the early stagnation during the stochastic search. A rational decision is to select the parameter ω according to the evaluated fitness of the search space. So, in this paper, a fuzzy procedure is proposed to calculate the parameter ω effectively. Thus, here, a Fuzzy Weighted SOA (FWSOA) algorithm is presented. The proposed FWSOA is detailed in the following paragraph.

We define an Average Fitness (AF) value of the search space at iteration t as an index of the quality of the obtained results of the algorithm so far. This AF index will be used as the input of the fuzzification part. Desired Fitness (DF) value of an objective function is the ideal solution of an engineering optimization problem and the Worst Fitness (WF) at the first iteration in the search space are used to form the sigmoid fuzzy membership function in the inference engine. The control rule as "If AF value is small, then parameter ω is small" is applied to the fuzzy inference system. Figure 3 illustrates the sigmoid fuzzy membership function.

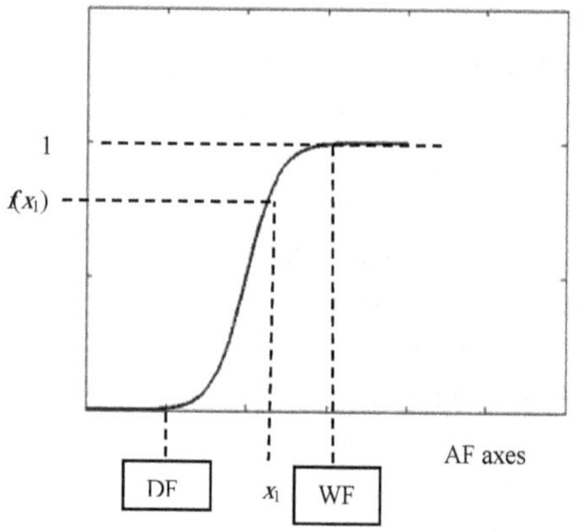

Fig. 3. Sigmoid membership function used in fuzzy inference system.

The output of the defuzzification part of the fuzzy inference system is the fuzzified weight ω. In the defuzzification part, ω will be extracted as a number in [0.1, 0.9] as follows:

$$\omega_t = 0.8 f(x_t) + 0.1 \quad (27)$$

$$f(x_t) = \frac{1}{1 + a e^{(b - x_t)}} \quad (28)$$

where f is the sigmoid function, x is the AF index at iteration t. Parameters a and b depend on DF and WF. The DF index has to be set for each optimization problem separately and WF index will be calculated at start of optimization process automatically.

3.4. FWSOA for simultaneous RPD and SVC placement

Here, there are follow steps in FWSOA employment for simultaneous RPD and SVC placement (Fig. 4):

Four mentioned algorithms are employed for simultaneous RPD and SVC placement problem and simulation results for IEEE 14 and 57-bus test systems will be presented.

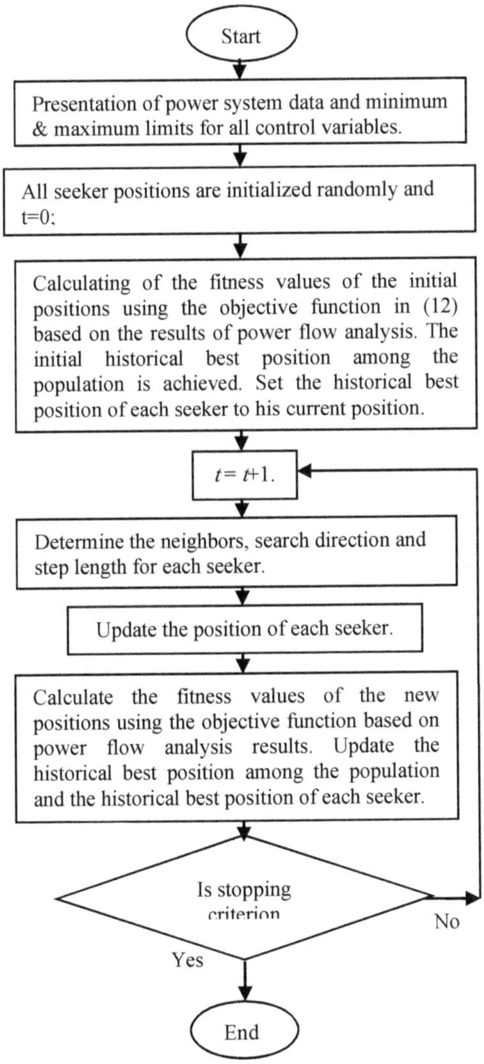

Fig. 4. The flowchart of FWSOA employment

4. SIMULATION RESULTS

4.1. IEEE 14- bus test system

The IEEE 14-bus test system has five generators, 20 transmission lines and three tap changing transformers. One capacitor placed at bus 9 [31]. One SVC also has been considered in the present study IEEE 14-bus test system. This is recommended that in this paper, SVCs are allocated to place in non-generator buses. Because the voltages of generator buses are regulated by excitation system. All variable limits have taken from [32]. The population size is 30 for IEEE 14-bus system, total 30 runs and the maximum generations of 25. Optimization parameters are w_1=0.6, w_2=0.2, w_3=0.2 [24], $Ploss_{min}=0.07$, $Ploss_{max}=0.2$, $\Delta V_{Lmin}=0$, $\Delta V_{Lmax}=1.5$, $VSM_{min}=0.05$, $VSM_{max}=2$, $\lambda_V=500$, $\lambda_Q=500$. Real power losses, voltage deviation and VSM are listed in Table 1. Table 2 includes all variable limitations that will be used for optimization process.

Table 1. Sub-objective function values, P_G and Q_G for IEEE 14- bus test system before optimization.

Real power losses (p.u.)	0.1339
Voltage deviation (p.u.)	0.0754
VSM(p.u.)	0.5489
$\sum P_G$ (MW)	272.3933
$\sum Q_G$ (MVAR)	82.4375

Table 2. Control variable limits for IEEE 14- bus test system.

Variable	Min variable	Max variable
Generator bus voltages (p.u.)	0.95	1.1
Q_{C9} (MVAR)	0	19
All taps (p.u.)	0.9	1.1
Q_{SVC} (MVAR)	-50	50

Best Function Value (BFV), Average Function Value (AFV) and Standard Deviation (STD) indices for IEEE 14- bus test system are listed in Table 3. These indices are obtained from 30 successive runs for each optimization algorithm. Real power losses *(Ploss)*, voltage deviation (ΔV_L) and *VSM* objectives for the best solution are listed in Table 4.

BFV, AFV and STD indices are very important. These parameters show the efficiency and robustness of an optimization algorithm in achieving global or near global optimal solution. BFV, AFV and STD indices of FWSOA are smaller than GA, PSO and SOA. From Table 3 it can be understood that FWSOA is better than GA, PSO and SOA in solving simultaneous RPD and SVC placement problem.

Table 3. Optimization indices by four algorithms for IEEE 14-bus test system.

Indices	GA	PSO	SOA	FWSOA
AFV	0.4600	1.2793e3	0.4542	0.4509
BFV	0.4097	0.4099	0.4091	0.4068
STD	0.0338	7.0038e3	0.0286	0.0271

Table 4. Sub-objective function values by four algorithms for IEEE 14-bus test system.

Indices	GA	PSO	SOA	FWSOA
Ploss	0.1317	0.1300	0.1254	0.1250
ΔV_L	0.0120	0.0271	0.0362	0.0416
VSM	0.5281	0.5289	0.5556	0.5621

The aim of the optimization is to further reduction in real power losses and voltage deviation and further increment in voltage stability margin. These objectives are listed in Table 4. However, increment or reduction in these objectives, depend quietly on the selection of weighted factors (w_i) in the overall objective function. So, these objectives may not validate the effectiveness of an optimization algorithm directly. For instance, by choosing the mentioned values for w_i for IEEE 14-bus test system, it is seen that ΔV_L becomes 0.0416 which is greater than GA, PSO and SOA, while Table 3 strictly represents the effectiveness of the FWSOA.

Figures 5 and 6 represent the voltage profile (before and after optimization) and convergence curve of FWSOA method for IEEE 14-bus system, respectively. Table 5 and Table 6 include control and dependent variables respectively, by FWSOA optimization method. Because of reduction in voltage deviation, voltage profile has been improved.

4.2. IEEE 57- bus test system

The IEEE 57- bus test system has seven generators, 80 transmission lines and 15 tap changer transformers. Three capacitors are placed at buses 18, 25, 53 [31]. Three SVCs are also considered in the present study for IEEE 57- bus test system. All variable limits have been taken from [25]. The population size is 60, total 30 runs and the maximum generations of 300 [24]. Optimization parameters are w_1=0.6, w_2=0.2, w_3=0.2, $Ploss_{min}=0.2$, $Ploss_{max}=0.5$, $\Delta V_{Lmin}=0$, $\Delta V_{Lmax}=1$, $VSM_{min}=0.05$, $VSM_{max}=0.4$, $\lambda_V=500$, $\lambda_Q=500$. All these limits are taken from [24]. Real power loss, voltage deviation and VSM, are in Table 7. Table 8 includes all variables that will be used for optimization process.

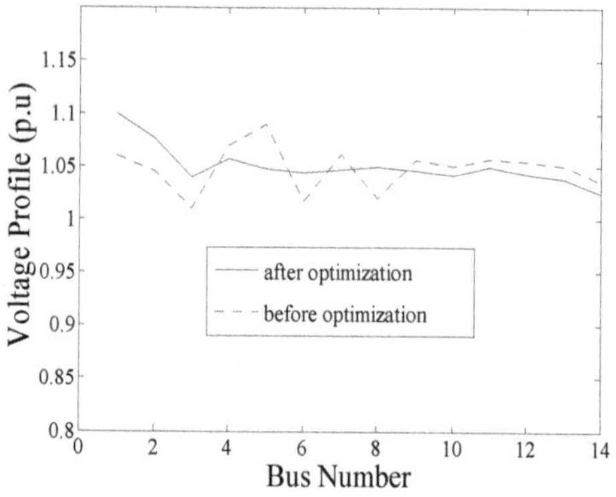

Fig. 5. Voltage profile for IEEE 14- bus test system before and after optimization.

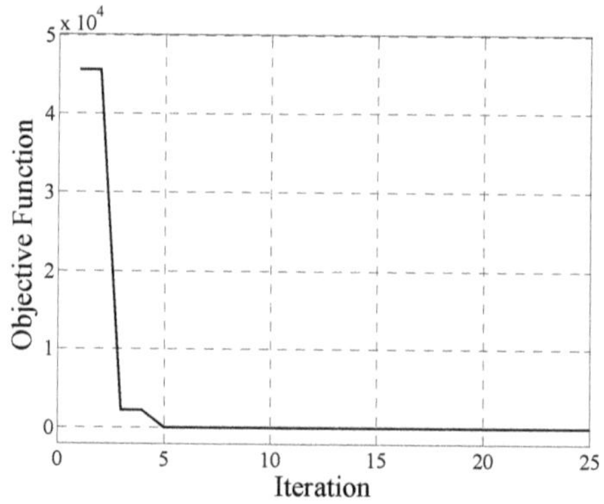

Fig. 6. Convergence curve of FWSOA method for IEEE 14-bus system

Table 5. Control variables of FWSOA optimization method for IEEE 14-bus system.

Control variables	Before optimization	After optimization
V_1	1.0600	1. 10
V_2	1.0450	1.08
V_3	1.0100	1.04
V_6	1.0177	1.06
V_8	1.0195	1.05
Q_{C9}	19	2.75
T_{4-7}	0.978	0.996
T_{4-9}	0.969	0.998
T_{5-6}	0.932	0.971
L_{SVC}	-	11
V_{SVC11}	-	1.05

Table 6. Dependent variables of FWSOA optimization method for IEEE 14-bus system.

$\sum Q_G$ (MVAR)	70.5607
Q_{SVC11}(MVAR)	4.3230

Table 7. Sub-objective function values, P_G and Q_G for IEEE 57- bus test system before optimization.

Real power losses (p.u.)	0.2786
Voltage deviation (p.u.)	0.0272
VSM (p.u.)	0.2101
$\sum P_G$ (MW)	1278.6638
$\sum Q_G$ (MVAR)	321.0800

Table 8. Control variable limits for IEEE 57- bus test system.

Variable	Min variable	Max variable
All voltages (p.u.)	0.94	1.06
Q_{C18} (MVAR)	0	10
Q_{C25} (MVAR)	0	5.9
Q_{C53} (MVAR)	0	6.3
All taps (p.u.)	0.9	1.1
Q_{SVCs} (MVAR)	-100	100

BFV, AFV and STD indices for IEEE 57- bus test system are listed in Table 9. These indices are obtained from 30 successive runs for each optimization algorithm. Real power losses *(Ploss)*, voltage deviation (ΔV_L) and *VSM* objectives for the best solution are listed in Table 10.

Table 9. Optimization indices by four algorithms for IEEE 57-bus test system.

Indices	GA	PSO	SOA	FWSOA
AFV	0.2171	0.4195	0.2193	0.2036
BFV	0.1972	0.2091	0.1829	0.1794
STD	0.0120	0.6536	0.0204	0.0117

Table 10. Sub-objective function values by four algorithms for IEEE 57-bus test system.

Indices	GA	PSO	SOA	FWSOA
Ploss	0.2494	0.2547	0.2437	0.2424
ΔV_L	0.0209	0.0197	0.0224	0.0255
VSM	0.2352	0.2325	0.2413	0.2432

BFV, AFV and STD indices are very important. These parameters show the efficiency and robustness of an optimization algorithm in achieving global or near global optimal solution. BFV, AFV and STD indices of FWSOA are smaller than GA, PSO and SOA. From Table 9 it can be understood that FWSOA is better than GA, PSO and SOA in solving simultaneous RPD and SVC placement problem.

The aim of the optimization is to further reduction in real power losses and voltage deviation and further increment in voltage stability margin. These objectives are listed in Table 10. However, increment or reduction in these objectives, depend quietly on the selection of weighted factors (w_i) in the overall objective function. So, these objectives may not validate the effectiveness of an optimization algorithm directly. For instance, by choosing the mentioned values for w_i for IEEE 57-bus test system, it is seen that ΔV_L becomes 0.0255 which is greater than GA, PSO and SOA, while Table 9 strictly represents the effectiveness of the FWSOA.

Figures 7 and 8 represent the voltage profile (before and after optimization) and convergence curve of FWSOA method for IEEE 57-bus system, respectively.

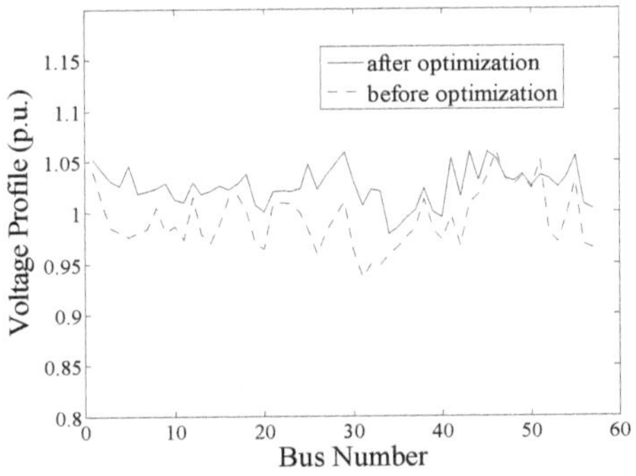

Fig. 7. Voltage profile for IEEE 57- bus test system before and after optimization.

Because of reduction in voltage deviation, voltage profile has been improved.

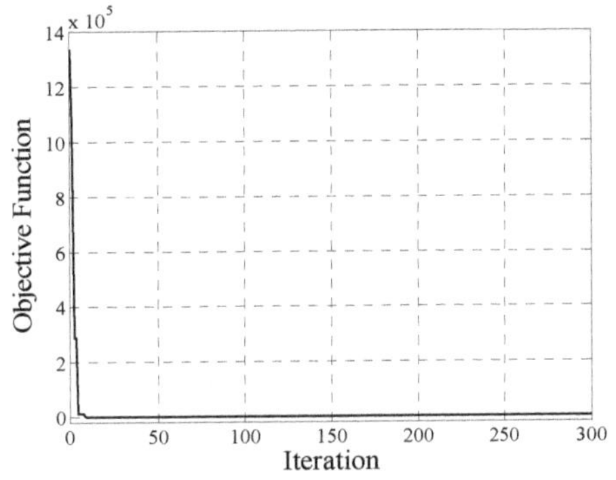

Fig. 8. Convergence curve of FWSOA method for IEEE 57-bus test system.

Tables 11 and 12 include control and dependent variables respectively, by FWSOA optimization method.

Table 11. Control variables by FWSOA optimization method for IEEE 57-bus test system.

Control variables	Before optimization	After optimization
V_1	1.04	1.05
V_2	1.01	1.04
V_3	0.985	1.03
V_6	0.98	1.03
V_8	1.005	1.05
V_9	0.98	1.02
V_{12}	1.015	1.02
Q_{C18}	10	7.52
Q_{C25}	5.9	5.90
Q_{C53}	6.3	6.30
$T_{4\text{-}18}$	0.97	1.02
$T_{4\text{-}18}$	0.978	0.991
$T_{20\text{-}21}$	1.043	1.02
$T_{24\text{-}26}$	1.043	0.999
$T_{7\text{-}29}$	0.967	0.968
$T_{32\text{-}34}$	0.975	0.927
$T_{11\text{-}41}$	0.955	0.900
$T_{15\text{-}45}$	0.955	0.964
$T_{14\text{-}46}$	0.9	0.983
$T_{10\text{-}51}$	0.93	0.974
$T_{13\text{-}49}$	0.895	0.974
$T_{11\text{-}43}$	0.958	0.949
$T_{40\text{-}56}$	0.958	0.999
$T_{39\text{-}57}$	0.98	0.962
$T_{9\text{-}55}$	0.94	0.960
L_{SVC}	-	50
L_{SVC}	-	49
L_{SVC}	-	46
V_{SVC50}	-	1.03
V_{SVC49}	-	1.04
V_{SVC46}	-	1.05

Table 12. dependent variables by FWSOA optimization method for IEEE 57-bus test system.

$\sum Q_G$ (MVAR)	211.8758
Q_{SVC50} (MVAR)	9.5144
Q_{SVCs49}(MVAR)	19.7419
Q_{SVCs46} (MVAR)	36.0037
$\sum Q_{SVCs}$ (MVAR)	65.2600

5. CONCLUSIONS

Voltage stability is very close to RPD in the power network. Thus, via optimization for included control variable settings in RPD problem, adequate and proper voltage stability margin, voltage deviation and real power losses reduction can be obtained. Therefore, in this paper, simultaneous RPD and SVC placement were investigated and optimization algorithms including GA, PSO, SOA and FWSOA were implemented to achieve these goals. The priority of this paper comparing with other researches is focusing on RPD with control variables including excitation settings of generators, tap positions of tap changing transformers, reactive power output of fixed capacitors and voltages with locations of SVCs in the network, simultaneously. Finally, by comparison, between performances of GA, PSO, SOA and FWSOA algorithms, the efficiency of the FWSOA based RPD and SVC placement approach in the present study was well verified.

REFERENCES

[1] M. Saravanan, S.M.R. Slochanal, P. Venkatesh and J. Prince-Stephen-Abraham, "Application of particle swarm optimization technique for optimal location of FACTS devices considering cost of installation and system loadability," *Electric Power System Research,* vol. 77 , no. 3-4, pp. 276-283, 2007.

[2] G.D. Galiana, "Assessment and control of the impact of FACTS devices on power system performance," *IEEE Transactions on Power Systems,* vol. 11, no. 4, pp. 1931-1936, 1996.

[3] H. Shayeghi and M. Ghasemi, "FACTS devices allocation using a novel dedicated improved PSO for optimal operation of power system,", *Journal of Operation and Automation in Power Engineering,* vol. 1, no. 2, pp. 124-135, 2013.

[4] R. Kazemzadeh, M. Moazen, R. Ajabi-Farshbaf and M. Vatanpour, STATCOM optimal allocation in transmission grids considering contingency analysis in OPF using BF-PSO algorithm, *Journal of Operation and Automation in Power Engineering,* vol. 1, no. 1, pp. 1-11, 2013.

[5] R. Sirjani, A. Mohamed and H. Shareef, "Optimal allocation of shunt Var compensators in power systems using a novel global harmony search algorithm," *International Journal of Electrical Power and Energy Systems,* vol. 43, no. 1, pp. 562–572, 2012.

[6] R. Minguez, F. Milano, R. Zarate-Minano and A. J. Conejo, "Optimal network placement of SVC devices," *IEEE Transactions on Power*

Systems, vol. 22, no. 4, pp 1851–1860., 2007.

[7] Sh. Udgir, S. Varshney and L. Srivastava, "Optimal placement and sizing of SVC for improving voltage profile of power system," *International Journal of Power System Operation and Energy Management,* vol. 1, no. 2, pp. 2231-4407, 2011.

[8] Y. Mansour, W. Xu, F. Alvarado and Ch. Rinzin, "SVC placement using modes of voltage instability," *IEEE Transactions on Power Systems,* vol. 9, no. 2, pp. 757 – 763, 1994.

[9] K. Sebaa, M. Bouhedda, A. Tlemcani and N. Henini, "Location and tuning of TCPSTs and SVCs based on optimal power flow and an improved cross-entropy approach," *International Journal of Electrical Power and Energy Systems,* vol. 54, pp. 536–545, 2014.

[10] I. Pisica, C. Bulac, L. Toma and M. Eremia, "Optimal SVC placement in electric power systems using a genetic algorithms based method," *Proceedings of the IEEE Power Technology Conference,* Bucharest, Romania, pp. 1-6, 2009.

[11] M.M. Farsangi, H. Nezamabadi-pour, S. Yong-hua and K.Y. Lee, "Placement of SVCs and selection of stabilizing signals in power systems", *IEEE Transactions on Power Systems,* vol. 22, no. 3, pp. 1061 - 1071, 2007.

[12] M. Gitizadeh, M. Shid-Pilehvar and M. Mardaneh, "A new method for SVC placement considering FSS limit and SVC investment cost," *International Journal of Electrical Power and Energy Systems* , vol. 53, pp. 900–908, 2013.

[13] T. Niknam, M.R. Narimani, R. Azizipanah-Abarghooee and B. Bahmani-Firouzi," Multi-objective optimal reactive power dispatch and voltage control: A new opposition-based self-adaptive modified gravitational search algorithm, "*IEEE Systems Journal,* vol. 7, no. 4, pp. 742-753, 2013.

[14] Ch-F Yang, G. G. Lai,Ch-H. Lee, Ch-T. Su and G. W. Chang, "Optimal setting of reactive compensation devices with an improved voltage stability index for voltage stability enhancement," *International Journal of Electrical Power and Energy Systems*, vol. 37, no. 1, pp. 50-57, 2012.

[15] A.A. Abou-El-Ela, M.A. Abido and S.R. Spea, "Differential evolution algorithm for optimal reactive power dispatch, "*Electric Power Systems Research,* vol. 81, no. 2,pp. 458-464, 2011.

[16] J. M. Ramirez, J.M. Gonzalez and T.O. Ruben, "An investigation about the impact of the optimal reactive power dispatch solved by DE," *International Journal of Electrical Power and Energy Systems,* vol. 33, no. 2, pp. 236-244, 2011.

[17] H.I. Shaheen, Gh.I. Rashed and S.J. Cheng, "Optimal location and parameter setting of UPFC for enhancing power system security based on differential evolution algorithm," *International Journal of Electrical Power and Energy Systems,* vol. 33, no. 1, pp. 94-105, 2011.

[18] A.H. Khazaliand and M. Kalantar, "Optimal reactive power dispatch based on harmony search algorithm," *International Journal of Electrical Power and Energy Systems,* vol. 33, no. 3, pp. 684-692, 2011.

[19] D. Devaraj and J. Preetha-Roselyn, "Genetic algorithm based reactive power dispatch for voltage stability improvement," *International Journal of Electrical Power and Energy Systems*, vol. 32, no. 10, pp. 1151-1156, 2010.

[20] K. Mahadevan and P.S. Kannan, "Comprehensive learning particle swarm optimization for reactive power dispatch," *Applied Soft Computing*, vol. 10, no. 2, pp. 641-652, 2010.

[21] B. Zhao, C.X. Guo and Y.J. Cao,"A multiagent-based particle swarm optimization approach for optimal reactive power dispatch," *IEEE Transactions on Power Systems*, vol. 20, no. 2, pp. 1070-1078, 2005.

[22] C. Zhang, M. Chen and C. Luo,"A multi-objective optimization method for power system reactive power dispatch," *Proceedings of the IEEE 8th World Congress on Intelligent Control and Automation,* pp. 6-10, Jinan, 2010.

[23] B. Zhao, C.X. Guo and Y.J.Cao,"An improved particle swarm optimization algorithm for optimal reactive power dispatch," *IEEE Power Engineering Society General Meeting* , pp. 272 - 279, 12-16 June 2005.

[24] Ch. Dai, W. Chen, Y. Zhu and X. Zhang, "Reactive power dispatch considering voltage stability with seeker optimization algorithm," *Electric Power Systems Research,* vol. 79, no. 10, pp. 1462-1471, 2009.

[25] S. Duman, Y. Sonmez, U. Guvenc and N. Yorukeren, "Optimal reactive power dispatch

using a gravitational search algorithm," *IET Proceedings on Generation, Transmission & Distribution,* vol. 6, no. 6, pp. 563-576, 2012.

[26] N. Karaboga, "Digital IIR filter design using differential evolution algorithm," *EURASIP Journal* of *Applied Signal Processing*, vol. *8*, no. pp. 1269-1276, 2005.

[27] L. dos-Santos-Coelho, B.M. Herrera, "Fuzzy identification based on a chaotic particle swarm optimization approach applied to a nonlinear Yo-yo motion system, "*IEEE Transactions on Industrial Electronics*, vol. 54, no.6 ,pp. 3234-3245, 2007.

[28] J. Kennedy and R. Eberhart, "Particle swarm optimization," *Proceedings of the IEEE International Conference on Neural Networks,* pp.1942-1948, 1995.

[29] D. Bratton and J. Kennedy," Defining a standard for particle swarm optimization," *Proceedings of the IEEE Conference on Swarm Intelligence Symposium,* pp. 120-127, Honolulu, 2007.

[30] K.F. Man, K.**S.** Tang, and **S.** Kwong, " Genetic algorithms: concepts and applications [in engineering design]," *IEEE Transactions on Industrial Electronics,* vol. 43, no. 5, pp. 519-534, 1996.

[31] R.D. Zimmerman, C.E. Murillo-Sanchez and D. Gan," *MATLAB Power System Simulation Package*," Version 4.1, Available at http://www.pserc.cornell.edu/matpower/2011.

[32] Q. Nana, M. Lixin, Sh. Guanhua and R. Youming, "An improved particle swarm optimization for reactive power optimization," *Proceedings of the IEEE 6th Joint International Information Technology and Artificial Intelligence Conference,* vol. 2, pp. 362-365, Chongqing, 2011.

An Improved Big Bang-Big Crunch Algorithm for Estimating Three-Phase Induction Motors Efficiency

M. Bigdeli[1,*], D. Azizian[2], E. Rahimpour[3]

[1]Department of Electrical Engineering, Zanjan Branch, Islamic Azad University, Zanjan, Iran
[2]Department of Electrical Engineering, Abhar Branch, Islamic Azad University, Abhar, Iran
[3]ABB AG, Power Products Division, Transformers, Bad Honnef, Germany

ABSTRACT
Nowadays, the most generated electrical energy is consumed by three-phase induction motors. Thus, in order to carry out preventive measurements and maintenances and eventually employing high-efficiency motors, the efficiency evaluation of induction motors is vital. In this paper, a novel and efficient method based on improved big bang-big crunch (I-BB-BC) algorithm is presented for efficiency estimation in the induction motors. In order to estimate the induction motor's efficiency, the measured current, the power factor and the input power are applied to the proposed method and an appropriate objective function is presented. The main advantage of the proposed method is efficiency evaluation of induction motor without any intrusive test. Moreover, a new effective and improved version of BB-BC algorithm is introduced. The presented modifications can improve the accuracy and speed of the classic version of algorithm. In order to demonstrate the capabilities of the proposed method, a comparison with other traditional methods and intelligent optimization algorithms is performed.

KEYWORDS: Efficiency estimation, Improved big bang-big crunch algorithm, Induction motor, Measurement.

1. INTRODUCTION

The electrical system as one of the main parts of the energy system is very important in any country such as industrial countries. This system shares the largest amount of energy comparing to other energy systems including oil, gas, coal, etc. Electrical motors use 70 to 75 percent of the total electricity which is consumed in industry. Thus, evaluation of induction motor's efficiency is an important issue for energy saving managements. Using the estimated efficiency of the induction motor, its performance can be judged and replacing the existing low efficiency motor by a high efficiency motor could be decided. In the recent years, several studies have been performed in this area which have some problems. The simplest and cheapest evaluation of efficiency is reading the motor's nameplate. This method assumes the efficiency to be constant and equal to the inscribed value on the plaque. This works well when the efficiency-load curve is smooth [1]. In Ref. [2], several intrusive methods have been explored to estimate the efficiency of the induction motor. Intrusive testing can be considered as a kind of interrupted testing. This procedure is done when the motor is removed from its normal operation mode. It usually requires no-load or blocked rotor tests. However, non-intrusive methods rely only on terminal voltage and current measurements while the motor is running. In these methods, to efficiency determination only requires values of the inputs to the motor, not the outputs. In simple terms, the non-intrusive method is performed for in-service motors. In the recent years, non-intrusive methods have been widely attended for these continuous applications [3-4].

*Corresponding author:
M. Bigdeli (E-mail: mehdi.bigdeli@iauz.ac.ir)

IEEE standard has introduced a proper measuring technique based on parameters identification of the equivalent circuit [5]. However, it requires rather intrusive measurements including no-load test, locked rotor test, DC test and the stray-load losses measurement, which are not possible in many cases. Another way for efficiency estimation is torque-based method. Torque-based methods are of two types: shaft torque measurement [1-2] and air-gap torque method [6]. In the shaft torque method, the shaft torque and the rotor speed are directly measured using sensors without the need to calculate the losses. The shaft torque measurement requires expensive equipments that may not be available in all areas and also, it is highly intrusive in efficiency estimation. In contrast to this, the air-gap torque method is a nonintrusive method for in-service motor-efficiency estimation using only motor terminal quantities and nameplate information. However, the no load test is avoided by the use of some empirical data. This causes lower estimation accuracy in comparison to the other well-known methods. In addition, the stator resistance shall be measured accurately at an operating temperature with a specific device. This causes extra cost, and installation concerns. The proposed method in Ref. [7] is the segregated losses method, which requires five types of losses measurements including: stator copper losses, rotor copper losses, core losses, stray-load losses and friction losses. The main disadvantage of this method is its intrusive measurements and the separation of different losses in machine. This technique involves separating rotor and reverse rotation tests for direct stray-load losses measurements. However, due to the need of performing two separate tests, it is not an appropriate method for efficiency determination of the induction motor. Some researchers used no-load tests to determine the motor efficiency in rated load [8-9]. The disadvantage of this approach is to separate the motor for no-load test. In Ref. [10], based on slip testing methods, motor efficiency is determined with a significant error (more than 10%). Ref. [11] has defined an instrument that could be used to measure the input power to determine the efficiency of an installed operating induction motor without removing power from the motor. The applied instrument is power analyzer which is very costly. So, it is not economical to use this method for small motors. In the recent years, the artificial intelligence-based methods are developed rapidly in the engineering fields [12-13] and several techniques have been proposed for efficiency evaluation of the induction motors [14-26]. Heuristics techniques employ the measured data such as current, voltage, input power, speed, and power factor. Using these measurements, parameters of the induction motor can be estimated and the efficiency of induction motor can be evaluated under different conditions. Genetic algorithm (GA) was one of the first employed heuristic methods in this field [14-20]; the main problem of GA is its premature convergence and low speed. Particle swarm optimization (PSO) was another employed algorithm that has higher speed rather than GA; but unfortunately it can conduct the estimation process to outside of the search space. Therefore, to obtain the final results, PSO must be implemented for several times [21-22]. In Refs. [23-26], the bacterial foraging algorithm (BFA) has been used to determine the motor efficiency. The main disadvantage of this method is its numerous parameters and lacking a reliable method for adjusting them.

Although the mentioned studies gave important results, such results are not efficient to estimate accurately the efficiency in induction motors without performing additional investigations. To address these shortcomings, here a new version of big bang-big crunch algorithm is proposed to overcome the above mentioned problems. This algorithm not only eliminates the mentioned problems of the previous methods but rather is very fast and there are a few numbers of parameters which are needed to be adjusted across the algorithm implementation.

Some important contributions of this work against past well-known works can be listed as follows:

- A new version of BB-BC algorithm is proposed for efficiency estimation.
- Every six equivalent circuit parameters of the induction motor are taken into the account.
- The efficiency is estimated and validated using the experimental data.
- The results of the proposed method are verified

against the results of past well-known works. According to the validation, the introduced method is more reliable than the methods that were presented in the previous researches.

- Three different fitness functions are applied to the proposed algorithm. The results related to these functions are discussed and compared with each other.

2. INDUCTION MOTOR'S EFFICIENCY ESTIMATION

2.1. Problem description

To estimate the efficiency of induction motor, multiple samples of input signals (current, speed, power factor and input power) should be available. So, at first the necessary tests have been carried out on a typical induction motor. Consequently, using the equivalent circuit model of induction motor in the steady-state condition, the objective function is created. In order to determine the efficiency of induction motor, three different objective functions are developed using the estimated and the measured data (current, power factor and input power). Afterwards, using various algorithms, equivalent circuit parameters are estimated in a way to minimize the difference between the measured and predicted values. Finally, using the estimated parameters and the input/output powers, the efficiency of an induction motor is calculated.

2.2. Machine equations

As previously mentioned, the equivalent circuit parameters are used for estimation the motor efficiency. Induction motor equivalent circuit is presented in Fig. 1.

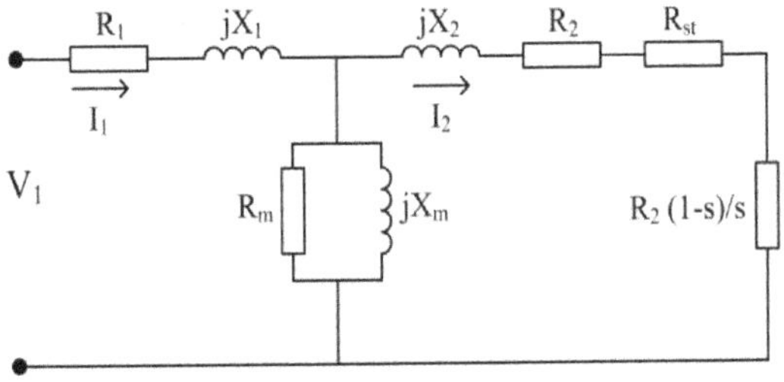

Fig. 1. Steady-state equivalent circuit of induction motor

where:

V_1: stator voltage (V)
I_1: stator current (A)
I_2: rotor current (A)
I_m: magnetizing current (A)
R_1: stator resistance (Ω)
R_2: rotor resistance referred to stator (Ω)
R_m: core loss resistance (Ω)
R_{st}: stray-load resistance (Ω)
X_1: stator leakage reactance (Ω)
X_2: rotor leakage reactance referred to stator (Ω)
X_m: magnetizing reactance (Ω)
S: slip
P_{in}: input power
P_{out}: output power

The stray-load losses could be modeled by adding an equivalent resistor to the equivalent circuit suggested by IEEE standard [5]. This resistance (R_{st}) can be calculated using the following equation:

$$R_{st} = \frac{0.018(1 - S_{fl})}{S_{fl}} \tag{1}$$

Using the equivalent circuit of Fig. 1, the admittances of stator and rotor can be expressed as follow:

$$Y_1 = \frac{1}{R_1 + jX_1} \tag{2}$$

$$Y_2 = \frac{1}{(R_2 / s) + jX_2 + R_{st}} \tag{3}$$

Similarly, the admittance of parallel branch in the equivalent circuit can be obtained from Eq. (4):

$$Y_m = \frac{1}{R_m} - \frac{j}{X_m} \tag{4}$$

At last, the total impedance of the equivalent circuit can be stated by the series-parallel combination of the above mentioned admittances as follows:

$$Z_1 = R_1 + jX_1 + \frac{1}{Y_2 + Y_m} \tag{5}$$

After calculating the admittances, efficiency of the motor can be calculated using the Kirchhoff's law as the following:

$$I_{1e} = \left| \frac{\overline{V_1 Y_1 (Y_2 + Y_m)}}{Y_1 + Y_2 + Y_m} \right| \tag{6}$$

$$I_{2e} = \left| \frac{\overline{V_1 Y_1 Y_2}}{Y_1 + Y_2 + Y_m} \right| \tag{7}$$

$$\cos\varphi_e = \frac{\mathrm{Re}(\overline{I_{1e}})}{\overline{I_{1e}}} \tag{8}$$

$$P_{ine} = 3(I_1^2 R_1 + I_2^2(R_2/s + R_{st}) + I_m^2 R_m) \tag{9}$$

$$P_{out\,e} = 3I_2^2 R_2(\frac{1-s}{s}) - P_{fw} \tag{10}$$

$$\eta = \frac{P_{out\,e}}{p_{ine}} \tag{11}$$

It should be noted that mechanical losses (or friction and windage losses), P_{fw}, are nearly constant from no-load to full-load. So, here it is taken as a constant percentage of the rated input power where P_{fw}=1.2%P_{in} as suggested by many motor efficiency estimation methods [1-6], [21-26].

2.3. Objective functions

To estimate the efficiency, in the first step the equivalent circuit parameters should be identified. To estimate the equivalent circuit parameters, different objective functions are introduced. The conventional way for this purpose is to estimate the unknown parameters to minimize the sum of squared errors between the calculated and measured results. Additionally, it must also be provided the constraints to be met. To achieve a comprehensive conclusion, different objective functions are defined. In past researches [23-26], current, power and torque have been used for this purpose. As a new work, in the current research a combination of these parameters is recommended to be used to construct the objective functions as follows:

$$F_1(x) = \sum_{j=1}^{n}\left(\frac{I_{1ej}}{I_{mj}} - 1\right)^2 + \sum_{j=1}^{n}\left(\frac{Pf_{ej}}{Pf_{mj}} - 1\right)^2 \tag{12}$$

$$F_2(x) = \sum_{j=1}^{n}\left(\frac{P_{inej}}{P_{inmj}} - 1\right)^2 + \sum_{j=1}^{n}\left(\frac{I_{1ej}}{I_{mj}} - 1\right)^2 + \sum_{j=1}^{n}\left(\frac{Pf_{ej}}{Pf_{mj}} - 1\right)^2 \tag{13}$$

$$F_3(x) = \sum_{j=1}^{n}\left(\frac{P_{inej}}{P_{inmj}} - 1\right)^2 + \sum_{j=1}^{n}\left(\frac{I_{1ej}}{I_{mj}} - 1\right)^2 \tag{14}$$

In the above equations, indices e and m are related to the estimated and measured values, respectively, and n is the sampling number.

2.4. Improved big bang-big crunch (i-bb-bc) algorithm

BB-BC algorithm is a recently developed method that relies on theory of the universe evolution [27]. BB-BC algorithm has been used in some problems such as optimal power flow [28-29]. BB-BC is developed from the Big Bang and the Big Crunch phases [27].

In the Big Bang phase, the first population is spread uniformly into search space. The second phase (Big Crunch) that computes a center of mass for population is a convergence operator. The center of mass could be calculated as follows [27]:

$$\vec{x}^c = \frac{\sum \vec{x}_i f_i^{-1}}{\sum f_i^{-1}} \tag{15}$$

where:

x_i: a member of population, and

f_i : fitness function's value.

Following the Big Crunch phase, new population must be generated for as the next Big Bang phase. The new generations must be spread around the center of mass by adding a normal random number as the following [28]:

$$\vec{x}_{new} = \vec{x}_c + \frac{rand \times (x_{\max} - x_{\min})}{k} \tag{16}$$

Where:

rand: a random number with normal distribution,

k: iteration number,

x_{min}: lower limit of the parameters, and

x_{max}: upper limit of the parameters.

In order to modify the performance of BB-BC method, here a modification is proposed for Eq. (16). This modification that is shown in Eq. (17) will be called as Improved Big Bang-Big Crunch (I-BB-BC) algorithm.

$$\vec{x}_{new} = \vec{x}_c + \frac{rand \times (x_{\max} - x_{\min})}{(f_{best})^p} \tag{17}$$

In this improved form the best fitness of individuals (f_{best}) is employed instead of iteration number (k); while the best fitness is increased, new populations will be spread nearer to center of mass. Note that the fitness function is declared as $f = 1/F$, where, F is the objective function that was introduced in Eqs. (12)-(14). Also, standard deviation of the normal random number must be adjusted for better results.

Following the new bang, center of mass should be recomputed and steps should be repeat till the ending condition is met. Fig. 2 shows the operation flowchart of the I-BB-BC algorithm.

3. EXPERIMENTAL MEASUREMENTS

As was mentioned before, for non-intrusive determination of the induction motor's efficiency, the losses should be determined. To estimate the parameters of the induction motor, experimental samples should be taken from the actual motor. As it is known, by the load torque changing, motor speed, current and power factor change. Sampling the measurements (current, power factor and speed) is performed from 25% to 100% of the full load. The reason for this kind of sampling is to increase the accuracy of optimization algorithms. Because a large number of samples help the algorithm to has a good performance in searching the problem domain. Fig. 3 shows the measurement circuit that was used in the laboratory. Tables 1 and 2 show the nameplate data and the measured values for a typical 2 hp induction motor, respectively.

4. PARAMETERS ESTIMATION AND EFFICIENCY EVALUATION RESULTS

To implement the algorithm, a program using the GUI/MATLAB is developed. Features of this program are applicable for any number of variables that are bounded by imposing constraints. To demonstrate the effectiveness of the proposed algorithm and compared it with other well-known algorithms, the problem is solved using various algorithms (GA, PSO and BB-BC) with different objective functions and the efficiency of induction motor have been identified. The results are given in the following.

4.1. Estimations using Eq. (12) as the objective function

Using Eq. (12), the motor parameters are estimated and the results are presented in Table 3. After estimating the parameters, motor efficiency was computed from Eq. (11) and the results are shown in Table 4. This Table also shows the output estimated errors (differences between the results of the heuristic algorithms and the actual values).

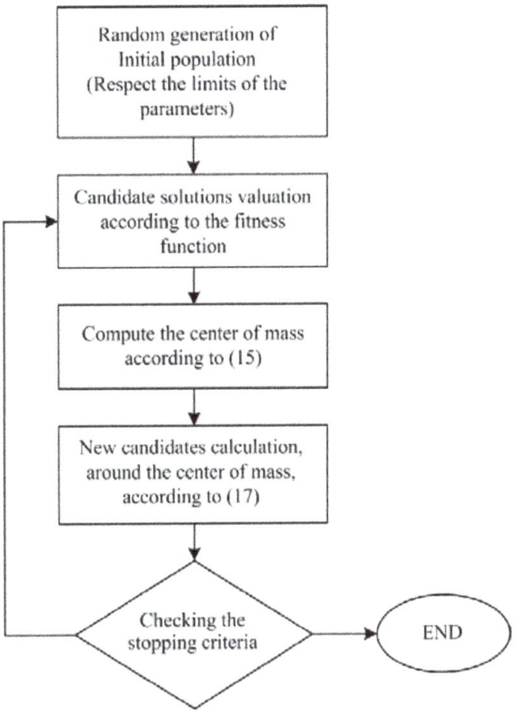

Fig. 2. I-BB-BC steps

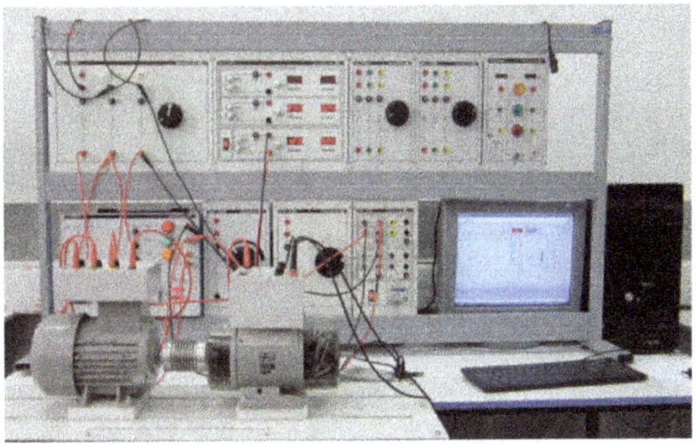

Fig. 3. Measurement setup in laboratory

Table 4 indicates that the I-BB-BC algorithm is close to the real values in three points (25%, 50% and 100% of full load) and only in 75% of the full load, the PSO algorithm has reached to a better result. To study the speed of the algorithms about achieving the optimal response, Fig. 4 shows the cost function reduction due to the number of iterations.

Table 1. Nameplate data for a typical 2hp induction motor

Brand	Siemens	Design class	B
Insulation Class	B	IP	55
I	4 A	Power Factor	0.76
Poles	4	Speed	1440 rpm
f	50 Hz	V_{L-L}	230/400 V
Connection	Y-d	Type	STC-3

Table 2. Measured values for a typical 2hp induction motor

Motor load (%)	I_1 (A)	PF	P_{in} (W)	P_{out} (W)	Efficiency (%)
25	2.71	0.412	738.9	382.01	51.70
50	3.01	0.573	1141.7	753.52	66.00
75	3.44	0.682	1555.6	1110.1	71.36
100	3.99	0.751	1999.5	1463.1	73.18

Table 3. The results of motor parameters estimation by various algorithms for objective function of Eq. (12)

Parameter (ohm)	GA	PSO	BB-BC	I-BB-BC
R_1	3.68	5.12	4.47	4.81
X_1	6.91	8.076	5.58	5.90
R_2	3.51	3.37	3.12	3.70
X_2	3.36	2.52	3.91	3.34
R_m	502.63	530.66	475.63	489.24
X_m	82.58	77.99	79.27	79.85

Table 4. The results of efficiency estimation by various algorithms for objective function of Eq. (12)

Motor load (%)	Efficiency (%)				Error (%)			
	GA	PSO	BB-BC	I-BB-BC	GA	PSO	BB-BC	I-BB-BC
25	55.74	53.92	53.45	52.18	7.82	4.29	3.38	0.93
50	69.53	67.44	66.69	66.54	5.35	2.18	1.04	0.81
75	74.52	72.47	72.95	73.06	4.43	1.56	2.22	2.38
100	76.17	74.01	74.13	73.49	4.09	1.11	1.29	0.42
Average					5.42	2.28	1.98	1.13

The obtained results show that the accuracy of the I-BB-BC is better than other algorithms.

4.2. Estimations using Eq. (13) as the objective function

In this objective function the input power, input power factor and input current are applied. In Tables 5 and 6, the results of the estimated parameters and the efficiency of the motor determined by different algorithms are given. Table 6 also shows the errors of efficiency estimation by different optimization algorithms.

As shown in Table 6, the mean error of BB-BC and I-BB-BC algorithms has increased with respect to the first objective function. However, when using this objective function for GA and PSO algorithms, the error is reduced.

Fig. 5 is a graph of cost function based on the number of iterations. It is clear that the convergence of BB-BC and PSO is almost identical. However, BB-BC convergence is faster than PSO and to achieve the optimal solution, the number of iterations is less than PSO needs.

Fig. 5 also shows that the convergence of I-BB-BC is rapid and the convergence of the GA is slow. In the low number of iterations, GA is not a good algorithm and for better performance, it needs to be increased the number of iterations. However, for better comparison, all the conditions are assumed to be equal in all cases.

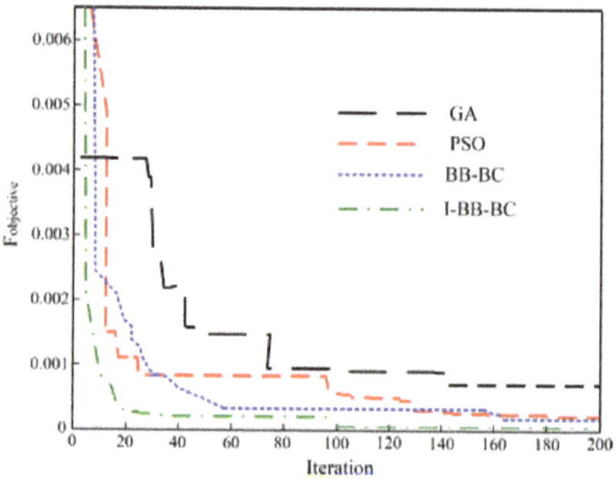

Fig. 4. Convergence characteristics of various algorithms for first objective function of Eq. (12)

Table 5. The results of motor parameters estimation by various algorithms for objective function of Eq. (13)

Parameter (ohm)	GA	PSO	BB-BC	I-BB-BC
R_1	3.79	3.58	3.48	3.55
X_1	7.15	4.85	4.77	5.02
R_2	3.49	3.82	3.97	3.88
X_2	3.92	4.61	4.46	4.51
R_m	512.78	471.14	466.95	475.04
X_m	82.66	82.69	82.23	82.76

4.3. Estimations using Eq. (14) as the objective function

To use the objective function of Eq. (14), only the input power and current are applied. Table 7 shows the results of different algorithms in estimating parameters of the induction motor.

Table 6. The results of efficiency estimation by various algorithms for objective function of Eq. (13)

Motor load (%)	Efficiency (%)				Error (%)			
	GA	PSO	BB-BC	I-BB-BC	GA	PSO	BB-BC	I-BB-BC
25	52.07	52.68	52.34	52.19	0.71	1.89	1.23	0.94
50	69.72	67.21	67.45	67.03	5.63	1.83	2.19	1.56
75	74.59	72.67	73.76	72.85	4.52	1.83	3.36	2.08
100	76.12	74.79	75.38	74.49	4.01	2.20	3.00	1.79
Average					3.71	1.93	2.44	1.59

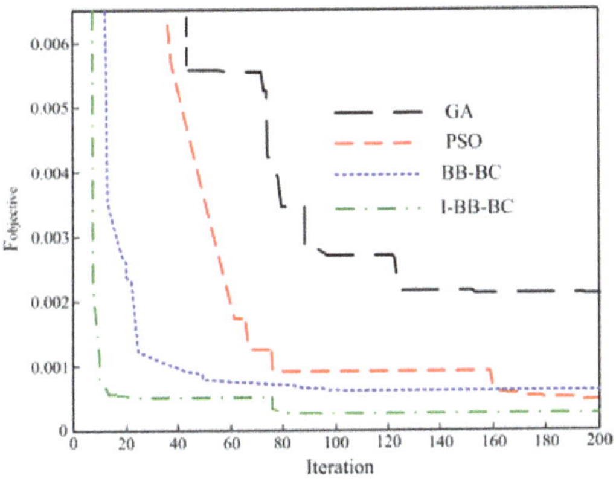

Fig. 5. Convergence characteristics of various algorithms for objective function of Eq. (13)

After estimating the parameters of the motor, it is also possible to calculate the efficiency as shown in Table 8. This Table also shows the estimated errors of the calculated efficiency with implementation of the third objective function. As seen, the resultant error by using this function is much less than other functions in the implementation of all four algorithms.

Figure 6 shows a graph of cost function reduction based on the number of iterations. It is clear that I-BB-BC algorithm has the lowest cost function and its convergence is very fast.

4.4. Validation against traditional methods and discussion

It is clear that despite the satisfactory results of the introduced algorithm in comparison with other heuristic algorithms, results of the proposed method should be compared with traditional methods. For this purpose, the results of efficiency evaluation using the introduced algorithm are compared to the classical methods (such as slip [10], current [6] and the equivalent circuit methods [11]). The obtained results are presented in Table 9.

Table 7. The results of motor parameters estimation by various algorithms for objective function of Eq. (14)

Parameter (ohm)	GA	PSO	BB-BC	I-BB-BC
R_1	4.12	4.15	4.55	4.19
X_1	4.33	4.54	4.67	4.58
R_2	3.80	3.81	3.75	3.78
X_2	3.08	4.19	5.08	4.92
R_m	514.48	495.60	487.36	497.43
X_m	81.37	81.99	81.42	81.59

Table 8. The results of efficiency estimation by various algorithms for objective function of Eq. (14)

Motor load (%)	Efficiency (%)				Error (%)			
	GA	PSO	BB-BC	I-BB-BC	GA	PSO	BB-BC	I-BB-BC
25	53.44	52.78	52.61	52.01	3.36	2.09	1.76	0.60
50	67.69	67.12	67.15	66.49	2.56	1.69	1.74	0.74
75	73.07	72.56	71.87	71.67	2.39	1.68	0.71	0.43
100	74.97	74.47	74.31	73.86	2.44	1.76	1.54	0.93
Average					2.68	1.80	1.43	0.68

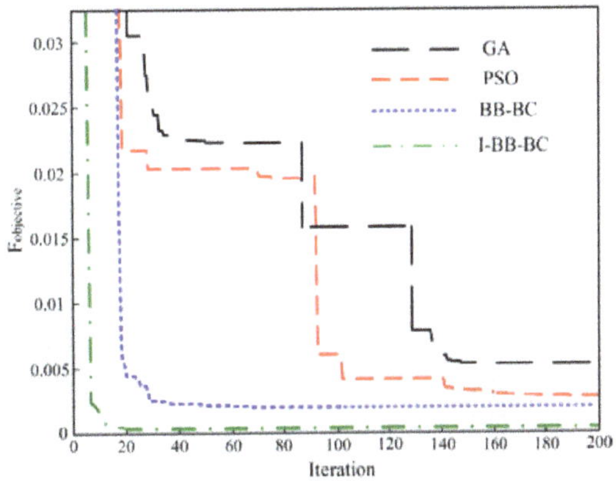

Fig. 6. Convergence characteristics of various algorithms for objective function of Eq. (14)

Table 9 shows that employing traditional methods for efficiency estimation during different loads (especially when the motor works with a load less than the rated load), causes many errors. The error of the equivalent circuit method is lower than the slip and the current methods. But as this method requires intrusive tests, it is not an appropriate method. The results of implementing various algorithms against different objective functions show that the objective function of Eq. (14) leads to the optimal solution with more accuracy and higher speed. As seen in Table 8, estimated efficiency by all algorithms is very close to the measured values and the estimated error in all algorithms is very lower than that for other objective functions.

Table 9. The results of efficiency estimation by traditional methods

Motor load (%)	Efficiency (%)			Error (%)		
	Slip method [10]	Equivalent circuit method [11]	Current method [6]	Slip method [10]	Equivalent circuit method [11]	Current method [6]
25	42.17	48.61	65.75	18.43	5.97	27.17
50	58.07	63.54	72.16	12.01	3.72	9.33
75	66.99	70.02	72.85	6.12	1.87	2.08
100	72.01	72.23	72.64	1.59	1.29	0.74

Among the mentioned traditional methods and the heuristic algorithms, I-BB-BC has smaller errors in efficiency estimation rather than other methods. Additionally, the comparison between efficiency computations for various heuristic algorithms is given in Table 10. From Table 10, it is clear that the average computation time of the I-BB-BC method is less than other methods. Meanwhile, the results (Figs. 4-6) show that the proposed algorithm can reach the lowest cost function in comparison with other algorithms. Therefore, the objective function of Eq. (14) employed with I-BB-BC algorithm is proposed as a suitable method for estimating the efficiency of induction motors.

Table 10. Comparison of computation time (second) for various algorithms

Algorithm	Objective function		
	Equation (12)	Equation (13)	Equation (14)
GA	4.98	5.28	5.12
PSO	4.27	4.48	4.39
BB-BC	4.18	4.23	4.13
I-BB-BC	3.33	3.42	3.28

In addition to very good results obtained in the estimation of efficiency, DC test which is usually performed in the stator resistance estimation was removed and the stator resistance was taken into account as a parameter that was estimated using the introduced method. In the past researches [11], [14-16], efficiency estimation is performed for motors with the same reactances of rotor and stator (class A); however, in this study, two reactances were not selected equal. So, the efficiency estimation could be applied to each motor with each insulation class.

5. CONCLUSION

In this paper, using an improved version of the I-BB-BC algorithm and employing different objective functions, the induction motor parameters were estimated. Then using the determined parameters, the motor's efficiency was calculated. Finally, to prove the capability of the introduced method, a comparison with other traditional and intelligent methods such as GA, PSO and BB-BC was performed. The comparison shows that the efficiency estimation by I-BB-BC algorithm has much less error than other methods and results are very close to the actual values. In addition, it quickly converges to optimal solution. So, according to the speed and accuracy of the proposed algorithm, it can be used as a reliable method for accurate determination of the efficiency in induction motors. In addition, the proposed method can be used to monitor the status of induction motors in the industrial areas.

ACKNOWLEDGEMENT

This research was undertaken in the framework of a research project supported by the Islamic Azad University, Zanjan Branch. So, the authors gratefully acknowledge Zanjan Branch of Islamic Azad University for their financial and spiritual supports.

REFERENCES

[1] J.S. Hsu, J. D. Kueck, M. Olszewski, D.A. Casada and P. J. Otaduy, "Comparison of induction motor field efficiency evaluation methods," *IEEE Transactions on Industry Applications*, vol. 34, no. 1, pp. 117-125, 1998.

[2] B. Lu, T.G. Habetler and R.G. Harley, "A survey of efficiency-estimation methods for in-service induction motors," *IEEE Transactions on Industry Applications*, vol. 42, no. 4, pp. 924-933, 2006.

[3] C.S. Gajjar, J.M. Kinyua, M.A. Khan and P.S. Barendse, "Analysis of a non-intrusive efficiency estimation technique for induction machines compared to the IEEE 112B and IEC 34-2-1 standards," *IEEE Transactions on Industry Applications*, vol. 51, no. 6, pp. 4541-4553, 2006.

[4] M. Chirindo, M.A. Khan and P.S. Barendse,

"Considerations for non-intrusive efficiency estimation of inverter-fed induction motors," *IEEE Transactions on Industrial Electronics*, Early Access, Published Online, 2015.

[5] IEEE standard test procedure for polyphase induction motors and generators, *IEEE Standard 112, IEEE Power Engineering Society*, New York, 1996.

[6] B. Lu, T.G. Habetler and R.G. Harley, "A nonintrusive and in-service motor-efficiency estimation method using air-gap torque with considerations of condition monitoring," *IEEE Transactions on Industry Applications*, vol. 44, no. 6, pp. 1666-1674, 2008.

[7] Y. El-Ibiary, "An accurate low cost method for determining electric motor's efficiency for the purpose of plant energy management," *IEEE Transactions on Industry Applications*, vol. 39, no. 4, pp. 12-19, 2003.

[8] A.G. Siraki, P. Pillay and P. Angers, "Full load efficiency estimation of refurbished induction machines from no-load testing," *IEEE Transactions on Energy Conversion*, vol. 28, no. 2, pp. 317-326, 2013.

[9] M. Al-Badri, P. Pillay and P. Angers, "A novel algorithm for estimating refurbished three-phase induction motors efficiency using only no-load tests," *IEEE Transactions on Energy Conversion*, vol. 30, no. 2, pp. 615-625, 2015.

[10] V. Dlamini, R. Naidoo, M. Manyage, "A non-intrusive method for estimating motor efficiency using vibration signature analysis," *International Journal of Electrical Power and Energy Systems*, vol. 45, no. 1, pp. 384-390, 2013.

[11] J.R. Holmquist and M. A. Rooks, "Richter practical approach for determining motor efficiency in the field using calculated and measured values," *IEEE Transactions on Industry Applications*, vol. 40, no. 1, pp. 242-248, 2004.

[12] E. Babaei and N. Ghorbani, "Combined economic dispatch and reliability in power system by using PSO-SIF algorithm," *Journal of Operation and Automation in Power Engineering,* vol. 3, no. 1, pp. 23-33, 2015.

[13] M. Sedighizadeh and M. Mahmoodi "Optimal reconfiguration and capacitor allocation in radial distribution systems using the hybrid shuffled frog leaping algorithm in the fuzzy framework," *Journal of Operation and Automation in Power Engineering*, vol. 3, no. 1, pp. 56-70, 2015.

[14] T. Phumiphak and C. Chat-Uthai, "Estimation of induction motor parameters based on field test coupled with genetic algorithm," in *Proceedings of the IEEE International Conference on Power System Technology*, pp. 1199- 1203, 2002.

[15] A. Charette, J. Xu, A. Ba-Razzouk, P. Pillay and V. Rajagopalan, "The use of the genetic algorithm for in-situ efficiency measurement of an induction motor," in *Proceedings of the IEEE International Conference on Power Engineering Society, Winter Meeting*, pp. 392-397, 2000.

[16] M. Cunkas and T. Sag, "Efficiency determination of induction motors using multi-objective evolutionary algorithms," *Advances in Engineering Software*, vol. 41, no. 2, pp. 255-261, 2010.

[17] P. Nangsue, P. Pillay and S.E. Conry, "Evolutionary algorithms for induction motor parameter determination," *IEEE Transactions on Energy Conversion*, vol. 14, no. 3, pp. 447-453, 1999.

[18] B. Lu, C. Wenping, I. French, K.J. Bradley and T.G. Habetler, "Non-intrusive efficiency determination of in-service induction motors using genetic algorithm and air-gap torque methods," in *Proceedings of the IEEE 42nd IAS Annual Meeting, International Conference on Industry Applications*, pp. 1186-1192, 2007.

[19] M. Al-Badri, P. Pillay and P. Angers, "A novel in situ efficiency estimation algorithm for three-phase IM using GA, IEEE method F1 calculations, and pretested motor data," *IEEE Transactions on Energy Conversion*, vol. 30, no. 3, pp. 1092-1102, 2015.

[20] I. Kostov, V. Vasil Spasov and V. Rangelova, "Application of genetic algorithm for determining the parameters of induction motors," *Technical Gazette*, vol. 16, no. 2, pp. 49-53, 2009.

[21] V.P. Sakthivel and S. Subramanian, "On-site efficiency evaluation of three-phase induction motor based on particle swarm optimization," *Energy*, vol. 36, no. 3, pp. 1713- 1720, 2011.

[22] C.P. Salomon, C. Wilson, E. Luiz, G. Lambert, E.L. Bonaldi, E. L. Levy, J. G. Borges, "Motor efficiency evaluation using a new concept of stator resistance," *IEEE Transactions on Instrumentation and Measurement*, vol. 64, no. 11, pp. 2908-2917, 2015.

[23] V.P. Sakthivel, R. Bhuvaneswari and S. Subramanian, "Non-intrusive efficiency estimation method for energy auditing and management of in service induction motor using bacterial foraging algorithm," *IET Electric Power Applications,* vol. 4, no. 8, pp. 579-590, 2010.

[24] V.S. Santos, P.R. Viego, J.R. Gomez, N.A. Lemozy, A. Jurado, E.C. Quispe, "Procedure for determining induction motor efficiency working under distorted

grid voltages," *IEEE Transactions on Energy Conversion*, vol. 30, no. 1, pp. 331-339, 2015.

[25] V.S. Santos, P.V. Felipe and J.G. Sarduy, "Bacterial foraging algorithm application for induction motor field efficiency estimation under unbalanced voltages," *Measurement*, vol. 46, no. 7, pp. 2232-2237, 2013.

[26] V.P. Sakthivel, R. Bhuvaneswari and S. Subramanian, "An accurate and economical approach for induction motor field efficiency estimation using bacterial foraging algorithm," *Measurement*, vol. 44, no. 4, pp. 674-684, 2011.

[27] O. K. Erol and I. Eksin, "A new optimization method: big bang–big crunch," *Advances in Engineering Software*, vol. 3, no. 7, pp. 106-111, 2006.

[28] S. Sakthivel, S.A. Pandiyan, S. Marikani and S.K. Selvi, "Application of big bang big crunch algorithm for optimal power flow problems," *The International Journal of Engineering and Science*, vol. 2, no. 4, pp. 41-47, 2013.

[29] S. Sakthivel, M. Gayathri and V. Manimozhi, "A nature inspired optimization algorithm for reactive power control in a power system," *International Journal of Recent Technology and Engineering*, vol. 2, no. 1, pp. 29-33, 2013.

Optimal Emergency Demand Response Program Integrated with Multi-Objective Dynamic Economic Emission Dispatch Problem

E. Dehnavi, H. Abdi*, F. Mohammadi

Department of Electrical Engineering, Engineering Faculty, Razi University, Kermanshah, Iran

ABSTRACT

Nowadays, demand response programs (DRPs) play an important role in price reduction and reliability improvement. In this paper, an optimal integrated model for the emergency demand response program (EDRP) and dynamic economic emission dispatch (DEED) problem has been developed. Customer's behavior is modeled based on the price elasticity matrix (PEM) by which the level of DRP is determined for a given type of customer. Valve-point loading effect, prohibited operating zones (POZs), and the other non-linear constraints make the DEED problem into a non-convex and non-smooth multi-objective optimization problem. In the proposed model, the fuel cost and emission are minimized and the optimal incentive is determined simultaneously. The imperialist competitive algorithm (ICA) has solved the combined problem. The proposed model is applied on a ten units test system and results indicate the practical benefits of the proposed model. Finally, depending on different policies, DRPs are prioritized by using strategy success indices.

KEYWORDS: Emergency demand response program, Dynamic economic emission dispatch, Imperialist competitive algorithm, Optimal incentive, Strategy success indices.

1. Introduction

Determining the optimal incentive in the incentive-based demand response programs (DRPs) should be based on a feasible and economical approach. Otherwise, it may impose a high additional cost at the supply side, create new peak when DRP ends [1], and decrease the network reliability [2].1 Due to the natures of the emergency DRP (EDRP) and dynamic economic emission dispatch (DEED) problems which focus on the demand side and supply side respectively, for a more comprehensive and effective investigation, integrating these two problems seems very useful. In other words, in the combined model, the fuel cost and emission are minimized and the optimal incentive is determined simultaneously. Modeling the customer's behavior based on price elasticity matrix (PEM) is one of the most feasible and powerful methods in this field [3-6].

EDRP and direct load control (DLC) are both voluntary incentive-based DRPs and there is no difference between their modeling. In other words these two programs have a same modeling as will be developed in the part two in this paper. Actually the ways of implementing of these two programs by independent system operator (ISO) is the main difference between them. In DLC, ISO directly controls some special loads which have possibility of being controlled remotely and this is not possible for all kind of loads. In EDRP, ISO motivates customers to reduce, interrupt, or shift their loads and it is possible for all kind of loads. On the other hand when EDRP is implemented, people have more social welfare in comparison to DLC. For the time-based DRPs like time of use (TOU), real time pricing (RTP), and critical peak pricing (CPP), instead of the optimal incentive, the optimal electricity price is determined during different periods.

*Corresponding author:
H. Abdi (hamdiabdi@razi.ac.ir).

A time-based DRP has been implemented in [7] to serve the power and heat demands of the customer with minimum cost. In the proposed DRP, the amount of responsive load can vary in different time intervals. The aim of the proposed DRP is to shift the load from high market price time intervals to the low market price time intervals. Actually they have presented the short-term hourly scheduling of industrial and commercial customers with cogeneration facilities, conventional power units, and heat-only units.

Real time pricing (RTP) program has been investigated in [8]. They have presented a combined scheduling and bidding algorithm for constructing the bidding curve of an electric utility that participated in the day-ahead energy markets.

In Ref. [9] EDRP and Interruptible/Curtail able (I/C) programs have been implemented in the unit commitment (UC) problem. Then, the effects of these two DRPs have been compared in the long-term UC problem with fuel constraints. The proposed methodology has been formulated as a mixed integer linear problem and implemented in GAMS environment.

A robust optimization approach has been proposed in Ref. [10] for decision making of electricity retailers. Meanwhile, considering the effect of DRP on total procurement cost, an optimal bidding strategy is proposed of electricity retailers with the time-based model of DRP in the electricity market. Also, it is considered that the consumers only participate in TOU programs. Moreover, rather than using the forecasted prices as inputs, the upper and lower limits of pool prices have been considered for the uncertainty modeling in their proposed model.

Up to now many works have been carried out based on the UC problem integrating with DRPs [11-15].

As mentioned above, the UC problem integrating with DRP has been investigated a lot. But, there are few works related to the economic dispatch problem integrating with the DRPs to appoint the optimal incentive or price and get the minimum generation cost. Y. Chen and J. Li [16] compared three formulations of the security constrained economic dispatch for facilitating participation of DRRs in the Midwest ISOs energy and the ancillary service market. They mainly focused on the interruptible loads [16]. A. Ashfaq et al. [17] presented a combined model of the economic dispatch problem integrating with demand side response. In their model, at the peak hours, the price signal is set by the generation company one hour ahead and sent to the residential area. They have neglected some constraints in the economic dispatch problem and like the pervious mentioned work the emission objective of the generating units has not been taken into account. Also, in their model just peak hours have been considered and it has not been applied to the whole day. N.I. Nwulu and X. Xia [18] investigated the game theory based DR integrating with the economic and environmental dispatch [18]. Game theory is the study of strategic decision making introduced by John von Neumann in 1928. Specifically, it is the study of mathematical models of conflict and cooperation between intelligent rational decision-makers. One of the main drawbacks of this theory is the difficulty of using it as a basis for estimation. In other words, after modeling, the model will not be so clear and in realistic systems may not be so helpful. Also, in Ref. [18] the valve-point loading effect and POZs have not been considered in their model. Also, their combined model is not so clear and by paying incentives to the customers, at the all hours of the operation, the demand is decreased which may not be always realistic, practical and economical. Also, this may not be based on the ISO point of view. In fact, customers who participate in DRPs can decrease or shift their demand during peak hours to off-peak hours. Actually they have neglected the shift-able loads.

Soft computing methods have higher capability of solving the non-linear multi-objective problems than the traditional methods and usually can optimally solve non-convex and non-smooth cost functions. Particle swarm optimization [19], gravitational search [20], artificial bee colony [21], harmony search [22], intelligent tuned harmony search (ITHS) [23], spiral [24], and imperialist competitive algorithm (ICA) [25-27] are some of these optimization algorithms. Among the mentioned optimization algorithms, ICA is a new one introduced in 2007 by E. A. Gargari [25]. It has a good performance in solving optimization problems in different areas such as DG planning, plate-fin heat exchangers design, template matching and electromagnetic problems [26, 27].

The major contributions of this paper are: (i) Integration of EDRP with the multi-objective DEED problem to schedule the online generators power output and determine the optimal incentive. (ii) The effectiveness of the final model is shown by applying it on the ten unit's test system in three different case studies. (iii) Investigation the effects of the EDRP-DEED model on the improvement of the load curve characteristics. (iv) Prioritizing of DRPs based on different policies by using the strategy success indices. Moreover, although valve point loading effect and POZ have been considered in DEED problem, but they have not been considered in an integrated model of DEED and DRPs. Actually addition of these constraints are the innovation of this paper.

The rest of this paper is organized as follows. In Section 2, the economic model of the price-based and incentive-based DRPs is developed based on PEM and the customer's benefit function. Formulation of the DEED problem is presented in Section 3. In Section 4 the optimal model through combining of DEED and EDRP including their constraints is developed. The characteristics of the test system are introduced in Section 5. Numerical simulation and results are presented in Section 6. Finlay, in Section 7 the conclusion is drawn.

2. ECONOMIC MODEL OF RESPONSIVE LOAD

To obtain the optimal consumption at the demand side, the elasticity is defined as the sensitivity of the demand respect to the price as Eq. (1) [2, 4, and 28].

$$E(t,t')=\frac{\rho_0(t')}{d_0(t)}\frac{\partial d(t)}{\partial \rho(t')}\quad \begin{cases}E(t,t')\le 0\, if\ t=t'\\ E(t,t')\ge 0\, if\ t\ne t'\end{cases} \tag{1}$$

where, E is the elasticity, $d(t)$ and $d_0(t)$ are the customer demands after implementing DRP and before it, during period t, $\rho(t')$ and $\rho_0(t')$ are the elasticity price and the initial electricity price during period t', respectively.

For 24 hours in a day, self and cross elasticity values can be given as a 24×24 matrix as Eq. (2).

$$\begin{bmatrix}\frac{\Delta d(1)}{d_0(1)}\\ \frac{\Delta d(2)}{d_0(2)}\\ \frac{\Delta d(3)}{d_0(3)}\\ \cdots\\ \frac{\Delta d(24)}{d_0(24)}\end{bmatrix}=\begin{bmatrix}E(1,1) & \cdots & E(1,24)\\ \vdots & \ddots & \vdots\\ E(24,1) & \cdots & E(24,24)\end{bmatrix}\times\begin{bmatrix}\frac{\Delta \rho(1)}{\rho_0(1)}\\ \frac{\Delta \rho(2)}{\rho_0(2)}\\ \frac{\Delta \rho(3)}{\rho_0(3)}\\ \cdots\\ \frac{\Delta \rho(24)}{\rho_0(24)}\end{bmatrix} \tag{2}$$

2.1. Modeling of single period elastic loads

In this case, the total revenue for the customers who participate in the DRPs will be calculated as Eq. (3) based on the hourly incentive rate. In other words, DRPs create a motivation for customers to reduce their consumption. The total payment given to the customers is as Eq. (3) [28].

$$INC(\Delta d(t))=inc(t)\times[\Delta d(t)] \tag{3}$$

where, $inc(t)$ is the amount of the incentive for reducing the consumption per $MW.h$, $\Delta d(t)$ is the amount of the reduced load.

Some programs consider a penalty for the customers who promise to participate in the DRP, but they don't (Eq. (4)). Most of the DRPs like EDRP and DLC are implemented voluntary. So, customers' are not penalized if they don't participate in DRPs (if they don't reduce or cut their consumption during peak hours). But, a DRP can be implemented mand-atorily which means that if customers' don't reduce their consumption during peak hours they will be penalized by an additional cost in their electricity bill. Some examples of mandatory programs are capacity market program (CAP) and interruptible / curtail-able (I/C) service programs.

$$PEN(\Delta d(t))=pen(t)\times\{IC(t)-[\Delta d(t)]\} \tag{4}$$

where, $IC(t)$ is the amount of the demand which the customer is responsive to reduce or shift.

Consumers who participate in DRP, increase their production benefit, decrease their consumption, and receive the reward from the system operator. Thus, the net-profit of the customer is as Eq. (5) which is related to the customer's income because of electricity consumption and producing their commodities.

$$NP(t)=B(d(t))-d(t)\rho(t)+INC(\Delta d(t))-PEN(\Delta d(t)) \tag{5}$$

where, B is the profit which customers obtain by consuming power.

To maximize the customer benefit, the derivative of Eq. (5) should be zero.

$$\frac{\partial NP}{\partial d(t)} = \frac{\partial B(d(t))}{\partial d(t)} - \rho(t) + \frac{\partial INC}{\partial d(t)} - \frac{\partial PEN}{\partial d(t)} = 0 \quad (6)$$

$$\frac{\partial B(d(t))}{\partial d(t)} = \rho(t) + inc(t) + pen(t) \quad (7)$$

As mentioned before, it is assumed that $B(d(t))$ is customer's benefit from the use of electricity during tth hour. Taylor series of B is given by Eq. (8).

$$B(d(t)) = B(d_0(t)) + \frac{\partial B(d_0(t))}{\partial d(t)}[d(t) - d_0(t)] + \frac{1}{2}\frac{\partial^2 B(d_0(t))}{\partial d^2(t)}[d(t) - d_0(t)]^2 \quad (8)$$

To obtain the optimal consumption by which the customers get the maximum profit, from Eq. (8):

$$B(d(t)) = B(d_0(t)) + \rho_0(t)[d(t) - d_0(t)] + \frac{1}{2}\frac{\rho_0(t)}{E(t,t)d_0(t)}[d(t) - d_0(t)]^2 \quad (9)$$

Differentiating:

$$\frac{\partial B(d(t))}{\partial d(t)} = \rho_0(t)\left(1 + \frac{d(t) - d_0(t)}{E(t,t)d_0(t)}\right) \quad (10)$$

By combining Eqs. (10) and (7), for the single-period model of the load can be obtained:

$$d(t) = d_0(t) \times \left(1 + \frac{\rho(t) - \rho_0(t) + inc(t) + pen(t)}{\rho_0(t)} E(t,t)\right) \quad (11)$$

2.2. Modeling of multi period elastic loads

Now, to consider shift-able loads in the, then we will have the multi period model as the following equation:

$$d(t) = d_0(t) \times \left\{1 + \sum_{\substack{t'=1 \\ t' \neq t}}^{24} E(t,t') \times \frac{[\rho(t') - \rho_0(t') + inc(t') - pen(t')]}{\rho_0(t')}\right\} \quad (12)$$

2.3. Load economic model

Finally, the combined model including the single and multi- period models of the load (considering curtail-able, interruptible, and shift-able loads) is given by Eq. (13).

$$d(t) = d_0(t) \times \left\{ \begin{array}{l} 1 + \frac{\rho(t) - \rho_0(t) + inc(t) - pen(t)}{\rho_0(t)} E(t,t) \\ + \sum_{\substack{t'=1 \\ t' \neq t}}^{24} E(t,t') \times \frac{[\rho(t') - \rho_0(t') + inc(t') - pen(t')]}{\rho_0(t')} \end{array} \right\} \quad (13)$$

Equation (11) is the single period elastic load model which considers just interruptible or curtail able (I/C) loads. Eq. (12) is for multi-period elastic load model which considers just shift-able loads. Eq. (13) is the combined model included both single and multi-periods models which consider both (I/C) and shift-able loads. In this paper the combined model is taken into account.

3. DYNAMIC ECONOMIC EMISSION DISPATCH FORMULATION

When the valve-point, loading effect is taken into account, the total fuel cost over the whole dispatch period is as follows.

$$\min \sum_{t=1}^{T} \sum_{i=1}^{Ng} F_i(P_{i,t}) \quad (14)$$

$$F_i(P_{i,t}) = a_i + b_i P_{i,t} + c_i P_{i,t}^2 + \left| d_i \sin\left(e_i\left(P_i^{min} - P_{i,t}\right)\right)\right| \quad (15)$$

where, a_i, b_i, c_i, d_i and e_i are the fuel cost coefficients of the ith unit, P_i^{min} is the minimum power generation, $P_{i,t}$ is the power output of the ith unit during the t-th time interval, N_g and T are the number of the generating units and the dispatch interval, respectively.

The atmospheric pollution caused by the fossil-fired generator contains carbon dioxide CO_2, nitrogen oxides NO_x, sulfur oxides SO_x, etc. The environmental objective is as follows.

$$\min \sum_{t=1}^{T} \sum_{i=1}^{Ng} E_i(P_{i,t}) \quad (16)$$

$$E_i(P_{i,t}) = \alpha_i + \beta_i P_{i,t} + \gamma_i P_{i,t}^2 + \eta_i \exp(\delta_i P_{i,t}) \quad (17)$$

where, $\alpha_i, \beta_i, \gamma_i, \eta_i$, and δ_i are coefficients of the emission issue for the i-th unit.

4. THE COMBINED MODEL OF EDRP INTEGRATED WITH THE DEED PROBLEM

The cost of implementing EDRP is as Eq. (18).

$$C_{EDRP}(t) = (d_0(t) - d(t))inc(t) \quad (18)$$

The multi-objective optimization problem can be changed to a single objective function using a penalty factor as follows:

$$\min TOF\left(P_{i,t}\right)=\sum_{t=1}^{T}\left\{\begin{array}{l}\omega_F\times\left[\sum_{i=1}^{N_g}\left\{\begin{array}{l}a_i+b_iP_{i,t}+c_i\left(P_{i,t}\right)^2\\+\left|d_i\sin\left(e_i\left(P_i^{min}-P_{i,t}\right)\right)\right|\end{array}\right\}\right]\\+\omega_F\times C_{EDRP}\left(t\right)\\+\omega_E\times\sum_{i=1}^{N_g}\left[pff\left(i\right)*\left(\begin{array}{l}\alpha_i+\beta_iP_{i,t}+\\\gamma_i\left(P_{i,t}\right)^2+\eta_i\exp\left(\delta_iP_{i,t}\right)\end{array}\right)\right]\end{array}\right\} \quad (19)$$

where,

$$\omega_F+\omega_E=1 \quad (20)$$

The first and second terms in Eq. (19) are cost functions ($) and the third term is emission (Ib). Therefore, to have same unit for cost and emission i.e. in dollar, the third term's unit should be changed from Ib to $. Therefore, that is why a price penalty factor (*pff*) is used. In other words, *pff* changes the unit of third term in Eq. (19) from Ib to $ [29].

$$pff\left(i\right)=\frac{F_i\left(P_i^{max}\right)}{E_i\left(P_i^{max}\right)}=\frac{a_i+b_iP_i^{max}+c_i\left(P_i^{max}\right)^2+\left|d_i\sin\left(e_i\left(P_i^{min}-P_i^{max}\right)\right)\right|}{\alpha_i+\beta_iP_i^{max}+\gamma_i\left(P_i^{max}\right)^2+\eta_i\exp\left(\delta_iP_i^{max}\right)} \quad (21)$$

4.1. Constraints

The DEED problem should satisfy the following equality and inequality constraints.

4.1.1. Power balance constraint

The total power output should be equal to the predicted load demand plus the total losses.

$$\sum_{i=1}^{N_g}P_i\left(t\right)=d\left(t\right)+P_L\left(t\right);t=1,\ldots T \quad (22)$$

where, $d(t)$ and $P_L(t)$ are the load demand and the power loss of transmission line at the t-th time interval. Generally $P_L(t)$ is calculated by Kron's loss formula, which can be given as Eq. (23).

$$P_L\left(t\right)=\sum_{i=1}^{N_g}\sum_{j=1}^{N_g}P_i\left(t\right)B_{ij}P_j\left(t\right) \quad (23)$$

where, $B_{i,j}$ is the power loss coefficient of the transmission network.

4.1.2. Incentive limits

For EDRP program, the incentives paid to the customers should be in a feasible range.

$$inc\left(t\right)^{min}\le inc\left(t\right)\le inc\left(t\right)^{max} \quad (24)$$

Referring to [32] $inc(t)^{min}$ and $inc(t)^{max}$ are usually considered to be $0.1\times\rho_0(t)$ and $10\times\rho_0(t)$, respectively.

4.1.3. Power generation limits

Generators power output is limited by its upper and lower generation limits.

$$P_i^{min}\le P_{i,t}\le P_i^{max}\quad i=1,2\ldots N_g \quad (25)$$

where, P_i^{min} and P_i^{max} are the lower and upper generation limits for the ith unit.

4.1.4. Prohibited operation zones constraint

In practice, generators should not work in some POZs. The main reason of this limitation is the vibration of the shaft bearing. So, generators should work in the feasible operating zones as given by Eq. (26).

$$\begin{cases}P_i^{min}\le P_i^t\le P_{i,1}^{l}\\P_{i,k-1}^{u}\le P_i^t\le P_{i,k}^{l};i=1,\ldots,Ng;t=1,\ldots,T\\P_{i,M}^{u}\le P_i^t\le P_i^{max}\end{cases} \quad (26)$$

where, $P^l_{i,k}$ and $P^u_{i,k}$ are the lower and upper limits of the kth POZ respectively, M is the number of POZs for the ith unit.

4.1.5. Generator ramp rate limits

Generator longevity is effectively influenced by the thermal stress. The increase and decrease rates of the generator power output are usually called the ramp-up and ramp-down, respectively. So, the operating range of the i-th unit is as Eq. (27).

$$\begin{cases}P_i^t-P_i^{t-1}\le UR_i\\P_i^{t-1}-P_i^t\le DR_i\end{cases} \quad (27)$$

where, UR_i and DR_i are the up-ramp and down-ramp limits of the i-th unit, respectively and are usually expressed in MW/h.

4.2. Solving the DEED-EDRP problem

In this part a general procedure for solving the DEED-EDRP problem by the population-based meta-heuristic algorithms is presented. In fact, the population includes some possible solutions of the optimization problem. The population size is determined by the number of possible solutions. The possible solutions in ICA are called countries, in PSO particles, in ABC artificial bees, etc. Also, every possible solution is called a candidate. In DEED-EDRP, every scheduled generating unit output at each hour comprises a component of the population. In other words, it is a candidate for DEED-EDRP optimization problem at each hour. The kth candidate (PG_k) at each hour is defined as Eq. (28).

$$PG_k = \left[P_{k,1}, P_{k,2}, \ldots, P_{k,j}, \ldots, P_{k,N_g}\right], k = 1,2..M' \quad (28)$$

where, PG_k is the current position of the kth vector, N_g is the number of generation units, M' is the population size, j is the generator number, and P_{kj} is the power output of the jth generation unit.

Constraint Eq. (22) can be handled by using a penalty term in Eq. (29). Thus, the evaluation function used in DEED-EDRP can be written as Eq. (29).

$$EF\left(P_{i,t}\right) = \sum_{t=1}^{T} \left\{ \begin{array}{c} \sum_{i=1}^{N_g} TOF\left(P_{i,t}\right) + \\ K_n abs\left(\sum_{i=1}^{N_g} P_{i,t} - P_{D,t} - P_{L,t}\right) \end{array} \right\} \quad (29)$$

where, K_n is the penalty factor, which is a positive real number. The amount of K_n at each hour increases with the algorithm iterations. If the constraint Eq. (22) is nonzero, the amount of the second term in Eq. (29) will be nonzero, too. In other words, a candidate which doesn't meet the constraint Eq. (22) will have a large evaluation function and more likely will be discarded. On the other hand, a candidate which meets the constraint Eq. (22) will have a relatively small evaluation function and consequently will be kept. K_n can be written as Eq. (30).

$$K_n = 1000 \times \sqrt{n}; n = 1, \ldots N_{iter} \quad (30)$$

where, N_{iter} is the maximum number of iterations at each hour.

To ensure meeting of constraints Eqs. (25) - (27), before calculating the evaluation function of each candidate by Eq. (29), the power generation outputs of each candidate should be in the acceptable ranges specified by constraints Eqs. (25) - (27). If a candidate meets the constraints Eqs. (25) - (27), its evaluation function will be determined by Eq. (29). Otherwise, its evaluation function will be penalized by a large number. For more information about ICA, refer to [25-27]. The solution method for EDRP-DEED problem can be summarized in some steps as following.

Step 1: Defining technical units' data, daily load demand, $\rho_0(t)$, PEM, load model, participation percentage (μ), and initial incentive (is set to zero by ISO). Moreover, in ICA optimization, the initial population which is a set of possible solutions of the EDRP-DED problem, is defined based on the pervious section.

Step 2: Increasing the amount of incentive by ISO and determining the hourly demand and total incentive. Also, generation costs are determined by the supply side in this step.

Step 3: Solving the EDRP-DED problem by ICA and determining the optimal generation power outputs which are announced to the supply side.

Step 4: Continuing the process from step 2 until the minimum cost of generation units is obtained and the optimal incentive is determined by ISO.

5. TEST SYSTEM

To show the correctness, features, and practical benefits of the proposed model, it is applied on the ten unit's test system. The test system is taken from [30, 31] with some modifications. The characteristics of the test system are given as Table 1-2. Also, the elements of PEM are like Table 3. The daily load curve is divided into the peak period (9 A.M. - 14 P.M. & 19 P.M. - 24 P.M.), the off-peak period (5 A.M. - 9 A.M. & 14 P.M. - 19 P.M), and the valley period (0 A.M.-5 A.M.). DR implementation potential (μ) is considered 20%. It means that 20 percent of the customers participate in the DRP. The initial electricity price (ρ_0)is as Fig. 1 [32]. Also, the transmission line losses coefficients of the ten-unit test system are as Eq. (31).

$$B = \begin{bmatrix} 49 & 14 & 15 & 15 & 16 & 17 & 17 & 18 & 19 & 20 \\ 14 & 45 & 16 & 16 & 17 & 15 & 15 & 16 & 18 & 18 \\ 15 & 16 & 39 & 10 & 12 & 12 & 14 & 14 & 16 & 16 \\ 15 & 16 & 10 & 40 & 14 & 10 & 11 & 12 & 14 & 15 \\ 16 & 17 & 12 & 14 & 35 & 11 & 13 & 13 & 15 & 16 \\ 17 & 15 & 12 & 10 & 11 & 36 & 12 & 12 & 14 & 15 \\ 17 & 15 & 14 & 11 & 13 & 12 & 38 & 16 & 16 & 18 \\ 18 & 16 & 14 & 12 & 13 & 12 & 38 & 16 & 16 & 18 \\ 19 & 18 & 16 & 14 & 15 & 14 & 16 & 15 & 42 & 19 \\ 20 & 18 & 16 & 15 & 16 & 15 & 18 & 16 & 19 & 44 \end{bmatrix} \times 10^{-6} \quad (31)$$

Table 1. Ten-unit test system characteristics (a)

Units	P_{max} (MW)	P_{min} (MW)	a_i ($/h)	b_i ($/MWh)	c_i ($/(Mh)^2h$)
1	455	150	1000	16.19	0.00084

2	455	150	970	17.26	0.00031
3	130	20	700	16.6	0.002
4	130	20	680	16.5	0.00211
5	162	25	450	19.7	0.00398
6	80	20	370	22.26	0.00712
7	85	25	480	27.74	0.00079
8	55	10	660	25.92	0.00413
9	55	10	665	27.27	0.00222
10	55	10	670	27.79	0.00173
Units	d_i *($/h)*	e_i *(rad/MW)*	***POZs*** *(MW)*		
1	40	0.0141	—		
2	60	0.0136	[185-210], [275-305], [410-420]		
3	30	0.0128	—		
4	20	0.0152	[25-40], [55-70], [75-85]		
5	20	0.0163	—		
6	30	0.0148	[30-45] , [50-65]		
7	30	0.0168	[35-50] , [55-70]		
8	32	0.0162	[15-25] , [35-45]		
9	25	0.0178	—		
10	33	0.0174	—		

Table 2. Ten-unit test system characteristics (b)

Units	α_i *(lb/h)*	β_i *(lb/MWh)*	γ_i *(lb/$(Mh)^2h$)*	η_i *(lb/h)*
1	42.8955	-0.5112	0.00460	0.25470
2	42.8955	-0.5112	0.00460	0.25470
3	40.2669	-0.5455	0.00680	0.24990
4	40.2669	-0.5455	0.00680	0.24800
5	13.8593	0.3277	0.00420	0.24970
6	13.8593	0.3277	0.00420	0.24970
7	330.0056	-3.9023	0.04652	0.25163
8	330.0056	-3.9023	0.04652	0.25163
9	350.0056	-3.9524	0.04652	0.25475
10	360.0012	-3.9864	0.04702	0.25475
Units	δ_i ***(1/MW)***	***UR (MW/h)***	***DR (MW/h)***	
1	0.01234	80	80	
2	0.01234	80	80	
3	0.01203	50	50	
4	0.01290	50	50	
5	0.01200	50	50	
6	0.01200	30	30	
7	0.01215	30	30	
8	0.01215	30	30	
9	0.01234	30	30	
10	0.01234	30	30	

Table 3. Self and cross elasticity values.

	Peak	**Off-peak**	**Valley**
Peak	-0.10	0.016	0.012
Off-peak	0.016	-0.10	0.01
Valley	0.012	0.01	-0.10

6. NUMERICAL SIMULATION AND RESULTS

In this paper, the cost based, emission based, and cost-emission based DEED integrated with EDRP through three different case studies are investigated. On the other hand, the effects of the elasticity values and incentives on the results are evaluated. The initial daily load demand is as Fig. 2.

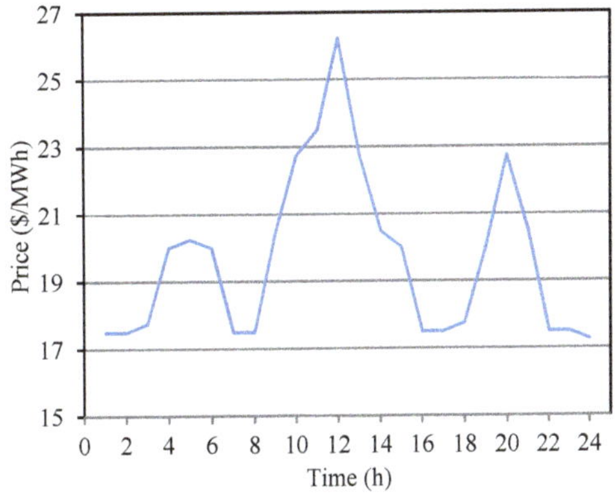

Fig. 1. Initial electricity price ($/MWh)

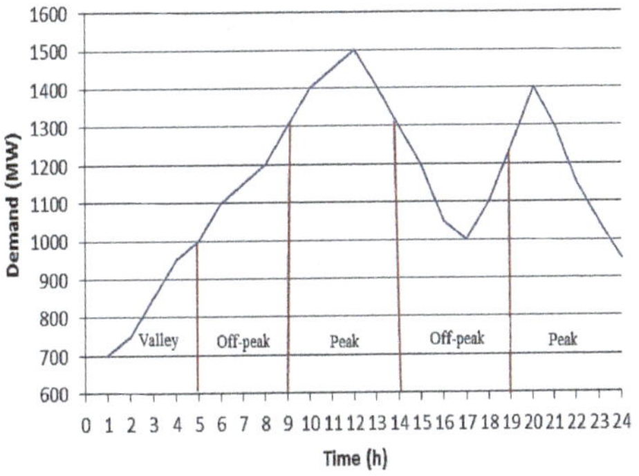

Fig. 2. Daily load curve

Three different groups with different values of PEM have been taken into account. In each case study, ten scenarios have been defined with different PEMs and incentives. Scenario 1 is the base case without implementing EDRP, scenarios 2-4 (group one with a PEM equals to E as Table 2) have incentives 4, 8, and 12 $/MWh, scenarios 5-7 (group two with PEM equals to 0.5× *E*) have incentives 4, 8, and 12 $/MWh, scenarios 8-10 (group three with PEM equals to 2×*E*) have incentives 4, 8, and 12 $/MWh, respectively.

To investigate the impacts of implementing EDRP on the load curve characteristics, some factors are defined as following. To evaluate the smoothness of the load curve, the load factor is defined as Eq. (32). Ideally, it is 100% which implies that at all hours of the operation the amount of demand is constant and does not change throughout the day.

$$Load - factor\% = 100 \times \left(\frac{\sum_{t=1}^{T} d(t)}{T \times d^{max}(t)} \right) \quad (32)$$

Peak-to-valley, peak-compensate, and deviation-of-peak-to-valley are the other important factors which are defined as Eqs. (33) - (35).

$$Peak - to - valley\% = 100 \times (\frac{d^{max}(t) - d^{min}(t)}{d^{max}ax(t)}) \quad (33)$$

$$Peak - compensate\% = 100 \times (\frac{d_0^{max}(t) - d^{max}(t)}{d_0^{max}(t)}) \quad (34)$$

$$Deviation - of - peak - to - valley\% = 100 \times \left(1 - \frac{d^{max}(t) - d^{min}(t)}{d_0^{max}(t) - d_0^{min}(t)} \right) \quad (35)$$

DRPs are usually prioritized by ISO to compare the performance value of the DR strategies. Therefore, in this paper to prioritize scenarios, the strategy index (SI) and strategy success index (SSI) are defined as Eqs. (36) and (37), respectively.

$$SI = \sum_{t=1}^{24} (St_1(t))^{W_1} \times (St_2(t))^{W_2} \times \ldots (St_k(t))^{W_k} \quad (36)$$

$$SSI = \frac{\sum_{i=1}^{N} SI(i)}{\sum_{i=1}^{N} SI(max)} \times 100 \quad (37)$$

where, $S_t(t)$ is the performance value of the scenario in the period t, W_k is the weighting for the k-th attribute, N is the total days of DRPs implementation, and SSI represents the normalized value of SI. The higher SSI represents the better profit.

6.1. Case study one: cost based DEED integrated with EDRP

In this case, the effects of implementing EDRP on the overall cost of the generation units are evaluated. So, ω_F and ω_E are considered to be 1 and 0, respectively. Results are shown in Table 4. In all scenarios, after implementing EDRP, the total cost reduces. Implanting EDRP imposes an additional cost (C_{EDRP}) which is paid as the incentive to the customers. But, the total cost which is sum of the cost of the generating units and the total incentive, reduces. Scenario 9 has the most reduction of the total cost by 17411.0128$ (679111.1848-661700.1720) and scenario 5 has the least one by 4014.4215 $ (679111.1848-675096.7633). On the other hand, the customer's benefit in each group increases with the incentive value and PEM and decreases with the generation cost of units. For example, scenario 10 has the most total incentive (35839.3653 $) and scenario 5 has the least one (995.5379$).

Table 4. Scenarios' performance in the case one

Scenario	Total generation cost ($)	Total incentive ($)	Total cost ($)
1	679111.1848	—	679111.1848
2	669700.6678	1991.0758	671691.7437
3	660272.7458	7964.3034	668237.0492
4	651999.4772	17919.6826	669919.1599
5	674101.2254	995.5379	675096.7633
6	669686.4824	3982.1517	673668.6341
7	665398.5145	8959.8413	674358.3558
8	661476.1561	3982.1517	665458.3078
9	645771.5652	15928.6068	661700.1720
10	635583.916	35839.3653	671423.2813

The optimal incentives for three different groups are determined as shown in Table 5. Also, in all scenarios total losses decreases, too. All characteristics of the load curve are improved for three groups as shown in Table 6.

Table 5. Groups' performance for the optimal incentives in the case one

Group	Optimal incentive ($/MWh)	Total generation cost ($)	Total incentive ($)
Base case	—	679111.1848	—
One	7.75	660626.9668	7474.3121
Two	10.32	665495.0939	6623.0299
Three	5.87	652631.6164	8590.4034
Group	Total cost ($)	Total power losses(MW)	
Base case	679111.1848	769.7731	—
One	668101.2789	739.4306	
Two	672118.1237	740.8757	
Three	661222.0198	693.7434	

Table 6. Load curve's characteristics in the case one

Group	Load factor	Peak to valley	Peak compensate	Deviation of peak to valley
Base case	76.39	53.33	—	—
One	80.15	49.11	7.48	14.80
Two	78.10	51.41	3.55	7.03
Three	81.33	46.46	11.34	22.77

The load curves before and after implementing EDRP for three different groups (for their optimal incentives) are shown in Fig. 3. Customers with the highest PEM have more willingness to reduce or shift their consumption during peak hours (group three of the customers) and vice versa (group two of the customers). Actually, by implementing the EDRP, the load curve smoothens which improves the network reliability.

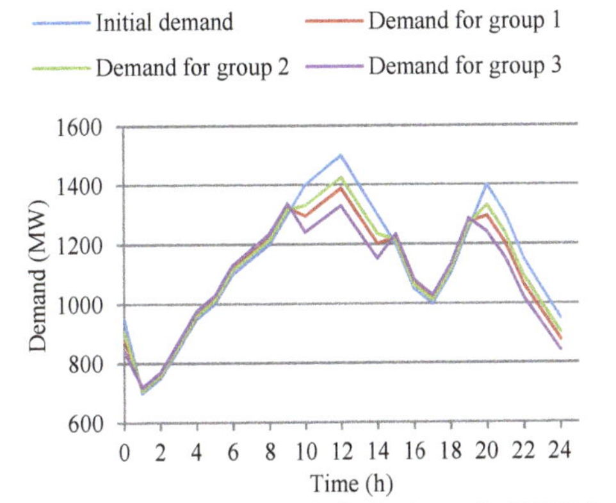

Fig. 3. The load curve before and after implementing EDRP for three different groups

6.2. Case study 2: emission-based DEED integrated with EDRP

In this case, the impact of implementing EDRP on reducing the emission is considered. It should be noted that in this case, the objective of implementing EDRP is 10 percent reduction of the initial emission. The initial emission before implementing EDRP is 49352.0551 Ib and in this case, the objective is to reduce it to 44416.8496 Ib. In this regard, ω_F and ω_E are considered to be 0 and 1, respectively. Results are shown in Table 7. Determining the optimal incentive is necessary from ISO point of view. From Table 7, it can be seen that the emission level of the generation units is decreased proportional to the elasticity of demand and the value of the incentive. For example, scenario 10 with larger PEM and incentive (2×E and 12 \$/MWh, respectively) has the most emission reduction by 11.05 % among the other ones.

6.3. Case study three: cost-emission based DEED integrated with EDRP

In this case, the cost-emission based DEED integrated with EDRP is investigated. Thus, there is a trade-off between the cost and the emission. In other words, ω_F and ω_E are both considered to be 0.5. But, depending on the system operator, different weights can be assigned too. Results are shown in Table 8. The objective function for the group three of customers reduces more. It is because of the fact that they have larger PEM which means that they have more willingness to reduce their consumption during the peak period or shift it to the valley or off-peak periods.

Table 7. Scenarios' performance in the case two

Sce.	Total emission (Ib)	Emission reduction%	Optimal incentive for 10 % emission reduction ($/MWh)	Total incentive ($) for 10 % emission reduction ($/MWh)
1	49352.0551	—	—	—
2	47892.3518	2.96	16.27	32941.4664
3	47203.6674	4.35		
4	45796.2286	7.21		
5	48956.6215	0.8	35.76	79566.9749
6	48528.7325	1.66		
7	47495.8976	3.76		
8	47294.4441	4.17	11.46	32686.3971
9	45554.8574	7.69		
10	43901.3077	11.05		

6.4. Prioritizing scenarios

Comparing the performance value of the scenarios using SSI coefficient is investigated in this part. For the best scenario, SSI is considered to be 100 % and for the other ones, it is calculated using Eqs. (36) and (37). Prioritizing of different scenarios based on different policies has been shown in Fig. 4. Here, reducing the total cost, emission, and objective function (case 1-3), energy reduction, load factor, and peak compensate are assumed as the important policies. As it is shown in Fig. 4, different policies have different priorities. In practice when there are some restrictions in implementing a higher priority scenario, ISO can choose another one with lower priority.

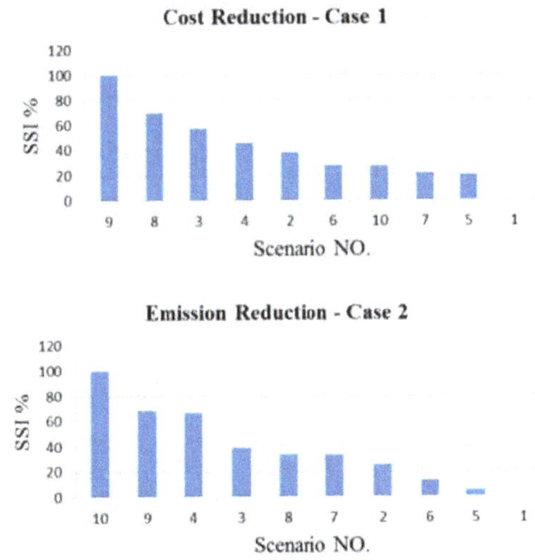

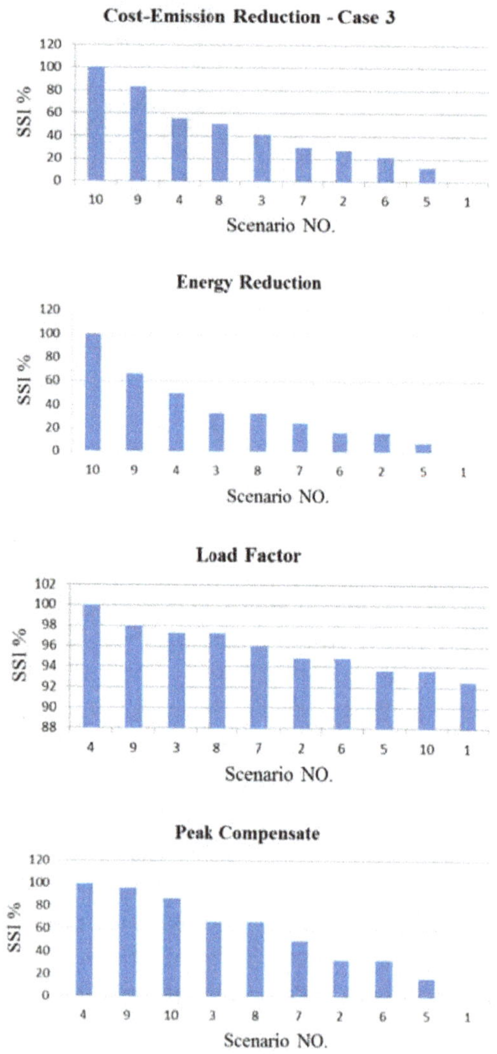

Fig. 4. Prioritizing of different scenarios based on different policies

Table 8. Scenarios' performance in the case three

Scenario	Total generation cost ($)	Total emission (lb)	Total incentive ($)
1	704386.6962	50752.9645	—
2	696378.1907	48965.6393	1991.0758
3	689097.4902	48083.3675	7964.3034
4	680312.5354	47426.8710	17919.6826
5	700936.9077	49775.3354	995.5379
6	695982.5925	49333.6351	3982.1517
7	691859.4736	48648.8467	8959.8413
8	689319.6233	48194.3079	3982.1517
9	673854.3426	46561.6077	15928.6068
10	655339.0089	45418.4736	35839.3653

Scenario	Objective function	Optimal incentive ($/MWh)
1	638283.5824	—
2	627284.3415	16.44
3	621368.4799	
4	615560.2359	
5	633206.8812	22.03
6	629264.0664	
7	625897.6729	
8	618088.7175	12.53
9	604602.4641	
10	597376.9502	

7. CONCLUSION

DRPs' capabilities such as the cost reduction and reliability improvement have made them into important priorities for the electricity markets. In this paper EDRP was integrated to the DEED problem to minimize the fuel cost and emission and appoint the optimal incentive simultaneously.

In fact, the combined model is a win-win situation for both the customers and the generation companies. It is because of the fact that implementing DRPs not only decreases the total generation cost and emission but also increases the customer's benefit and network reliability. By applying the proposed model on the ten-unit system, some analyses were carried out to investigate the impacts of some important factors such as the elasticity values and incentives on the results. Also, prioritizing of different scenarios based on different policies was presented by using strategy success indices. Results showed the effectiveness and practical benefits of the proposed model. In the future work, the price-based DRPs such as the time of use (TOU) program will be integrated with the DEED problem. In this model instead of the optimal incentive, the optimal price during different periods is determined.

APPENDIX A

Generators' optimal scheduling for case 1 (before and after implementing EDRP).

ACKNOWLEDGMENT

This work is sponsored by Iran Energy Efficiency Organization (IEEO) (SABA).

Table 9. Optimal generators power output (case 1, without implementing EDRP)

Hour	P_1(MW)	P_2(MW)	P_3(MW)	P_4(MW)	P_5(MW)	P_6(MW)	P_7(MW)	P_8(MW)	P_9(MW)	P_{10}(MW)
1	198.3231	241.684	31.63968	94.59544	27.19408	25.49634	26.8687	45	10.26797	10
2	150	247.7146	49.83891	110.7146	34.63525	65	27.30336	45	21.64562	10
3	230	167.7146	99.51208	110.4738	83.34554	65	53.15977	28.24109	16.64621	10.49327
4	310	222.6107	68.41978	85	101.9436	77.48763	33.43801	45	15.70465	10

5	390	267.0335	52.33952	85	101.6804	47.48763	30.61284	15	10.19548	24.39218
6	433.8736	347.0335	50.11793	92.06238	56.89958	28.54348	33.46109	45	27.28848	15.57493
7	446.6911	345.5086	75.58416	112.9852	56.08574	26.88016	32.65191	31.60081	40.43256	13.77038
8	366.6911	425.5086	98.37808	103.086	106.0857	65	29.89463	13.56929	15.49595	10
9	367.0852	430.1813	118.431	89.70195	156.0857	68.29093	29.53957	45	10	24.07745
10	447.0852	420.5234	128.4045	123.2035	133.2943	65	70	30.52909	10	16.76485
11	455	443.6234	130	111.8638	136.884	65	70	12.54559	33.12652	40.32668
12	455	449.4167	130	126.3022	162	65.24816	71.82764	45	35.87504	10.32668
13	452.9055	389.4062	129.8111	119.1658	125.3109	65	76.56733	50.54382	13.80363	21.79765
14	455	400.4528	114.9359	85	80.23671	65	70.1974	34.32804	10	24.65694
15	375	442.6851	64.9359	122.8244	50.6497	73.36889	52.6237	12.72644	15.00224	24.50812
16	301.8589	377.1026	114.9359	85.98152	35.88778	65	30.40196	32.65795	19.08779	12.50596
17	381.8589	329.3905	64.9359	70.42312	29.41574	65	30.78129	12.12719	19.96428	20.55156
18	405.0488	394.5353	95.00446	46.83145	26.91788	69.63742	51.6138	13.84193	13.9593	12.51913
19	455	436.1718	46.09915	96.83145	76.91788	65	70	12.6892	18.38966	11.26751
20	455	437.4784	96.09915	124.533	116.9341	77.66695	70	45	10	12.62963
21	455	394.2677	94.5559	102.0265	79.42167	74.46078	70	15	12.40555	42.62963
22	375.745	397.9068	74.50345	89.71224	77.12804	49.73815	70	10.98237	12.08122	23.45051
23	392.2455	374.1333	62.13671	46.03439	78.53448	25.33387	50.02141	26.8977	10.26565	11.73639
24	312.2455	307.1181	108.4851	72.7076	28.53448	65	30.08306	11.00563	11.13948	24.57502

Table 10. Optimal generators power output (case 1, group 1, incentive 7.75 $/MWh, after implementing EDRP)

Hour	P_1(MW)	P_2(MW)	P_3(MW)	P_4(MW)	P_5(MW)	P_6(MW)	P_7(MW)	P_8(MW)	P_9(MW)	P_{10}(MW)
1	159.8035	226.6205	85.82839	48.51468	57.2006	22.95584	26.69639	46.18369	21.64581	21.47983
2	239.8035	179.4035	60.54318	85	44.8248	29.9754	29.40787	53.08889	18.88343	28.18016
3	299.2614	259.4035	31.77494	85	43.34917	27.70434	50.01592	30.67103	36.29844	11.24974
4	379.2614	182.563	64.44217	85	93.34917	49.67918	34.08422	45.07673	23.30656	22.96594
5	299.2614	262.563	114.4422	85	125.3316	24.51648	32.31975	50.47622	22.14551	15.03279
6	379.2614	320.7022	70.89777	108.1699	75.33159	65	26.909	31.35368	48.05633	15.96953
7	371.5328	400.7022	50.6717	99.26182	84.44543	65	32.46259	49.30955	27.98502	14.20571
8	426.4736	382.9169	51.28612	110.4204	96.25642	65	25.51265	45	30.0347	16.4447
9	455	452.0681	101.2861	98.48865	46.25642	65	27.24451	32.80284	55	24.35663
10	455	425.0122	102.4829	85	96.25642	46.83228	31.781	34.70053	25.72526	32.87751
11	417.6214	400.1585	112.6041	127.3472	146.2564	65	30.68051	52.05293	20.68154	10.01162
12	430.0328	422.7976	127.0438	97.44613	142.8555	66.77055	27.93882	49.91953	31.27763	35.84971
13	350.0328	408.3643	129.9616	111.1635	162	65	28.89328	45	12.19888	20.03089
14	430.0328	328.3643	97.95841	93.85765	112	72.77545	31.50753	15	21.81147	33.06842
15	357.5289	408.3643	125.3128	102.6092	72.24709	69.51408	34.69267	45	21.79396	10.99171
16	323.6045	375.7817	116.0473	85	27.09675	65	32.01686	15	20.62362	28.75407
17	271.9672	438.765	56.47937	85.94941	34.48164	46.23864	54.02472	11.16455	17.65936	20.58893
18	351.9672	358.765	106.4794	104.2435	84.48164	28.46595	31.19196	45	21.08438	10
19	389.0167	438.765	126.9681	107.086	44.83316	65	70	15	31.66874	13.99587
20	455	430.5116	89.23892	119.4467	25.21213	65	84.02351	26.13844	31.05067	10
21	420.2483	355.867	107.5968	124.5094	75.21213	46.34431	70	14.22062	10.2291	12.23036
22	384.2155	349.7336	108.9581	74.50942	25.21213	21.9332	52.42813	30.218	17.73097	26.37444
23	304.2155	269.7336	97.37275	112.9782	75.21213	24.17103	34.51549	14.37614	39.02653	20.92095
24	224.2155	234.2253	79.80722	122.3652	125.2121	24.19863	28.64941	27.8196	17.39608	11.38794

REFERENCES

[1] H. Falsafi, A. Zakariazadeh and Sh. Jadid, "The role of demand response in single and multi-objective wind-thermal generation scheduling: A stochastic programming," *Energy*, vol. 64, pp. 853-867, 2013.

[2] M. Joung and J. Kim, "Assessing demand response and smart metering impacts on long-term electricity market prices and system reliability," *Applied Energy*, vol. 101, pp. 441-448, 2013.

[3] A.K. David and Y.C. Lee, "Dynamic tariffs theory of utility-consumer interaction," *IEEE Transactions on Power System*, vol. 4, pp. 904-911, 1989.

[4] A.K. David and Y.Z. Li, "Effect of inter-temporal factors on the real time pricing of electricity," *IEEE Transactions on Power System*, vol. 1, pp. 44-52, 1993.

[5] N. Venkatesan, J. Solanki and S. Kh. Solanki, "Residential demand response model and impact on voltage profile and losses of an electric distribution network," *Applied Energy*, vol. 96, pp. 84-91, 2012.

[6] M. Parvania, M. Fotuhi-Firuzabad and M. Shahidehpour, "Optimal demand response aggregation in wholesale electricity markets," *IEEE Transactions on Smart Grid*, vol. 4, pp. 1957-1965, 2013.

[7] M. Alipour, K. Zare and B. Mohammadi-Ivatloo, "Short term scheduling of combined heat and power generation units in the presence of demand response programs," *Energy*, vol. 71, pp. 289-301, 2014.

[8] M. Kazemi, B. Mohammadi-Ivatloo and M. Ehsan, "Risk constrained strategic bidding of Gencos considering demand response," *IEEE Transactions on Power Systems*, vol. 30.1, pp. 376-384,

2015.
[9] M. M. Sahebi, E.A. Duki, M. Kia, A. Soroudi and M. Ehsan, "Simultaneous emergency demand response programming and unit commitent programming in comparison with interruptible load contracts," *IET Generation, Transmission & Distribution*, vol. 6.7, pp. 605-611, 2012.
[10] S. Nojavan, B. Mohammadi-Ivatloo and K. Zare, "Optimal bidding strategy of electricity retailers using robust optimization approach considering time of use rate demand response programs under market price uncertainties," *IET Generation, Transmission & Distribution*, vol. 9.4, pp. 328-338, 2015.
[11] M. Parvania and M. Fotuhi Firuzabad, "Demand response scheduling by stochastic SCUC," *IEEE Transactions on Smart Grid*, vol. 1, pp. 89-98, 2010.
[12] F. H. Magnago, J. Alemany and J. Lin, "Impact of demand response resources on unit commitment and dispatch in a day-ahead electricity market," *International Journal of Electrical Power and Energy Systems*, vol. 68, pp. 142-149, 2015.
[13] H.R. Arasteh, M. Parsa Moghaddam, M.K.Sheikh-El-Eslami and A. Abdollahi, "Integrating commercial demand response resources with unit commitment," *Electrical Power and Energy Systems*, vol. 51, pp. 153-161, 2013.
[14] J. Aghaei and M.I. Alizadeh. "Robust n-k contingency constrained unit commitment with ancillary service demand response program," *IET Generation, Transmission & Distribution*, vol. 8, pp. 1928-1936, 2014.
[15] Ch. Zhao, J. Wang, J.P. Watson and Y. Guan, "Multi-stage robust unit commitment considering wind and demand response uncertainties," *IEEE Transactions on Power Systems*, vol. 28, pp. 2708-2717, 2013.
[16] Y. Chen and J. Li. "Comparison of security constrained economic dispatch formulations to incorporate reliability standards on demand response resources into Midwest ISO co-optimized energy and ancillary service market," *Electric Power Systems Research*, vol. 81, pp. 1786-1795, 2011.
[17] A. Ashfaq, S. Yingyun and A. Zia Khan, "Optimization of economic dispatch problem integrated with stochastic demand side response," in Proceedings of the *IEEE International Conference on Intelligent Energy and Power Systems*, pp. 116-121, 2014.
[18] N. I. Nwulu and X. Xia, "Multi-objective dynamic economic emission dispatch of electric power generation integrated with game theory based demand response programs," *Energy Conversion and Management*, vol. 89, pp. 963-974, 2015.
[19] H. Khorramdel, B. Khorramdel, M. T. Khorrami and H. Rastegar, "A multi-objective economic load dispatch considering accessibility of wind power with here-and-now (hn) approach," *Journal of Operation and Automation in Power Engineering*, vol. 2, pp. 49-59, 2014.
[20] Sh. Jiang, Zh. Ji and Y. Shen, "A novel hybrid particle swarm optimization and gravitational search algorithm for solving economic emission load dispatch problems with various practical constraints," *International Journal of Electrical Power & Energy Systems*, vol. 55, pp. 628-644, 2014.
[21] D. C. Secui, "A new modified artificial bee colony algorithm for the economic dispatch problem," *Energy Conversion and Management*, vol. 89, pp. 43-62, 2014.
[22] L. Wang and L.P. Li, "An effective differential harmony search algorithm for the solving non-convex economic load dispatch problems," *International Journal of Electrical Power & Energy Systems*, vol. 44 pp. 832-843, 2013.
[23] A. Hatefi and R. Kazemzadeh, "Intelligent tuned harmony search for solving economic dispatch problem with valve-point effects and prohibited operating zones," *Journal of Operation and Automation in Power Engineering*, vol. 1, pp. 84-95, 2013.
[24] L. Benasla, A. Belmadani and M. Rahli, "Spiral optimization algorithm for solving combined economic and emission dispatch," *International Journal of Electrical Power & Energy Systems*, vol. 62, pp. 163-174, 2014.
[25] A. Gargari, "Imperialist competitive algorithm: An algorithm for optimization\ inspired by imperialistic competition," in Proceedings of the *IEEE Congress on Evolutionary Computation*, pp. 4661-4667, 2007.
[26] B. Mohammadi-ivatloo, A. Rabiee, A. Soroudi and M. Ehsan, "Imperialist competitive algorithm for solving non-convex dynamic economic power dispatch," *Energy*, vol. 44, pp. 228-240, 2012.
[27] R. Roche, L. Idoumghar, B. Blunier, and A. Miraoui. "Imperialist competitive algorithm for dynamic optimization of economic dispatch in power systems," *Springer-Verlag Berlin Heidelberg*, vol. 7401, pp. 217-228, 2012,
[28] H. Aalami, M. Parsa Moghadam and G. R. Yousefi, "Modeling and prioritizing demand response programs in power markets," *Electric Power System Research*, vol. 80, pp. 426-435, 2010.
[29] N. Pandita, A. Tripathia, Sh. Tapaswia and M. Panditb, "An improved bacterial foraging algorithm for combined static/dynamic environmental economic dispatch," *Applied Soft Computing*, vol. 12, pp. 3500-3513, 2012.
[30] A. Abdollahi, M. Parsa Moghaddam, M. Rashidinejad and M. K. Sheikh-El-Eslami, "Investigation of economic and environmental-driven demand response measures incorporating

UC," *IEEE Transactions on Smart Grid*, vol. 3, pp. 12-25, 2012.

[31] R. Zhang, J. Zhou, L. Mo, Sh. Ouyang and X. Liao, "Economic environmental dispatch using an enhanced multi-objective cultural algorithm," *Electric Power Systems Research*, vol. 99, pp. 18-29, 2013.

[32] Staff Report, "Assessment of demand response and advanced metering," *FERC*, Available: http://www.FERC.gov Dec. 2008.

14

Stochastic Multiperiod Decision Making Framework of an Electricity Retailer Considering Aggregated Optimal Charging and Discharging of Electric Vehicles

A. Badri*, K. Hoseinpour Lonbar

Faculty of Electrical Engineering, Shahid RajaeeTeacher Training University, Tehran, Iran

ABSTRACT

This paper proposes a novel decision making framework for an electricity retailer to procure its electric demand in a bilateral-pool market in presence of charging and discharging of electric vehicles (EVs). The operational framework is a two-stage programming model in which at the first stage, the retailer and EV aggregator do their medium-term planning. Determination of retailer's optimum selling price and the amount of energy that should be purchased from bilateral contracts are medium-term decisions that are made one month prior to real-time market. At the second stage, market agents deal with their activities in the short-term period. In this stage the retailer may modify its preliminary strategy by means of pool market option, interruptible loads (ILs), self-scheduling and EVs charging and discharging (V2G). Thus, a bi-level programming is introduced in which the upper sub-problem maximizes retailer profit, whereas the lower sub-problem minimizes the aggregated EVs charging and discharging costs. Final decision making is obtained in this stage that may be considered as a day-ahead market, keeping in mind the medium-term decisions. Due to the volatility of pool price and uncertainties associated with the consumers and EVs demand, the proposed framework is a mixed integer nonlinear stochastic optimization problem; therefore, Monte Carlo Simulation (MCS) is applied to solve it. Furthermore, a market quota curve is utilized to model the uncertainty of the rivals and obtaining retailer's actual market share. Finally, a case study is presented in order to show the capability and accuracy of the proposed framework.

KEYWORDS: Aggregator, Bilateral, Decision making, Electric vehicle, Interruptible load, Retailer, Self-production.

1. INTRODUCTION

Role of retailer in electricity market is highlighted more, because a large number of consumers due to lack of familiarity with the market rules, cannot play active role in that. Retailers take part in power markets by procuring energy from the bilateral and the pool markets and by selling energy to their consumers at fixed prices during a specific medium-term period. Because of volatility of pool price, the retailer is exposed to the uncertainties of the pool price and demand of consumers. On the other hand, the costs of multiple options at the medium-term period are generally greater than the average pool price. Therefore, the retailer faces a trade-off between different purchasing options of electricity [1].

There are noticeable literatures describing retailer role in the electricity markets. In [2] a stochastic based decision-making framework for an electricity retailer is proposed in which the retailer determines the sale price of electricity to the consumers based on TOU rates. The proposed framework in [3] is modeled in the form of a multi-objective framework to simultaneously maximize retailers' profit and minimize selling prices to clients. The work addressed in [4] includes a stochastic medium-term planning for an electricity retailer considering objective functions of expected profit and downside risk for determining the selling price offered to the

*Corresponding author:
A. Badri (E-mail:ali.badri@srttu.edu)

consumers and the optimal quantity of forward contracts. Reference [5] provides a novel technique based on Information Gap Decision Theory (IGDT) to assess different strategies for a retailer under unstructured pool price uncertainty. All of above mentioned models in [3-6] have focused on retailers' medium-term planning that may be inaccurate. In [6], a Non-dominated Sorting Genetic Algorithm-II (NSGA-II) based approach is presented for conversion of multi-objective function into an equivalent single objective function; however impacts of bilateral contracts and pool price volatilities are not taken into account. A reliability assessment model in presence of micro grids is represented in [7]. Although the model considers distribution load uncertainties but it mostly investigates system reliability enhancement. Electricity procurement for a large consumer from the pool market and forward contracts is reported in [8,9] and [11]. Although, a mixed pool-forward market is represented in [8-10]; however the problem is discussed from electricity consumers and retailers' perspectives are not considered.

Recently due to environmental issues and customer preferences a great attention has been paid to electric vehicles (EVs). Thus, increasing deployment of the EVs in the power system needs an agent responsible for aggregating of large EV fleets and controlling their charging and discharging process. In the electricity market environment, this agent is popularly referred to as EV aggregation agent [11], or in short, EV aggregator.

In this paper, a stochastic programming approach [11] is presented for an electricity retailer who procures its demand in a mixed bilateral-pool market. The retailer load consists of conventional loads and flexible EV loads. In fact, EV aggregator is in charge of controlling EVs charging and discharging process and the retailer supplies its flexible demand as well as other conventional consumers. Accordingly, a two stage operational framework is presented in which at the first stage, the retailer and aggregator do their medium-term planning that is made one month prior to real-time market.

At the second stage, the retailer and EV aggregator deal with their activities in the short-term period. In this stage, the retailer may modify its preliminary strategy by means of different sources such as pool market, interruptible loads, self-scheduling and EVs charging and discharging (entitled, vehicle to grid (V2G)) strategies, keeping in mind the medium-term decisions. Subsequently, a bi-level programming approach is adopted to solve the decision-making problem in which the upper sub-problem maximizes retailer profit, whereas the lower sub-problem minimizes the aggregated EVs charging and discharging costs. Due to the volatility of pool price and uncertainties associated with the consumers and EVs demand, the proposed framework is a mixed integer nonlinear stochastic optimization problem; therefore, Monte Carlo Simulation (MCS) is applied to solve it;. Furthermore, a market-quota curve is utilized to model the uncertainty of the rivals and obtaining retailer's actual market share.

The rest of this paper is organized as follows: the proposed market framework in terms of retailer's medium-term and short-term strategies in pool and bilateral markets as well as EV aggregator model are presented in Section 2. The case study is provided in Section 3 and finally Section 4 represents the conclusion.

2. FORMULATION OF PROPOSED MARKET FRAMEWORK

It is assumed the retailer buys electricity from the wholesale market and sells it back to conventional consumers and EV aggregator based on TOU rates for a specified period. Following, a detailed structure of the proposed multi-period decision making model is presented that allows a retailer and EV aggregator to determine the optimal strategy in the medium and short-term programming.

2.1. Medium-term planning framework

The medium-term program is a stochastic program in which uncertain parameters are modeled through scenario generation. The retailer's medium-term program is to determine the selling price and the quantity of power that should be purchased from bilateral contract as well as an approximate estimate of EVs aggregated demand.

2.1.1. Scenario generation

As previously mentioned, the pool price and consumer's demand are uncertainties in the medium-term program. The pool price uncertainty is modeled by mean–variance model of historical data, and it is assumed that the pool price distribution around the expected value is normal. We assume that the retailer has a forecast of the expected demand of conventional consumers, $\bar{\lambda}_{tj}^{P}$. Also, we consider that the amount of load demand is highly dependent on pool prices; therefore, after generating each pool price scenario $\lambda_{tj}^{P}(\omega)$, conventional load demands can be generated as a function of pool market prices that is calculated as below [12]:

$$d_{tj}(\omega) = \bar{d}_{tj}(1 + \xi \frac{\lambda_{tj}^{P}(\omega) - \bar{\lambda}_{tj}^{P}}{\bar{\lambda}_{tj}^{P}}) \quad (1)$$

$$\forall \omega \in \Omega_{\omega}, \forall t \in \{v, s, p\}, \forall j = j1, \ldots, j7$$

where, ξ is a parameter that depends on the relationnship between the pool price and the demand of conventional consumers. In this paper, we assume ξ = -0.1. It is notable that advanced methods for forecasting such as scenario generation, scenario reduction and model building could be easily used for medium-term planning strategy. Afterwards, the retailer generates scenarios for EVs load demand as well. The most effective factors on the EVs load demand are home departure time, daily travelled distance and home arrival time. Besides, road traffic condition, driving habits, battery capacity and its charger efficiency should also be considered. In medium-term, in order to generate MCS random samples, some of EVs related data are used to obtain corresponding probability density functions (PDFs). Non-Gaussian PDFs are suggested to create EVs random variables due to the better approximation of these functions. The Weibull PDF has been selected as the most appropriate function for departure time (d_k) of EVs as bellow:

$$f_{d_h}(\mathrm{h}) = \frac{\beta}{\alpha}(\frac{h}{\alpha})^{(\beta-1)} e^{-(\frac{h}{\alpha})^{\beta}} \qquad h > 0 \quad (2)$$

Also, to model daily travelled distance (tr_d) and arrival time (a_k), a type III Generalized expected value (Gev) PDF is selected. These functions are presented in Eqs. (3) and (4):

$$f_{tr_d}(\mathrm{t}) = \frac{1}{\sigma_{tr_d}}(1 + \mathrm{k}_{tr_d}\frac{(\mathrm{d} - \mu_{tr_d})}{\sigma_{tr_d}})^{-(1+\frac{1}{k_{tr_d}})} e^{-(1+\mathrm{k}_{tr_d}\frac{(\mathrm{d}-\mu_{tr_d})}{\sigma_{tr_d}})^{\frac{1}{k_{tr_d}}}} \quad (3)$$

$$f_{a_h}(\mathrm{t}) = \frac{1}{\sigma_{a_h}}(1 + \mathrm{k}_{a_h}\frac{(\mathrm{d} - \mu_{a_h})}{\sigma_{a_h}})^{-(1+\frac{1}{k_{a_h}})} e^{-(1+\mathrm{k}_{a_h}\frac{(\mathrm{d}-\mu_{a_h})}{\sigma_{a_h}})^{\frac{1}{k_{a_h}}}} \quad (4)$$

Required power for full-charge of EV battery in each day, is equal to the difference between its battery capacity and initial state of charge, when EV comes back from its last daily trip. This statement can be expressed as follows:

$$charge_{j}^{EV} = Cap_{bat} - SOC_{0}^{EV} \quad (5)$$

The SOC_{0}^{EV} of EVs depends on several factors such as daily travelled distance and battery capacity. Hence, SOC_{0}^{EV} can be derived as:

$$SOC_{0}^{EV} = 100 - \frac{tr_d}{C_{eff} \times Cap_{bat}} \times 100 \quad (6)$$

where, C_{eff} is the efficiency coefficient of the EV which depends on the traffic conditions and driving patterns as well as converter efficiency.

Here, EV aggregator generates total EVs demand scenarios. Subsequently, it estimates EVs aggregated power in the valley, shoulder and peak hours of EVs fleet using the following model:

$$Min: \sum_{EV=1}^{N'_{EV}} \sum_{j} \sum_{h=a_h}^{d_h} (P_{hj,EV}^{P}(\omega) . (1 + \xi \frac{\lambda_{hj}^{P}(\omega) - \bar{\lambda}_{hj}^{P}}{\bar{\lambda}_{hj}^{P}})) \quad (7)$$

s.t.

$$P_{hj,EV}(\omega) = \frac{CH_{hj,EV}(\omega)}{\eta} \qquad \forall \omega, \forall EV, \forall h, \forall j \quad (8)$$

$$\sum_{h=a_h}^{d_h} (CH_{hj,EV}(\omega)) + SOC_{0}^{EV}(\omega) = Cap_{bat} \quad \forall \omega, \forall EV, \forall h, \forall j \quad (9)$$

$$SOC_{h,j}^{EV}(\omega) = SOC_{h-1,j}^{EV}(\omega) + CH_{hj,EV}(\omega) \quad \forall \omega, \forall EV, \forall h, \forall j (10)$$

$$CH_{\min} \leq CH_{hj,EV}(\omega) \leq CH_{\max} \qquad \forall \omega, \forall EV, \forall h, \forall j (11)$$

$$SOC_{\min} \leq SOC_{h,j}^{EV}(\omega) \leq SOC_{\max} \qquad \forall \omega, \forall EV, \forall h, \forall j (12)$$

$$P_{hj,EV}^{P}(\omega) = P_{hj,EV}(\omega) \qquad \forall \omega, \forall EV, \forall h, \forall j (13)$$

In which, Eq. (8) represents the amount of required power for charging EVs battery in ωth scenario during hour h in day j. Based on Eq. (9) EVs battery should be fully charged within charging time [a_k d_k]. The charging state of the battery at the end of interval h considering charging power at that interval is given in Eq. (10). Moreover, constraints associated with variables of optimization problem are represented in the Eqs. (11)- (12). The power balance for each EV in ωth scenario during hour h in day j can be expressed as Eq. (13).

After implementation of the above mentioned optimization problem, the aggregated EVs fleet required power during valley, shoulder and peak hours can be derived as follows:

$$d_{tj}^{Fleet}(\omega)=\sum_{EV=1}^{N_{EV}}\sum_{h\in t}\mathrm{P}_{hj,EV}(\omega)\quad \forall\omega,\forall t\in\{v,s,p\},\forall j \quad (14)$$

By means of market-quote curve which will be explained in the next subsection, retailer's selling price and the percentage of consumers demand supplied by the retailer can be obtained.

2.1.2. Formulation of medium-term strategy

The profit objective function of the retailer in this stage can be formulated as follows:

$$Max: Exp\left(\sum_{j=1}^{7}\sum_{t}\begin{pmatrix}\mathrm{Rev}_{tj}^{mid-term}(\omega)\\ -COST_{tj}^{B,mid-term}(\omega)-COST_{tj}^{P,mid-term}(\omega))\end{pmatrix}\right) \quad (15)$$
$$\forall\omega\in\Omega_{\omega},\ \forall t\in\{v,s,p\},\ \forall j=j1,\ldots,j7$$

where, the first term is the retailer's revenue from selling to the consumers while the second and third terms are the cost of purchasing from bilateral contract and pool market, respectively. The individual parts of the above function can be explained as follows:

2.1.2.1. Setting retailer selling price

The relationship between the actual demand supplied to the consumers and the price offered by the retailer is proposed through a step wise market-quota curve. This curve represents retailer's market share among its other rivals.

A market-quota curve with three blocks is shown in Fig. 1. From mathematical perspective, the market-quota curve for consumers during period t in day j and ωth scenario of MCS can be formulated as follows:

$$D_{tj}(\omega)=\sum_{i=1}^{N_i}(\overline{D}_{tj,\mathrm{i}}^{D}(\omega)+\overline{D}_{tj,\mathrm{i}}^{Fleet}(\omega)).v_{\mathrm{t,i}} \quad (16)$$
$$\forall\omega,\forall t,\forall j, i=1,\ldots,N_i$$

$$\lambda_t^R=\sum_{i=1}^{N_i}\lambda_{t,i}^R \qquad \forall t \quad (17)$$

$$\overline{\lambda}_{\mathrm{t,i-1}}^R v_{\mathrm{t,i}}\le\lambda_{\mathrm{t,i}}^R\le\overline{\lambda}_{\mathrm{t,i}}^R v_{\mathrm{t,i}} \qquad \forall t,\forall i \quad (18)$$

$$\sum_{i=1}^{N_i}v_{\mathrm{t,i}}=1 \qquad v_{\mathrm{t,i}}\in\{0,1\} \quad (19)$$

Where, $\overline{D}_{tj,\mathrm{i}}^{D}(\omega)$ and $\overline{D}_{tj,\mathrm{i}}^{Fleet}(\omega)$ are percentage of $d_{tj}(\omega)$ and $d_{tj}^{Fleet}(\omega)$, respectively.

The revenue obtained from selling to the consumers (conventional consumers and EVs loads) is equal to the product of the selling price and power supplied by the retailer.

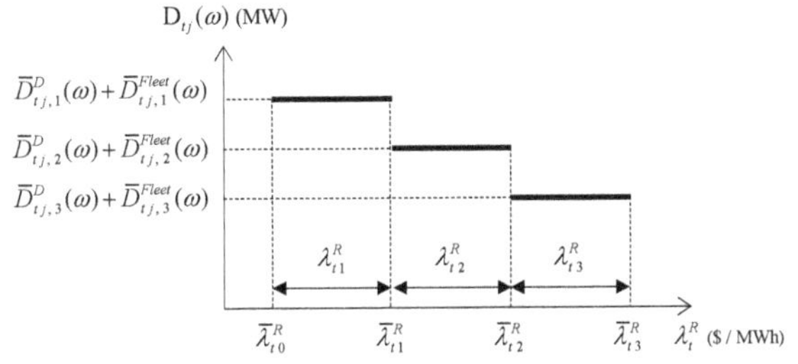

Fig. 1. Retailer market-quota curve [13].

Due to the stochastic behavior of consumer demands, the retailer's obtained revenue would be a random variable; Thus, the corresponding medium-term revenue from selling to the consumers during period t in day j and in ωth scenario can be formulated as:

$$\mathrm{Rev}_{tj}^{mid-term}(\omega)=D_{tj}(\omega)\,\lambda_t^R \qquad \forall\omega,\forall t,\forall j \quad (20)$$

$$\mathrm{Rev}_{tj}^{mid-term}(\omega)=\sum_{i=1}^{N_i}(\overline{D}_{tj,\mathrm{i}}^{D}(\omega)+\overline{D}_{tj,\mathrm{i}}^{Fleet}(\omega)).v_{\mathrm{t,i}}.\lambda_{\mathrm{t,i}}^R \quad (21)$$
$$\forall\omega,\forall t,\forall j$$

2.1.2.2. The cost of bilateral contracts

In bilateral contracts a maximum and a minimum bound of the purchased power is defined for each period. In the maturity period, if consumed power violates from these bounds, the retailer incurs a penalty. In the medium-term program, the quantity of power that is procured from bilateral contract during valley, shoulder and peak hours for a week is defined. Accordingly, the cost of bilateral contract throughout the time horizon of one week in ωth scenario during period t in day j is given as:

$$COST_{tj}^{\mathrm{B,mid-term}}(\omega)=P_{tj}^{B}(\omega)\lambda_{tj}^{B} \qquad \forall\omega,\forall t,\forall j \quad (22)$$

2.1.2.3. The cost of pool market

Besides the bilateral contract, the retailer may also procure its demand from the pool market. Due to the volatility of pool prices, the retailer faces uncertainty while offering in the medium-term market. In order to consider this uncertainty, a stochastic programming should be considered for retailer's medium-term planning. Therefore, purchasing cost from the pool market in ωth scenario can be formulated as:

$$COST^{P,mid-term}_{tj}(\omega) = P^{P}_{tj}(\omega)\lambda^{P}_{tj}(\omega) \quad \forall\,\omega, \forall\,t, \forall\,j \qquad (23)$$

The medium-term model is a stochastic mixed integer non-linear problem (MINLP) due to the randomness of pool prices and consumer demands.

2.2. Short-term planning framework

A bi-level programming approach is proposed to solve the decision making problem faced by the retailer and EV aggregator in the short-term problem; the upper sub-problem maximizes retailer profit whereas the lower sub-problem minimizes aggregator charging cost. On the other hand, in the lower level, the EV aggregator agent is responsible for optimal scheduling of its EVs battery charging and discharging process. In this way EV owners would pay less and also the retailer can adhere its plan thereby reduces its imbalance costs. Following the lower and upper levels of short-term optimization problem are addressed.

2.2.1. EVs aggregator model in the lower sub-problem

In this stage, aggregator should supply EVs charging loads. Furthermore, aggregator may set up bilateral contracts with retailer in order to supply a part of its required demand through V2G concept, when the price of pool market and other options are high. It is assumed that the EVs demand is responsive to the selling price offered by the retailer and is scheduled by the aggregator. In the lower level of short-term optimization problem, the objective function of the aggregator is to minimize the total cost over the scheduling time horizon. Mathematically, this objective function in day j can be formulated as:

$$Min: \sum_{EV=1}^{N_{EV}} \sum_{h\in t=a_h}^{d_h} (p^{S}_{hj,EV}\cdot\lambda^{R}_{t} - p^{P}_{hj,EV}\cdot\lambda^{Pei}_{t}) \qquad \forall j \quad (24)$$

s.t.

$$p^{S}_{hj,EV} = \frac{ch_{hj,\mathrm{EV}}}{\eta} \qquad \forall EV, \forall h, \forall j \quad (25)$$

$$p^{P}_{hj,EV} = dch_{hj,\mathrm{EV}}\cdot\eta \qquad \forall EV, \forall h, \forall j \quad (26)$$

$$\sum_{h=a_t}^{d_t} (\mathrm{ch}_{hj,\mathrm{EV}} - \mathrm{dch}_{hj,\mathrm{EV}}) + soc^{EV}_{0} = Cap_{bat} \quad \forall EV, \forall h, \forall j \quad (27)$$

$$soc^{EV}_{h,j} = soc^{EV}_{h-1,j} + ch_{hj,\mathrm{EV}} - dch_{hj,\mathrm{EV}} \qquad \forall EV, \forall h, \forall j \quad (28)$$

$$ch_{\min} \le ch_{hj,\mathrm{EV}} \le ch_{\max} \qquad \forall EV, \forall h, \forall j \quad (29)$$

$$dch_{\min} \le dch_{hj,\mathrm{EV}} \le dch_{\max} \qquad \forall EV, \forall h, \forall j \quad (30)$$

$$soc_{\min} \le soc^{EV}_{h,j} \le soc_{\max} \qquad \forall EV, \forall h, \forall j \quad (31)$$

Where all mentioned constraints are previously described except for Eq. (26) that shows useful power for discharging EV batteries.

2.2.2. Retailer profit model in the upper sub-problem

In the upper level of short-term optimization problem, the retailer seeks to maximize its short-term profit during each day keeping in mind the medium-term decisions. The complete formulation of the upper level is given as:

$$\begin{aligned} Max:\ & \mathrm{Rev}^{Selling}_{j} + \mathrm{Rev}^{Fleet}_{j} + \mathrm{Rev}^{IL}_{j} - COST^{\mathrm{B,short-term}}_{j} \\ & - COST^{\mathrm{B,penalty}}_{j} - COST^{P,short-term}_{j} - COST^{G}_{hj} - COST^{Fleet}_{j} \end{aligned} \qquad (32)$$

This objective function consists of two main parts, the revenue obtained by the retailer from selling to both consumers and the pool market as well as the cost of buying from various options. Different parts of the objective function can be explained as follows:

2.2.2.1. Cost of bilateral contract

It is assumed that in the medium-term a Contract for Difference (CFD) agreement is signed between wholesale market and the retailer in such a way that the difference between the bilateral contract price and actual pool price is equally split between two sides. The proportion of the difference could be changed by negotiating between two sides. Based on CFD, in the short-term program the bilateral contract cost can be formulated as:

$$\begin{aligned} COST^{\mathrm{B,short-term}}_{j} &= \sum_{t}\sum_{h\in t}\Big(\lambda^{B}_{tj} + \frac{\lambda^{P}_{tj} - \lambda^{B}_{tj}}{2}\Big)\mathrm{p}^{B}_{hj} \\ &= \sum_{t}\sum_{h\in t}\frac{1}{2}\lambda^{B}_{tj}p^{B}_{hj} + \sum_{h=1}^{24}\frac{1}{2}\lambda^{P,est}_{hj}p^{B}_{hj} \qquad \forall j \end{aligned} \qquad (33)$$

On the other hand, in the short-term period, retailer should pay penalties due to under or over-consumption of bilateral contracts that are signed in the medium-term. The total power consumed from bilateral contract during period t in day j is:

$$\sum_{h\in t} p^{B}_{hj} = \sum_{m=1}^{5} x_{\mathrm{m},tj} \qquad \forall t, \forall j \quad (34)$$

where, $X_{m,tj}$ is an auxiliary variable for penalty

calculations of bilateral contracts in the short-term stage. Fig. 2 shows the penalties that are incurred for under or over-consumption of a bilateral contract related to period t in day j.

The penalty constraints can be expressed mathematically as:

$$0 \le x_{1,tj} \le 0.8\,\overline{P}_{tj}^{B} \qquad \forall t, \forall j \quad (35)$$

$$0 \le x_{2,tj} \le 0.1\,\overline{P}_{tj}^{B} \qquad \forall t, \forall j \quad (36)$$

$$0 \le x_{3,tj} \le 0.2\,\overline{P}_{tj}^{B} \qquad \forall t, \forall j \quad (37)$$

$$0 \le x_{4,tj} \le 0.1\,\overline{P}_{tj}^{B} \qquad \forall t, \forall j \quad (38)$$

$$0 \le x_{5,tj} \le M \qquad \forall t, \forall j \quad (39)$$

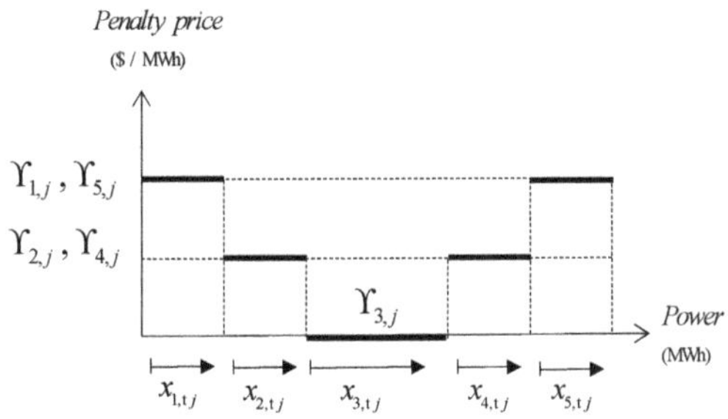

Fig. 2. Penalty function for under or over-consumption of bilateral contracts.

In the above constraints, $X_{m,tj}$ (m=1 … 5) is a variable representing the power consumed within block m. With this approach under or over-consumption penalties can be expressed in cumulative blocks and formulated as follows:

$$COST_{tj}^{B,\text{Penalty}} = \begin{pmatrix} \Upsilon_{1,t}\,(0.8\,\overline{P}_{tj}^{B} - x_{1,tj}) \\ +\Upsilon_{2,t}\,(0.1\,\overline{P}_{tj}^{B} - x_{2,tj}) \\ +\sum_{m=3}^{5} \Upsilon_{m,t}\, x_{m,tj} \end{pmatrix} \qquad \forall t, \forall j \quad (40)$$

Note that, $\overline{P}_{tj}^{B}$ is computed in the medium-term program and is a sufficiently large constant, e.g. $2\overline{P}_{tj}^{B}$.

Finally, total penalty cost considering all bilateral contracts throughout the time horizon of day j is:

$$COST_{j}^{B,\text{penalty}} = \sum_{t} COST_{tj}^{B,penalty} \qquad \forall j \quad (41)$$

2.2.2.2. Retailer self-production

Some retailers hedge against risk of pool market by owning some self-production utilities such as distributed generators. It is considered that self-production facility can only supply a part of the retailer's demand, which is a realistic assumption. Thus, the retailers face a trade-off between the cost of self-production and the cost of other options. In the short-term program the self-produced power at hour h of day j, P_{hj}^{G} is given as:

$$P_{hj}^{G} = u_{hj}\, P^{G,\min} + \sum_{f=1}^{F} P_{fhj} \qquad \forall h, \forall j \quad (42)$$

$$0 \le P_{fhj} \le u_{hj}\, P_{f}^{\max} \qquad \forall h, \forall j, \forall f \quad (43)$$

The retailer production cost may be implemented using an approximate piecewise linear function curve as shown in Fig. 3. Considering DG corresponding costs, the aggregated self-production cost during hour h in day j is thus obtained as:

$$COST_{hj}^{G} = u_{hj} C^{fix} + y_{hj} C^{su} + z_{hj} C^{shd} + \sum_{f=1}^{F} \ell_f\, P_{fhj} \qquad \forall h, \forall j \quad (44)$$

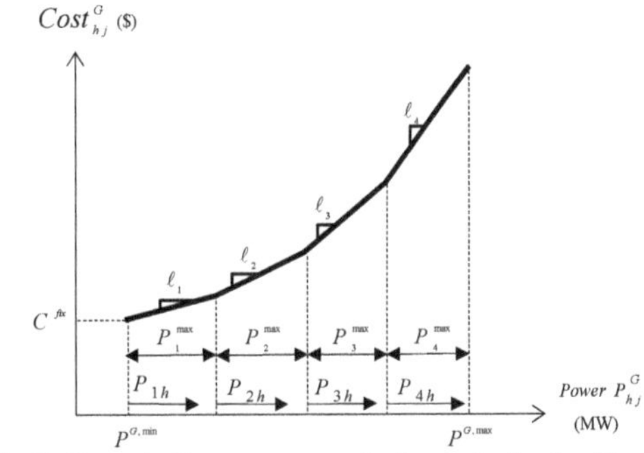

Fig. 3. Piecewise linear convex production cost using four blocks [9].

Subsequently, self-production constraints in terms of minimum up time and down time as well as ramp-up and ramp-down constraints are expressed as Eqs. (45)- (48):

$$\left[X_{(h-1)}^{on} - T^{on}\right].\left[u_{(h-1)j} - u_{hj}\right] \ge 0 \qquad \forall h, \forall j \quad (45)$$

$$\left[X_{(h-1)}^{off} - T^{off}\right].\left[u_{hj} - u_{(h-1)j}\right] \ge 0 \qquad \forall h, \forall j \quad (46)$$

$$P_{h+1,j}^{G} - P_{h,j}^{G} \le R^{UP} \qquad \forall h, \forall j \quad (47)$$

$$P_{h,j}^{G} - P_{h+1,j}^{G} \le R^{DN} \qquad \forall h, \forall j \quad (48)$$

The relationship between binary variables used to model the ON and OFF status of the self-production facility should meet the following constraints to avoid conflicting situations:

$$u_{hj} - u_{(h-1)j} = y_{hj} - z_{hj}\ ,\ y_{hj} + z_{hj} \le 1 \qquad \forall h, \forall j \quad (49)$$

On the other hand, self-produced power can be locally consumed or sold to the pool in the short-term program during hour h in day j as shown in Eq. (50):

$$P_{hj}^{G} = P_{hj}^{consume} + P_{hj}^{Sell} \qquad \forall h, \forall j \quad (50)$$

In order to avoid simultaneous buying and selling

power during hour h in day j at an identical price, the following constraints can be used in the content of short-term program:

$$p_{hj}^{P} \leq k_{hj} L_{hj} \quad , \quad k_{hj} \in \{0,1\} \qquad \forall h, \forall j \quad (51)$$

$$P_{hj}^{Sell} \leq (1-k_{hj}) P^{G,\max} \qquad \forall h, \forall j \quad (52)$$

Where L_{kj} is total load of consumers during hour h in day j with interrupted load. It is notable that based on Eq. (51), retailer can consume the obtaining power of self-production facility, but the output power should be less than or equal to the actual demand of consumers. On the other hand, based on Eq. (52) retailer can sell the obtaining power of self-production facility to the pool, however, the output power should be less than or equal to the maximum power output of the self-production.

2.2.2.3. Retailer strategy with EV aggregator agent

In the short-term program, the revenue of retailer from selling to EVs fleet during hours h of type t in day j is:

$$\mathrm{Rev}_{j}^{Fleet} = \sum_{EV=1}^{N_{EV}} \sum_{h \in t} p_{hj,\mathrm{EV}}^{S} \lambda_{t}^{R} \qquad \forall j \quad (53)$$

On the other hand, the cost of purchasing from the EVs fleet through V2G capability can be formulated as:

$$COST_{j}^{Fleet} = \sum_{EV=1}^{N_{EV}} \sum_{h \in t} p_{hj,\mathrm{EV}}^{P} \lambda_{t}^{Pei} \qquad \forall j \quad (54)$$

where, $p_{hj,EV}^{S}$ and $p_{hj,\mathrm{EV}}^{P}$ are obtained by EV aggregator through short-term lower sub-problem.

2.2.2.4. Participating in the pool market

As the medium-term stage, in the short-term program the retailer may also procure its residual demand from the pool market. Thus, purchasing cost from the pool in the short-term program for each day j is computed as:

$$COST_{j}^{P,short-term} = \sum_{h=1}^{24} p_{hj}^{P} \lambda_{hj}^{P,est} \qquad \forall j \quad (55)$$

Also, it is assumed that the retailer can sell back its excess self-produced power to the pool market. Therefore, selling revenue from the pool for each day j is computed as:

$$\mathrm{Rev}_{j}^{Selling} = \sum_{h=1}^{24} \lambda_{hj}^{P,est} P_{hj}^{Sell} \qquad \forall j \quad (56)$$

2.2.2.5. Presence of IL contracts

In addition to the self-production and V2G capabilities of EVs, ILs can be utilized by retailers as a risk management tool against volatility of pool prices. In this paper, two types of IL contracts are considered. In the first case the consumers pay λ_{t}^{R} for their loads; but if in case of emergency, the retailer is forced to interrupt the consumers loads, a penalty λ_{t}^{Fine} ($\lambda_{t}^{Fine} \rangle \lambda_{t}^{R}$) should be paid to the consumers. In the second case, an IL contract has been signed between two sides and consumers pay a reduced price for their loads, $\lambda_{t}^{\mathrm{Re}duce}$ but do not receive any additional pecuniary compensation in case of interruption. For this type of consumers $\lambda_{t}^{Fine} = 0$ [13]. Mathematically, the revenue of retailer from selling to the conventional consumers considering IL contract in the short-term program can be formulated as:

$$\mathrm{Rev}_{j}^{IL} = \begin{pmatrix} \sum_{h \in t} (D_{1,hj} - p_{1,hj}^{IL}) . \lambda_{t}^{R} \\ + \sum_{h \in t} (D_{2,hj} - p_{2,hj}^{IL}) . \lambda_{t}^{\mathrm{Re}duce} \\ - \sum_{h \in t} p_{1,hj}^{IL} . \lambda_{t}^{Fine} \end{pmatrix} \qquad \forall j \quad (57)$$

where, the first and second terms are associated with the net revenue of retailer from selling to consumers of type 1 and 2, respectively. Also, the cost of interrupting consumers of type 1 is given in the third term. Eventually, the short-term power balance at hour h in day j can be expressed as:

$$\begin{pmatrix} \sum_{EV=1}^{N_{EV}} p_{hj,EV}^{P} + P_{hj}^{consume} \\ + p_{hj}^{P} + p_{hj}^{B} \end{pmatrix} = \begin{pmatrix} \sum_{EV=1}^{N_{EV}} p_{hj,EV}^{S} + D_{1,hj} \\ + D_{2,hj} - p_{1,hj}^{IL} - p_{2,hj}^{IL} \end{pmatrix} \forall h, \forall j \quad (58)$$

It is notable that, self-production facility, V2G and IL options, not only meet a part of the retailer's demand, but also their implementation in the short-term program is more realistic. The short-term framework is a bi-level optimization problem. EMP (Extended Mathematical Programming) solver in the GAMS software is used to solve this model. Here, in order to simulate the uncertainty of the mathematical model a large number of parameters related to uncertainties of the spot market price $\lambda_{tj}^{P}(\omega)$, the conventional consumers demand $d_{tj}(\omega)$ and EVs fleet demand $d_{tj}^{Fleet}(\omega)$ are randomly produced by MCS. Subsequently, for each of these uncertainties, the optimization problem is solved to get $P_{tj}^{B}(\omega)$, $P_{tj}^{P}(\omega)$ *and* λ_t^R. Finally, the expected values of all allocation schemes are computed for the decision variables. These decisions are imposed as boundary constraints in the short-term program.

3. CASE STUDY

To show the efficiency of the proposed framework, a case study is performed based on a typical electricity retailer data in Nord Pool Market [14]. Decision making time horizon is one month prior to real-time market for medium-term and one day prior to real-time market for short-term. In this paper, one week time horizon is considered for numerical analysis. According to employed TOU pricing, the valley period is defined at hours 1-7, hours 11-13 and 17-21 are peak periods and the remaining hours are considered as shoulder period. It is assumed that the selling price at each hour is fixed. Moreover, the retailer pays a fine $\lambda_t^{Fine} = 1.15\,\lambda_t^R$ per unit of interruption to conventional consumers of type 1 and offers a 7% discount in the selling price, $\lambda_t^{Reduce} = 0.93\,\lambda_t^R$ for consumers of type 2. Also, it is supposed that, 80% and 20% of consumers are type 1 and type 2, respectively, and the retailer can interrupt maximum 30% of each customer's load. Fig. 4 shows expected demand data of conventional consumers, $\bar{d}_{tj}$, for valley, shoulder and peak hours of a sample week days. Subsequently, Fig. 5 illustrates retailer's market-quota curve with 100 points (100 steps) for each valley, shoulder and peak periods.

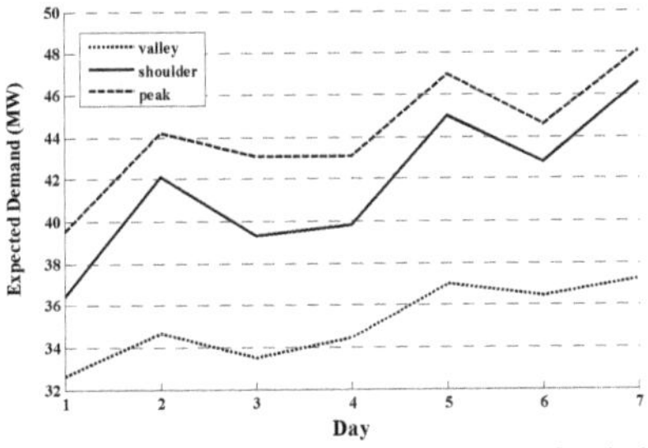

Fig. 4. Expected demand data during each hour of period t (MW)

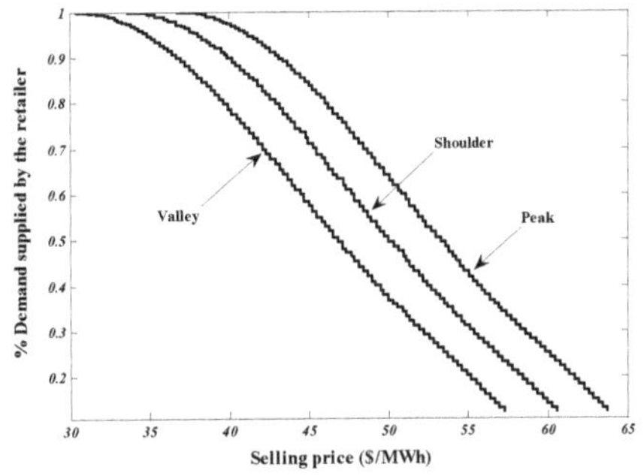

Fig. 5. Market-quota curves data with 100 intervals

Table 1, provides the mean and standard deviation of existing pool prices for one week time horizon that are obtained from historical data of Nord Pool spot market. Penalty data of bilateral contracts for under-consumption and over-consumption is given in Table 2.

Table 1. Mean and standard deviation of pool price data

	valley	shoulder	peak
Mean (μ)	40.86	44.68	48.08
Standard deviation (δ)	1.20	4.33	6.68

Data associated with self-production unit in terms of unit technical data and cost coefficients is also represented in Tables 3-5.

Table 2. Penalty data of bilateral contracts($\Upsilon_{3,t} = 0$) $/MWh

Penalty slope	Valley	Shoulder	Peak
$\Upsilon_{1,t}$	3	4	5
$\Upsilon_{2,t}$	1.5	2	3
$\Upsilon_{4,t}$	1.5	2	3
$\Upsilon_{5,t}$	3	4	5

Table 3. Technical characteristics of the self-production facility

$P^{G,\min}$ (MW)	$P^{G,\max}$ (MW)	Ramp-up limit, (MW/h)	Ramp-down limit (MW/h)	Minimum up/down time (h)
3	12	3	3	2

Table 4. Fixed, shut-down and start-up costs ($).

Fixed	Shut-down	Start-up
200	100	150

Table 5. Piecewise linear cost for self-production unit.

Block	Block size (MW)	Cost ($/MWh)
1	2.5	29
2	2.5	38
3	2.5	45
4	1.5	55

In this paper considering 10% penetration level an introduction of 6000 EVs has been estimated in the retailer area. The parameters of the non-Gaussian fitted PDFs associated with EVs random parameters are given in Table 6.

Table 6. Parameters of the fitted PDFs.

Datasets	The suggested PDF
d_t	$\alpha = 7.67$ $\beta = 21.38$
tr_d	$k_{tr_d} = -0.05$ $\mu_{tr_d} = 17.65$ $\sigma_{Ntr_d} = 7.12$
a_t	$k_{tr_d} = -0.06$ $\mu_{tr_d} = 17.27$ $\sigma_{Ntr_d} = 0.84$

EV battery parameters such as Cap_{bat}, η and C_{eff} are 24 kW, 90% and 2km/kWh, respectively. Also, it is assumed that, maximum charging and discharging power in each hour is equal to the 10% of EV battery capacity. Price and conventional demand forecasting data during the time horizon of one week are given in [15].

In the medium-term problem MCS with 200 iterations is employed to obtain the expected values of pool and bilateral procurement levels, as well as optimal selling prices. Based on the market-quota curve the percentage of total demand supplied by the retailer for both conventional and aggregated EVs load in the valley, shoulder and peak hours are 34%, 38% and 40%, respectively. Consequently, expected retail selling prices for valley, shoulder and peak hours are 50.732, 52.899 and 55.673 ($/MWh), respectively.

Expected weekly power procurement from bilateral contract, $\overline{P}_{tj}^{B}$ (MW), is represented in Table 7. Also, in Fig. 6 the probability distribution function of expected profit in the medium-term problem is shown that is obtained from 200 MCS scenarios. As shown in Fig. 6, the expected value and standard deviation of the profit in the medium-term problem are 28364.5 $ and 2078.27 $, respectively. As it is obvious from Fig. 6, the probability density function of expected profit is approximately close to the normal distribution function.

To analyze the impact of each option on the retailer's optimal bidding strategy in the short-term problem, five cases are considered as follows:

- Case 1: The retailer procures its demand only from pool market.
- Case 2: The retailer procures its demand from both pool market and bilateral contract. Also, in this case retailer will sign IL contracts with conventional consumers.
- Case 3: Same as case 2; however, in this case a contract is signed between retailer and EV aggregator for buying its own V2G contribution.
- Case 4: Same as case 3; however, in this case the retailer utilizes its self-production unit just for self-consumption.
- Case 5: Same as case 4; however, in this case the retailer can also sell its excess self-production power to the pool market.

Table 7. Expected power procurement from bilateral contract $\overline{P}_{tj}^{B}$ (MW).

Day	Valley	Shoulder	Peak
j1	51.61	61.75	63.56
j2	59.36	73.19	79.01
j3	54.93	69.84	72.78
j4	58.93	65.58	70.41
j5	61.49	76.70	64.48
j6	59.00	69.14	76.26
j7	57.79	84.95	67.94

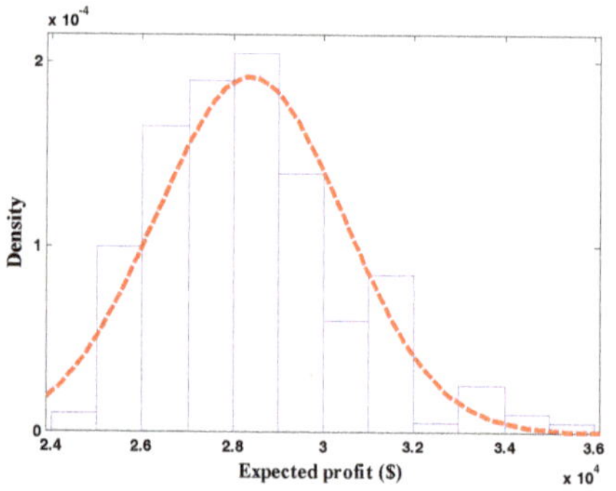

Fig. 6. The probability distribution function of expected profit in the medium-term program

Retailer's profits in individual days of the week are represented in Table 8. As shown increases in pool prices in seventh day, causes a profound reduction in all cases and as indicated in the first case it is noticeable due to its force to buy just from the pool market. However, in other cases the considered options will prevent loss of reduction, nevertheless, the total profit is drastically reduced. In cases 3, 4 and 5, the retailer hedges against the risk of pool market with self-production facility and uses a part of its self-production power for self-consumption. Moreover, the retailer has IL contracts and can also procure some of its demand from V2G capability of EVs in hours that pool price is high. In these cases retailer's strategy within these hours would be beneficial.

Table 8. Retailer's profit in different cases in days of the week ($).

Day	Case 1	Case 2	Case 3	Case 4	Case 5
1	3782.38	3691.82	3691.82	3691.82	3685.17
2	2000.13	2178.75	2218.99	2280.23	2444.27
3	3899.97	3743.24	3743.24	3743.24	3733.33
4	3725.76	3585.72	3585.72	3585.72	3582.53
5	2651.76	2694.16	2712.10	2641.89	2753.50
6	3758.51	3619.73	3619.73	3394.52	3389.06
7	-1528.4	-24.17	294.44	2073.01	3542.85
Sum	18290.09	19489.28	19866.08	21410.47	23130.73

Percentages of various power source contributions to meet retailer's weekly demand for cases 4 and 5 are illustrated in Fig. 7. As indicated in case 5 the self-production quota in final demand is reduced in comparison to case 4. This is due to retailer incentive to sell back its excess self-production to the pool market considering corresponding high prices. As a result more percentage of the demand would be provided by other options. It is noticeable that despite existing high pool prices the retailer still prefers to sell its power while repurchasing from the pool market.

The amount of interrupted loads and penalties that are incurred for under or over-consumption of bilateral contracts are represented in Table 9. As it is obvious, penalty of bilateral contract in case 3 is less than case 2; however in case 4, the use of various options such as self-production, ILs and V2G capability of EVs, prevents over-consumption of bilateral contracts, thus reduces penalties incurred for these contracts. Also, use of self-production facility for self-consumption, causes profound reductions in interrupted loads and bilateral contract penalties in cases 4 and 5. Note that in case 5, due to selling power the retailer may be obligated to over-consume from bilateral contracts. Consequently, incurred penalty due to over-consumption of bilateral contracts as well as IL contracts are relatively increased.

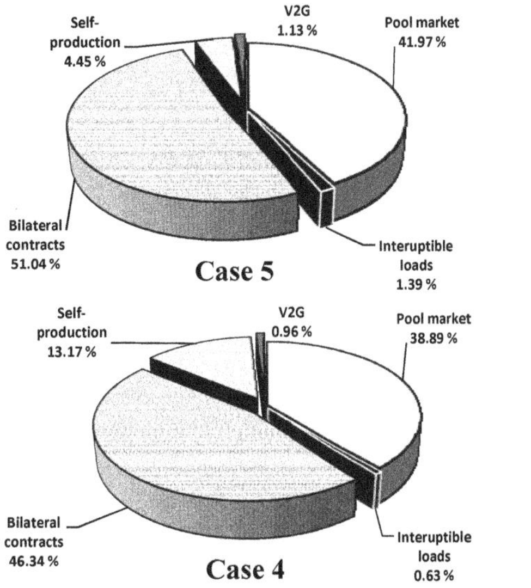

Fig. 7. Percentages of various power source contributions during whole week for cases 4 and 5 (MW).

Table 10 represents the detailed status of the retailer in terms of generation, demand, revenue and cost in case 5. Note that in this Table, the net profit is calculated as the total revenue obtained from selling power to the pool, conventional loads and EVs minus the costs of all options. Fig. 8 provides the optimal sharing of retailer's procurement options to meet the corresponding demand in Case 5 for a representative day. It can be observed that the retailer employs its self-production facility for self-consumption within hours 11,13 and 21-22, while benefits from selling power to the pool during hours 11,12 and 14-20. Note that in these periods retailer prefers to procure its demand from bilateral contracts as well as V2G capability of EVs rather than participating in the pool market. It is notable that the retailer has employed IL contracts in these hours since the amount of pool price is drastically high.

Table 9. the amount of interrupted load and penalties associated with bilateral contracts in different cases ($).

CASE	1	2	3	4	5
Penalty of bilateral contracts ($)	-	229.18	148.73	0	286.68
IL (MW)	-	36.38	36.38	18.17	40.40

Table 10. Numerical results of the case 5.

Gen.	Total power generated	396.5 MW
	Percentage self-consumed	32.5 %
	Percentage sold	67.5 %
Demand	Demand supplied	2897.0 MW
	Percentage of self-production and consumption	4.45 %
	Percentage of purchased power from pool	51.04 %
	Percentage of procured power from bilateral contracts	41.97 %
	Percentage of procured power from V2G capability of EVs	1.13 %
	percentage of interrupted loads (ILs)	1.39 %
Revenue	selling to the pool	16689.3 $
	selling to the conventional consumers	138371.6 $
	selling to the EVs	12041.1 $
	Total revenue	167102.0 $

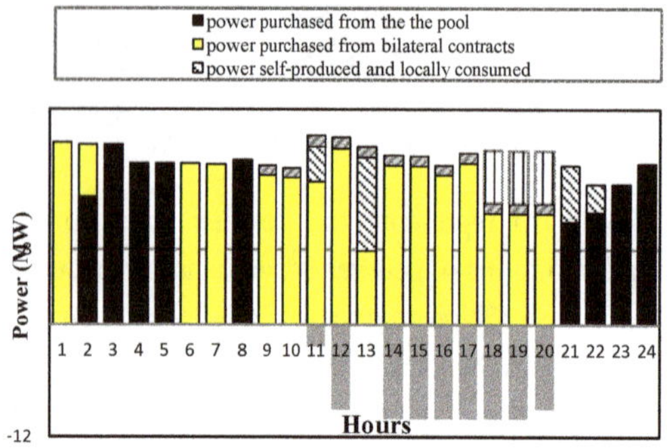

Fig. 8. Mix of electricity sources for a typical working day.

Furthermore, during hours 1,6,7 and 9-10 pool prices are much higher than corresponding bilateral prices; thus within these hours power can be bought exclusively from bilateral contracts. Nevertheless, the retailer provides its required power just from pool market during hours 3-5, 8 and 23-24 due to its relatively cheap prices. On the other hand, during hours 2 and 21-22, the pool prices are greater than other options. Nonetheless, the retailer procures a part of its demand from the pool. The penalties incurred for over-consumption of bilateral contracts are the main reason of this behavior.

The role of aggregator is to coordinate EVs charging in order to minimize corresponding charging costs. This in turn reduces retailer's aggregated cost due to shifting a part of demand to off-peak periods. Furthermore, V2G capability of EVs may affect retailer's profit.

Totally one can conclude that the retailer does not rely heavily on single electric power resource and procures its demand from different resources. Among which pool and bilateral markets are usually considered as the most reliable markets due to their relatively high certainty and firm structure. Subsequently, self-production, IL and some other options are in the next priority. Also, due to the flexible power procurement framework, penalties of bilateral contract associated with under-consumptions or over-consumptions are comparatively very small (see Table 10). Moreover, without self-production facility, the net profit of retailer would be decreased by 14.10 %. As a result the self-production is a tool that acts as a hedge against the volatility of pool prices. Finally, it is worth to note if neither the self-production facility, nor other options (e.g. IL, bilateral contracts and V2G capability of EVs) are available, the net profit would be drastically decreased by 20.92 % that is noticeable from the retailer's point of view.

4. CONCLUSIONS

In this paper, a multi-stage and multi-period framework for the decision making of an electricity retailer and EV aggregator is proposed. A bi-level programming approach is adopted to solve the decision making problem in the short-term problem; in which the upper sub-problem maximizes retailer revenue whereas the lower sub-problem minimizes the EVs charging costs. This paper provides a model that allows a retailer to procure optimally its power using different power supply options such as bilateral contracts, ILs programs, V2G capabilities of EVs and self-production facility in order to hedge against market risks. The appropriate use of these options allows significant increment in retailer's profit in comparison with buying exclusively from the pool. Due to the volatility of pool price and uncertainties associated with the consumers and EVs demand, a mixed integer nonlinear stochastic optimization problem is utilized to solve the problem. Numerical results show the capability and economic advantages of the proposed model for the retailer.

NOMENCLATURE

Sets and numbers:

t	Set of periods in the valley (v), shoulder (s) and peak (p) hours
h	Set of hours
j	Set of days

ω	Index for scenarios of MCS
Ω_{ω}	Set of scenarios in MCS
EV	Index for each EV
N'_{EV}	Total number of EVs in the aggregator's area
N_{EV}	Number of EVs supplied by EV aggregator
f	Number of blocks in the self-production function
i	Index of blocks in market-quote curve
N_{i}	Number of blocks in market-quote curve

Parameters:

$\lambda^{P}_{tj}(\omega)$	Pool price in ωth scenario during period t in day j
$\overline{\lambda}^{P}_{tj}$	Mean value of pool prices during period t in day j
$\lambda^{P}_{hj}(\omega)$	Hourly pool price in ωth scenario during hour h in day j
$\lambda^{R}_{t,i}$	Selling price associated with ith block of market-quote curve in period t
$\overline{\lambda}^{R}_{t,i}$	Upper bound of the ith interval of market-quote curve in period t
λ^{B}_{tj}	Bilateral contract price during period t in day j
λ^{B}_{hj}	Hourly bilateral contract price during hour h in day j
λ^{Pei}_{t}	Payment price to the EV aggregator by the retailer due to V2G capability during period t
$\lambda^{P,est}_{hj}$	Short-term pool market price forecast during hour h in day j
ℓ_{f}	Price of fth block of self-production function
$d_{tj}(\omega)$	Consumer's demand in ωth scenario during period t in day j
$\overline{d}_{tj}$	Expected demand of consumers during period t in day j
$\mathrm{SOC}^{EV}_{0}(\omega)$	Initial state of charge of EV battery in ωth scenario
soc^{EV}_{0}	Initial state of charge of EV battery
$P^{\mathrm{G,min}}, P^{\mathrm{G,max}}$	Minimum and maximum output power of self-production unit
$P^{\max}_{f}$	Maximum output power of fth block of self-production function
$\Upsilon_{m,\mathrm{t}}$	Slope of mth block during period t associated with penalty function of bilateral contract
C^{fix}, C^{su}, C^{shd}	Fixed start-up and shut-down costs of self-production facility
T^{on}, T^{off}	Up-time and down-time of the self-production facility
R^{UP}, R^{DN}	Ramp-up and ramp-down rates of self-production facility
$D_{1,hj}$, $D_{2,hj}$	Short-term forecasted demands of consumer types 1 and 2 during hour h in day j

Variables:

$P^{P}_{hj,EV}(\omega)$	Required power of EV that would be traded in the pool market in ωth scenario during hour h in day j
$P_{hj,EV}(\omega)$	Required power for charging EVs batteries in ωth scenario during hour h in day j
$CH_{hj,\mathrm{EV}}(\omega)$	Useful charging power for each EV in ωth scenario during hour h in day j
$\mathrm{SOC}^{EV}_{h,j}(\omega)$	State of charge of EV battery at the end of interval h in day j in ωth scenario
$d^{Fleet}_{tj}(\omega)$	Total EVs fleet required power in ωth scenario during period t in day j
$D_{tj}(\omega)$	Total demand supplied by the retailer to the consumers in ωth scenario during period t in day j
$\overline{D}^{D}_{tj,\mathrm{i}}(\omega), \overline{D}^{Fleet}_{tj,\mathrm{i}}(\omega)$	Power associated with ith block of the market-quote curve during period t in day j and in ωth scenario for conventional consumers and EVs fleet, respectively
$P^{B}_{tj}(\omega)$	Medium-term power purchased from bilateral contracts during period t in day j and in ωth scenario
$P^{P}_{tj}(\omega)$	Medium-term power purchased from the pool market during period t in day j and in ωth scenario
$p^{S}_{hj,EV}$	Short-term EVs required charging power during hour h of type t in day j
$p^{P}_{hj,\mathrm{EV}}$	Short-term power purchased from the EVs during hour h in day j
$ch_{hj,\mathrm{EV}}$	Short-term useful charging power of EVs batteries during hour h in day j
$dch_{hj,\mathrm{EV}}$	Short-term discharging power of EVs batteries during hour h in day j
$soc^{EV}_{h,j}$	State of charge of EV battery at the end of interval h in day j
p^{B}_{hj}	Short-term power purchased from bilateral contract during hour h in day j
p^{P}_{hj}	Short-term power purchased from pool market during hour h in day j
$\overline{P}^{B}_{tj}$	The expected value of purchasing power from bilateral contract during period tin day j
P^{G}_{hj}	Power self-produced at hour h of day j
P_{fhj}	Self-produced power associated with fth block of self-production function during hour h in day j
P^{Sell}_{hj}	Power self-produced and sold to the pool during hour h in day j
$P^{consume}_{hj}$	Power self-produced and locally consumed at hour h in day j
$p^{IL}_{1,hj}$, $p^{IL}_{2,hj}$	Interrupted load from consumer types 1 and 2 in the short-term stage during hour h in day j
λ^{R}_{t}	Selling price offered by the retailer to the conventional consumers and EV aggregator in period t
$v_{\mathrm{t,i}}$	Binary variable that is equal to 1 if the selling price offered by the retailer to consumers belongs to block i of the market-quota curve, and 0 otherwise
u_{hj}	Binary variable that is equal to 1 if unit is committed during hour h and 0 otherwise
y_{hj}	Binary variable that is equal to 1 if the unit starts up at the beginning of hour h in day j and 0 otherwise
z_{hj}	Binary variable that is equal to 1 if the unit shuts down at the beginning of hour h in day j and 0 otherwise
k_{hj}	Binary variable that is equal to 1 if power is bought from the pool and 0 if it is sold to the pool during hour h in day j
X^{on}_{h}, X^{off}_{h}	Number of continuously on (off) time

hours of self-production unit up to the hour h

REFERENCES

[1] S.J. Deng and S.S. Oren, "Electricity derivatives and risk management," *Energy*, vol. 31, pp. 940-953, 2006.

[2] A. Hatami, H. Seifi and M.K. Sheikh-El Eslami, "A stochastic-based decision-making framework for an electricity retailer: time-of-use pricing and electricity portfolio optimization," *IEEE Transactions on Power Systems*, vol. 26, no. 4, pp. 1808-1816, 2011.

[3] A. Ahmadi, A.R. Heidari and A. Esmaeel Nezhad, "Benders decomposition and normal boundary intersection method for multi-objective decision making framework for an electricity retailer in energy markets," *IEEE Systems Journals*, vol. 99, pp. 1- 10, 2014.

[4] A. Ahmadi, M. Charwand and J. Aghaei, "Risk-constrained optimal strategy for retailer forward contract portfolio," *International Journal of Electrical Power and Energy Systems*, vol. 53, pp. 704-713, 2013.

[5] M. Charvand and Z. Moshavash, "Midterm decision-making framework for an electricity retailer based on Information Gap Decision Theory," *International Journal of Electrical Power & Energy Systems*, vol. 63, pp. 185-195, 2014.

[6] J.. Moshtagh, S. Ghasemi, "Optimal distribution system reconfiguration using non-dominated sorting genetic algorithm (NSGA-II)," *Journal of Operation and Automation in Power Engineering*, vol. 1, no. 1, pp. 12-21, 2013.

[7] M Allahnoori, Sh. Kazemi, H. Abdi, and R. Keyhani, "reliability assessment of distribution systems in presence of microgrids considering uncertainty in generation and load demand," *Journal of Operation and Automation in Power Engineering*, vol. 2, no. 2, pp. 113-120, 2014.

[8] A.J. Conejo, J.J. Fernandez-Gonzalez and N. Alguacil, "Energy procurement for large consumers in electricity markets," *IEE Proceeding on Generation, Transmission and Distribution*, vol. 152, no. 3, pp. 357-364, 2005.

[9] A.J. Conejo and, M. Carrion, "Risk-constrained electricity procurement for a large consumer," *IEE Proceedings on Generation, Transmission and Distribution*, vol. 153, no. 4, pp 407- 413, 2006.

[10] M. Carrion, A.B. Philpott, A.J. Conejo and J.M. Arroyo, "A stochastic programming approach to electric energy procurement for large consumers," *IEEE Transactions on Power Systems*, vol. 22, no. 2, pp. 744-754, 2007.

[11] J. Tomi and W. Kempton, "Using fleets of electric-drive vehicles for grid support," *Journal of Power Sources*, vol. 168, no. 2, pp. 459-468, 2007.

[12] M. Carrion, J.M. Arroyo and A.J. Conejo, "A bilevel stochastic programming approach for retailer futures market trading," *IEEE Transactions on Power Systems*, vol. 24, no. 3, pp. 1446-1556, 2009.

[13] M. Nazari and A.A. Foroud, "Optimal strategy planning for a retailer considering medium and short-term decisions," *International Journal of Electric Power and Energy Systems*, vol. 45, pp. 107-116, 2013.

[14] Available: http://www.nordpoolspot.com/Market-data1/.

[15] H.M. Ghadikolaei, E. Tajik, J. Aghaei and M. Charwand, "Integrated day-ahead and hour-ahead operation model of discos in retail electricity markets considering DGs and CO_2 emission penalty cost," *Applied Energy*, vol. 95, pp. 174-185, 2012.

15

Low Voltage Ride Through Enhancement Based on Improved Direct Power Control of DFIG under Unbalanced and Harmonically Distorted Grid Voltage

A. R. Nafar Sefiddashti[1], G. R. Arab Markadeh[2*], A. Elahi[1], R. Pouraghababa[3]

[1]Department of Electrical Engineering, Shahrekord University, Shahrekord, Iran

[2]Member of Center of Excellence for Mathematics, Department of Electrical Engineering, Shahrekord University, Shahrekord, Iran

[3]Esfahan Regional Electrical Company

ABSTRACT

In the conventional structure of the wind turbines along with the doubly-fed induction generator (DFIG), the stator is directly connected to the power grid. Therefore, voltage changes in the grid result in severe transient conditions in the stator and rotor. In cases where the changes are severe, the generator will be disconnected from the grid and consequently the grid stability will be attenuated. In this paper, a completely review of conventional methodes for DFIG control under fault conditions is done and then a series grid side converter (SGSC) with sliding mode control method is proposed to enhance the fault ride through capability and direct power control of machine. By applying this controlling strategy, the over current in the rotor and stator windings will totally be attenuated without using additional equipments like as crowbar resistance; Moreover, the DC link voltage oscillations will be attenuated to a great extent and the generator will continue operating without being disconnected from the grid. In addition, the proposed method is able to improve the direct power control of DFIG in harmonically grid voltage condition. To validate the performance of this method, the simulation results are presented under the symmetrical and asymmetrical faults and harmonically grid voltage conditions and compared with the other conventional methods.

KEYWORDS: Doubly-fed induction generator, Fault ride through capability, Sliding mode control, Series compensator.

1. INTRODUCTION

In recent years, wind energy has made the most progress among other energy resources. The more development of wind energy utilization in electric power systems, the poorer performance of the wind farms in the case of fault. The reason is that the disconnection of the wind turbine from the grid causes instability. When a voltage drop occurs in the grid, it affects the connection point of the grid and generator and causes a sudden voltage drop in the stator which consequently increases the rotor current and the DC link voltage of back to back converter [1].

In such circumstances, the generator may be disconnected from the grid. This may; furthermore, leads to the frequency and voltage control problems. Therefore, it is recommended that the wind turbines stay connected and actively help in the stability of the power system during the fault occurrence. In other words, wind turbine should have the fault ride through capability.

With the increasing penetration of wind plants in the power systems to ensure the quality of the grid, several grid regulations have been established. These regulations define the performance limit of wind turbines connected to the grid in terms of power factor, frequency range, voltage variations and the fault ride through capability. To meet the grid regulations, the wind turbines should be capable of neutralizing the additional currents which are the

*Corresponding author:
G. R. Arab Markadeh (arab-gh@eng.sku.ac.ir)

result of the transient voltage drop and continue power transmission to the grid during the short circuit. Fig. 1 presents the grid regulations, according to the voltage changes for different countries [2].

There are many ways to protect and prevent damage to the wind turbine. In order to prevent excessive current damage to the rotor side converter, the generator rotor circuit will be blocked with a series short circuit resistor and the rotor side converter (RSC) which is called the crowbar [3-6]. Since the rotor side converter is di sconnected at the time of the fault, it makes impossible to control the reactive power.

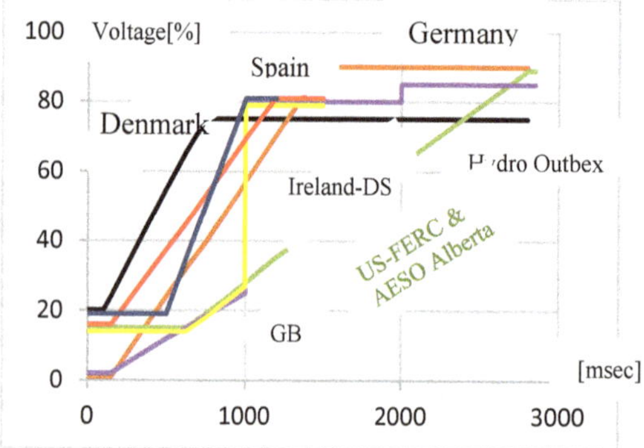

Fig. 1. The voltage through drop requirement and the grid regulations for different countries [2]

Each of the above mentioned references applied different controlling method to control the RSC and the grid side converter (GSC). The oscillations in the rotor and stator current are the result of the DC component of the stator flux in the case of a fault [7]. Therefore, the demagnetization method is applied to control RSC in Ref. [7]. So if the DC component of the rotor current reduces, the stator linkage can be demagnetized and then the oscillation of the current will be attenuated. As well as, a crowbar along with a hysteresis controller is used for controlling. Since there was no reactive power control at the time istantof RSC disconnection, the transient time to get to the desired performance will be reduced by the reduction of the crowbar connection duration time. While in other references, the crowbar is still active at the total fault time or even some time after that.

In Ref. [8] the application of a series resistor in the stator or rotor circuit is proposed to limit the machine currents at fault time. The vector control is used in this reference to control the converters; however, the DC link voltage is not controlled.

Utilizing a series compensator with the stator widingsis another strategy which helps to enhance the performance compared with the other methods [9-13]. In this method, the stator voltages of the disturbed lines will be refined by this compensator and the generator continues operating without getting disconnected from the grid. In Ref. [9], the RSC and GSC are controlled based on vector control method with the aim of controlling the active and reactive powers and limiting the currents in the case of an asymmetric fault. As well as, aproportional+resonant controller is used for the series converter. This structure has to tolerate the total power through of the grid because of the existence of a series converter in the grid. Furthermore, this controlling method is not robust to the parameters' changes because of the application of PI controllers; therefore, it cannot maintain the system stability in the case of fault occurrence.

In Ref. [10], the vector control method is applied for both converters to control the active and reactive power and the DC link voltage. As well as, the series converter is used to inject the desired voltage to the grid voltage in order to balance the stator voltages.

In Ref. [11], the control structure of the back to back converters is the same as [10], but an adaptive fuzzy PI controller is used for series converter. In Ref. [12], using a supercapasitor energy storage system is used to store the rotor energy in fault conditions and return it to the network whenever needed.

In most of the above mentioned references, the DC link voltage regulation and the rotor current limitation were not both concurrently controlled in the case of the symmetric and asymmetric faults.

In Ref. [13], a nine-switchesinverter with two independent output voltage is proposed to be used instead of GSC and series converter. It is known that multi-output inverters increase the voltage stress of DC link capacitor and the switching losses because of the higher switching frequency in comparison with the conventional inverters.

Another method for the rotor current limitation in fault conditions is proposed in Ref. [14], in which the additional energy of the wind turbine is stored in the rotor moment of inertia by increasing the rotor speed with the pitch angle control; however, it is impossible due to the mechanical system's time constant.

A sensitivity analysis approach integrated with a novel hybrid approach combining wavelet transform, particle swarm optimization and an adaptive-network-based fuzzy inference system known as Wavelet-ANFIS-PSO is proposed in Ref. [15] to acquire the optimal control of DFIG based wind generation. In Ref. [16], a multi-objective economic load dispatch model is developed for the system consisting of both thermal generators and wind turbines. Using two optimization methods, sequential quadratic programming and particle swarm optimization, the system are optimally scheduled. The objective functions are total emission and a total profit of units.

In Ref. [17] the sliding mode control method without decomposition of negative and positive sequences of unbalanced variables is used for control of active and reactive power under unbalanced and harmonic distorted grid voltage, but stator voltages and currents and rotor currents are still unbalanced and DC link voltage is not controlled properly.

In Ref. [18] a DC-capacitor current control method is investigated for a grid-side converter to eliminate the negative impact ofunbalancedgrid voltage on the DC-capacitor. Although Vdc fluctuations is reduced, the oscillation and unbalance of other variables still exist.

In this paper, a series compensator, placed in the DFIG stator, will be used to control the rotor and stator currents in the case of symmetric and asymmetric fault. Using this compensator, there is no need for a crowbar. Moreover, for the robustness of the controller structure respect to the changes in the machine parameters, the sliding controller will be used to control the series compensator converter and RSC and GSC converters. This compensator is directly fed by the DC link between the two RSC and GSC converters and at the same time, it has the fault ride through capability in the case of symmetric and asymmetric faults. Due to the rapid response rate of the controller, there would not be any oscillations in the rotor and stator currents in the case of different types of fault. Furthermore, the DC link voltage regulation and control of active and reactive power and torque is well achieved.

The present paper has the following structure. In Section II, Introducing different DFIG structures with additional equipment in fault ride through. In Section III, the generator has been modeled. In Section IV, all three converters have been investigated using sliding mode control. In Section V, simulation results using this method of control are compared with vector control method presented in Ref. [9] and finally, the general conclusion of this paper is presented in the last Section.

2. Fault Ride Through Control Structures

As it was stated in the previous parts, different topologies exist for the fault ride through control. In one of these structures, while a fault happens in the grid, the rotor side converter will be disconnected and the rotor windings will have a short circuit with a series of external resistors known as a crowbar. An example of this structure is shown in Fig. 2 [4]. In many papers, the three-phase crowbar structure with different controlling methods is simultaneously used by RSC and GSC for the fault ride through [5, 6].

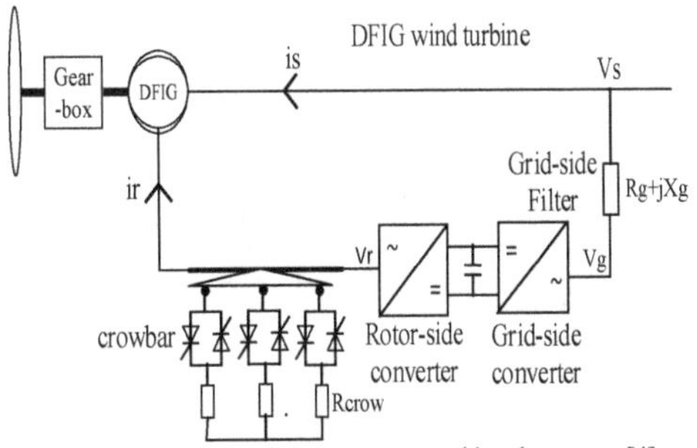

Fig. 2. Crowbar three-phase structure and its placement [4]

To reduce the rating power of crowbar switches, the various structures of crowbar connections are presented in Ref. [19]. In this reference, the allowable voltage of each switch and the current of each switch in the Delta connection will be reduced to 0.577 by the use of Wye connection (Fig. 3).

Another topology proposed in Ref. [20] for the fault ride through is a switching resistive load which is paralleled with the DC link capacitor. This structure prevent the DC link voltage to reach to an unauthorized value in the fault condition (Fig. 4).

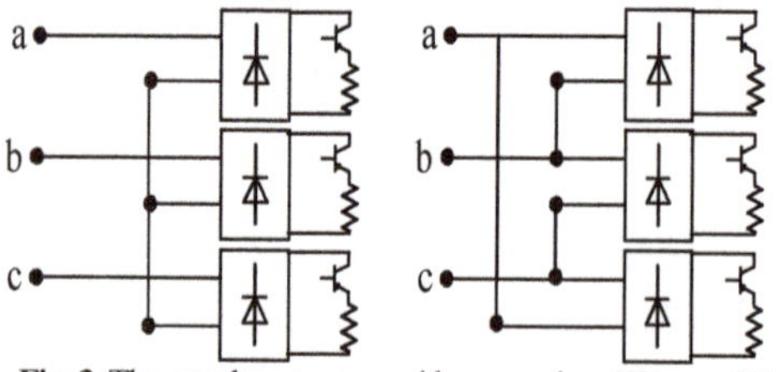

Fig. 3. The crowbar structure with connection a)Wye and b) Delta [19]

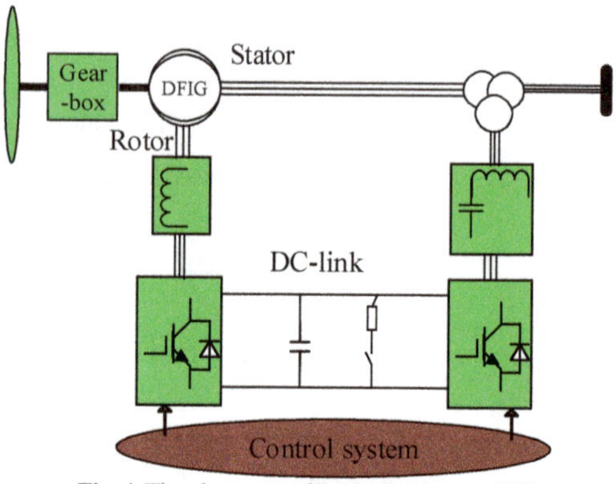

Fig. 4. The placement of DC brake chopper [20]

Another structure, proposed in Ref. [21], is the simultaneous use of DC brake chopper and crowbar. As shown in Fig. 5, this strategy will be capable of limiting the rotor current and the DC link voltage to an authorized limit.

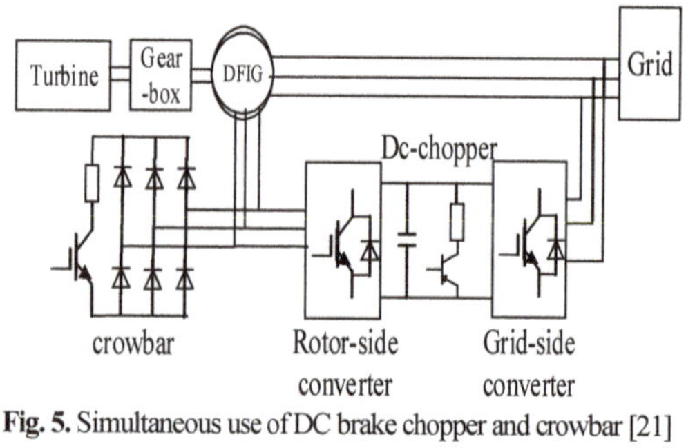

Fig. 5. Simultaneous use of DC brake chopper and crowbar [21]

In Ref. [8], limiting the stator and rotor currents in the fault condition has been achieved through the application of a controllable resistor in the stator circuit. As presented in Fig. 6, when the bidirectional switch is turned off, the resistor enters the circuit and is able to limit the rotor and stator currents.

One of the other methods for the fault ride through control is using a series compensator to inject a voltage in the stator circuit in fault condition [9], [22]. This structure in some research works known as dynamic voltage restore (DVR) and is depicted in Fig. 7. In [9], the DC link capacitor of DVR is an independent capacitor while in Ref. [22], a common capacitor is used for RSC, GSC and DVR.

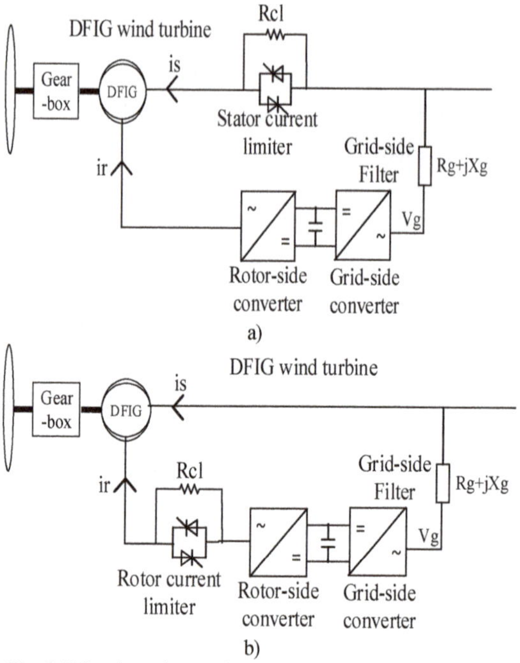

Fig. 6. Using the series crowbar structure in a) stator, b) rotor [8]

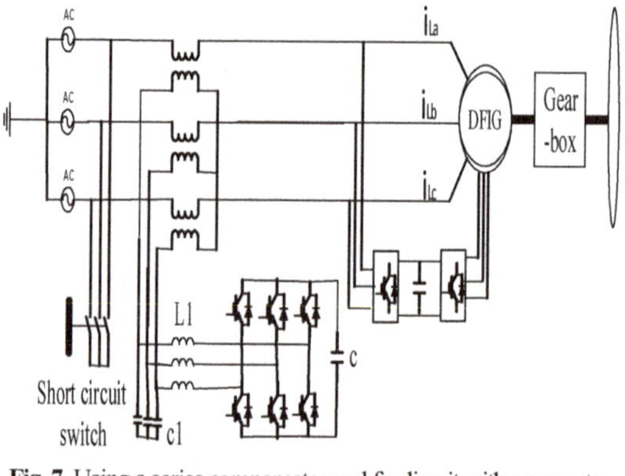

Fig. 7. Using a series compensator and feeding it with a separate capacitor [9]

Besides the structures which are based on the series compensator, another structure has been proposed in Ref. [12] in which not only the series compensator is fed by the DC link capacitor, but also another storage capacitor or battery, parallel with the DC link capacitor is also considered to storge the additional energy absorbed from the RSC in the fault condition. In the other words, any electrical generated energy in fault condition will be stored in this supercapacitor and after fault clearing, the energy will be transferred to the grid. Fig. 8 presents the structure of this compensator.

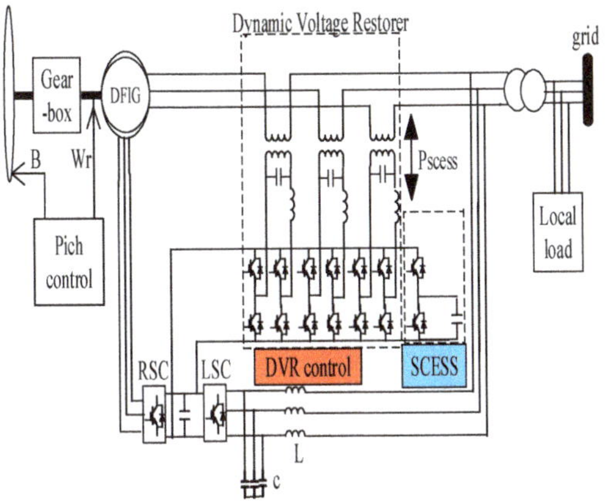

Fig. 8. The series compensator along with the energy storage capacitor (battery) [12, 24]

3. MATHEMATICAL MODEL

The equivalent circuit of a DFIG in the stationary reference frame is presented in Fig. 9. In this circuit, all the rotor parameters are transferred to the stator side. The rotor and stator voltage equations are as follows.

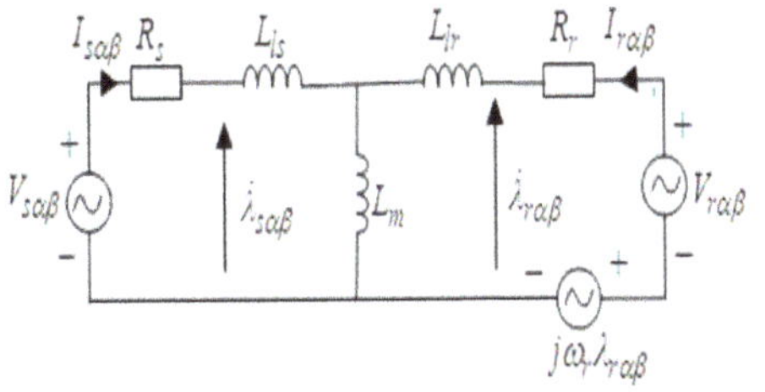

Fig. 9. The equivalent circuit of the induction machine in the stationary reference frame

$$\vec{v}_{s\alpha\beta} = R_s\vec{i}_{s\alpha\beta} + \frac{d\vec{\Psi}_{s\alpha\beta}}{dt} \tag{1}$$

$$\vec{v}_{r\alpha\beta} = R_r\vec{i}_{r\alpha\beta} + \frac{d\vec{\Psi}_{r\alpha\beta}}{dt} j\omega_r\vec{\Psi}_{r\alpha\beta} \tag{2}$$

where R_s and R_r are stator and rotor resistances.

The flux linkage of the stator and rotor will be obtained from the Eqs. (3) and (4).

$$\vec{\Psi}_{s\alpha\beta} = L_s\vec{i}_{s\alpha\beta} + L_m\vec{i}_{r\alpha\beta} \tag{3}$$

$$\vec{\Psi}_{r\alpha\beta} = L_r\vec{i}_{r\alpha\beta} + L_m\vec{i}_{s\alpha\beta} \tag{4}$$

where L_s, L_r and Lm are the stator, rotor and the magnetizing inductances respectively.

By instituting the Eqs. (3) and (4) in Eqs. (1) and (2) and the calculation of the space phasor of the stator currents, we have:

$$\frac{d\vec{i}_{s\alpha\beta}}{dt} = \frac{1}{L_r'}[L_r\vec{V}_{s\alpha\beta} - L_m\vec{V}_{r\alpha\beta} - R_sL_r\vec{i}_{s\alpha\beta} \tag{5}$$

$$-j\omega_rL_m(L_r\vec{i}_{r\alpha\beta} + L_m\vec{i}_{s\alpha\beta}) + L_mR_r\vec{i}_{r\alpha\beta}]$$

In the above equation, $L_r' = L_sL_r - L_m^2$.

Equation (6) presents the relation between active and reactive power of the stator with machine variables.

$$\begin{cases} P_s = \frac{3}{2}(v_{\alpha s}i_{\alpha s} + v_{\beta s}i_{\beta s}) \\ Q_s = \frac{3}{2}(v_{\beta s}i_{\alpha s} - v_{\alpha s}i_{\beta s}) \end{cases} \tag{6}$$

4. CONTROL STRUCTURE

4.1. DFIG vector control

As it was mentioned in section one, one of the ways for the fault ride through is the application of the series compensator on the grid. In Ref. [9], a vector control method has been used to control the active and reactive powers of the stator. Furthermore, to control the DC link voltage and the GSC reactive power, the vector control method is used. The series converter control has been achieved using a proportional+ resonant controller. To control the series converter, first the voltage phase of the grid will be identified through a phase lock loop by which the positive and negative sequences of the grid voltage will be obtained. Now, the series converter sets the amount of the injected voltage to each phase through the obtained values. By this method, a safe and normal voltage is achievable. Since the PI controller is used for controlling in this reference, this controller is not tolerant of the parameter's changes. The controlling block diagram used is shown in Appendix C.

4.2. SMC of the RSC

RSC is generally used to control the active and reactive power of the stator and the rotor speed. In this paper, the active and reactive powers are controlled by RSC.

Considering the controlling variables of Ps and Qs for RSC, the sliding surface will be considered with integral type.

$$\begin{aligned} S_{P_S} &= e_{P_S} + k_{P_S}\int e_{P_S}\,dt\,, e_{P_S} = P_S^* - P_S \\ S_{Q_S} &= e_{Q_S} + k_{Q_S}\int e_{Q_S}\,dt\,, e_{Q_S} = Q_S^* - Q_S \end{aligned} \tag{7}$$

k_{P_S} and k_{Q_S} are positive constants. Moreover, e_{P_S} and e_{Q_S} are the errors of active and reactive powers of the stator. By the derivation of the sliding surface with respect to time in Eq. (7), we have:

$$\begin{aligned} \dot{S}_{P_S} &= \dot{P}_S^* - \dot{P}_S + k_{P_S}(P_S^* - P_S) \\ \dot{S}_{Q_S} &= \dot{Q}_S^* - \dot{Q}_S + k_{Q_S}(Q_S^* - Q_S \end{aligned} \tag{8}$$

$\dot{P}_S$ and $\dot{Q}_S$ can be obtained based on Eq. (6) as follows:

$$\begin{cases} \dfrac{dP_s}{dt} = \dfrac{3}{2}(i_{\alpha s}\dfrac{dv_{\alpha s}}{dt} + i_{\beta s}\dfrac{dv_{\beta s}}{dt} + v_{\alpha s}\dfrac{di_{\alpha s}}{dt} + v_{\beta s}\dfrac{di_{\beta s}}{dt}) \\ \dfrac{dQ_s}{dt} = \dfrac{3}{2}(i_{\alpha s}\dfrac{dv_{\beta s}}{dt} + v_{\beta s}\dfrac{di_{\alpha s}}{dt} - i_{\beta s}\dfrac{dv_{\alpha s}}{dt} - v_{\alpha s}\dfrac{di_{\beta s}}{dt}) \end{cases} \tag{9}$$

In the case of a balanced voltage source, we have:

$$\begin{cases} V_{\alpha s} = V_m \cos(\omega_s t) \\ V_{\beta s} = V_m \sin(\omega_s t) \end{cases} \tag{10}$$

Therefore, the voltage deviation respect to time obtained as fallows:

$$\begin{cases} \dfrac{d}{dt}V_{\alpha s} = -\omega_s V_m \sin(\omega_s t) = -\omega_s V_{\beta s} \\ \dfrac{d}{dt}V_{\beta s} = \omega_s V_m \cos(\omega_s t) = \omega_s V_{\alpha s} \end{cases} \tag{11}$$

By separating Eq. (5) to real and imaginary parts and substituting Eq. (11) in Eq. (9) can be rewritten as:

$$\begin{aligned} \frac{d}{dt}\begin{pmatrix} P_s \\ Q_s \end{pmatrix} &= \frac{3L_m}{2L_r'}(\begin{pmatrix} -V_{\alpha s} & -V_{\beta s} \\ -V_{\beta s} & V_{\alpha s} \end{pmatrix}\begin{pmatrix} V_{\alpha r} \\ V_{\beta r} \end{pmatrix} \\ &+ \begin{pmatrix} R_r V_{\alpha s} - \omega_r L_r V_{\beta S} & R_r V_{\beta s} + \omega_r L_r V_{\alpha s} \\ R_r V_{\beta s} + \omega_r L_r V_{\alpha S} & -R_r V_{\alpha s} + \omega_r L_r V_{\beta S} \end{pmatrix}\begin{pmatrix} i_{\alpha r} \\ i_{\beta r} \end{pmatrix} \\ &+ \begin{pmatrix} \dfrac{L_r L_r'}{L_m}(V_{\alpha s}^2 + V_{\beta s}^2) \\ 0 \end{pmatrix}) + \begin{pmatrix} -\dfrac{R_s L_r}{L_r'} & -(\omega_r L_m^2 + \omega_s) \\ -(\omega_r L_m^2 + \omega_s) & -\dfrac{R_s L_r}{L_r'} \end{pmatrix}\begin{pmatrix} P_s \\ Q_s \end{pmatrix} \end{aligned} \tag{12}$$

By substituting Eq. (12) in Eq. (8) the $\dot{\boldsymbol{S}}$ yields:

$$\frac{d}{dt}\begin{pmatrix} S_{PS} \\ S_{QS} \end{pmatrix} = -\frac{d}{dt}\begin{pmatrix} P_s \\ Q_s \end{pmatrix} + \begin{pmatrix} k_{PS}(P_s^* - P_s) \\ k_{QS}(Q_s^* - Q_s) \end{pmatrix} = F + DV_{r\alpha\beta} \tag{13}$$

Finally the control input can be expressed as:

$$\begin{pmatrix} V_{\alpha r} \\ V_{\beta r} \end{pmatrix} = -D^{-1}(F + \begin{pmatrix} k_{PS1} sign(S_{PS}) \\ k_{QS1} sign(S_{QS}) \end{pmatrix}) \tag{14}$$

4.3. SMC of the GSC

The controlling objectives of the grid side converter control are: 1) DC link voltage control and 2) the grid side reactive power control. To design the GSC controller, first a sliding surface, as Eq. (15), should be considered to attain the controlling objectives.

$$\begin{aligned} S_{V_{dc}} &= \dot{e}_{V_{dc}} + \lambda_1 e_{V_{dc}} + \lambda_2 \int e_{V_{dc}}\, dt, \\ e_{V_{dc}} &= V_{dc}^* - V_{dc} \\ S_{Q_g} &= e_{Q_g} + \lambda_{Q_g} \int e_{Q_g}\, dt, \\ e_{Q_g} &= Q_g^* - Q_g \end{aligned} \tag{15}$$

By the derivation of the sliding surface, we will have:

$$\begin{aligned} \dot{S}_{V_{dc}} &= -\ddot{V}_{dc} - \lambda_1 \dot{V}_{dc} + \lambda_2(V_{dc}^* - V_{dc}) \\ \dot{S}_{Q_g} &= \dot{Q}_g^* - \dot{Q}_g + \lambda_{Q_g}(Q_g^* - Q_g) \end{aligned} \tag{16}$$

λ_1, λ_2 and λ_{Q_g} are positive constants. The electrical equations of this converter in the stationary reference frame are as follows:

$$CV_{dc}\frac{dV_{dc}}{dt} = P_g - P_r \tag{17}$$

In the above equation,

$$P_g = \frac{3}{2}(e_{n\alpha} i_{g\alpha} + e_{n\beta} i_{g\beta}) \tag{18}$$

$$Q_g = \frac{3}{2}(e_{n\beta} i_{g\alpha} - e_{n\alpha} i_{g\beta}) \tag{19}$$

$$\vec{e}_n = \vec{U}_{gn} + R_g \vec{i}_g + L_g \frac{d\vec{i}_g}{dt} \tag{20}$$

P_g and Q_g are the active and reactive power of the grid, respectively and $\vec{e}_n$ is the voltage space phasor of the grid. The relation between the voltages of each inverter base and the two-axial frame of $\alpha\beta$ is as follows:

$$\overbrace{\begin{bmatrix} V_{g\alpha} \\ V_{g\beta} \end{bmatrix}}^{V_g} = M_2 M_1 U_g \tag{21}$$

$$M_1 = \frac{1}{3}\begin{bmatrix} 2 & -1 & -1 \\ -1 & 2 & -1 \\ -1 & -1 & 2 \end{bmatrix},\; M_2 = \frac{2}{3}\begin{bmatrix} 1 & \dfrac{-1}{2} & \dfrac{-1}{2} \\ 0 & \dfrac{\sqrt{3}}{2} & -\dfrac{\sqrt{3}}{2} \end{bmatrix}$$

In the above relation, M_1 is the transformation matrix between U_r and the rotor winding voltage, M_2 is the Clark transformation matrix.

By the derivation of Eqs. (18) and (19) and separating the real and imaginary sequences of the space phasor derivation of the grid current from the relation Eq. (20) and instituting in Eq. (16), we have:

$$\overbrace{\begin{bmatrix} \dot{S}_{V_{dc}} \\ \dot{S}_{Q_g} \end{bmatrix}}^{\dot{S}_{V_{dc}Q_g}} = \overbrace{\begin{bmatrix} F_{V_{dc}} \\ F_{Q_g} \end{bmatrix}}^{F_{V_{dc}Q_g}} - \overbrace{\frac{3}{2L_g}\begin{bmatrix} \dfrac{-e_{n\alpha}}{CV_{dc}} & \dfrac{-e_{n\beta}}{CV_{dc}} \\ -e_{n\beta} & e_{n\alpha} \end{bmatrix}}^{K}\begin{bmatrix} v_{g\alpha} \\ v_{g\beta} \end{bmatrix} \tag{22}$$

In relation Eq. (22), $F_{V_{dc}Q_g}$ is a function of state variables, the entrance reference values and their derivations; however, it is not related the controlling entrances of $v_{g\alpha}$ and $v_{g\beta}$. By instituting Eq. (21) in Eq. (22), we have:

$$\dot{S}_{V_{dc}Q_g} = F_{V_{dc}Q_g} - \overbrace{K \underbrace{M_2 M_1}_{M}}^{C_{V_{dc}Q_g}} U_g \qquad (23)$$

Since $\dot{S}_{V_{dc}Q_g}$ varies with time, a new sliding surface is used to obtain the control rule which is related to the previous sliding surface ($S_{V_{dc}Q_g}$) as follows [25, 26]:

$$S^*_{V_{dc}Q_g} = C^+_{V_{dc}Q_g} S_{V_{dc}Q_g} \qquad (24)$$

In this relation, $C^+_{V_{dc}Q_g}$ is the Moor-Penrose pseudo-reverse matrix for $C_{V_{dc}Q_g}$. Similar to the rotor side controller, the control rule for the grid side parallel converter is:

$$U_g = \mathrm{u}_0 \operatorname{sgn}(S^*_{V_{dc}Q_g}) \qquad (25)$$

4.4. SMC of the SGSC

The main objective in the series converter control is the injection of the balanced or imbalanced voltage (or both) to the grid so that the stator voltages and as a consequence, stator and rotor currents get balanced. A balanced voltage in the stator reduces the oscillations of the electromagnetic torque and other under control variables. Therefore, the following sliding surfaces can be defined:

$$\begin{cases} s_{v_\alpha} = \lambda_{v_\alpha} \int e_{v_\alpha}\, dt \;, e_{v_\alpha} = v^*_{inj_\alpha} - v_{inj_\alpha} \\ s_{v_\beta} = \lambda_{v_\beta} \int e_{v_\beta}\, dt \;, e_{v_\beta} = v^*_{inj_\beta} - v_{inj_\beta} \end{cases} \qquad (26)$$

λ_{v_α} and λ_{v_β} are the positive constants. Assuming that $R_{gs} = 0$, the relation between the injected voltage in line and the voltage of each inverter leg in the stationary reference frame will be obtained from Eq. (27).

$$\begin{aligned} V_{inj_\alpha} &= V_{conv_\alpha} \\ V_{inj_\beta} &= V_{conv_\beta} \end{aligned} \qquad (27)$$

The reference voltages for the series converter will be obtained from the following relations:

$$\begin{cases} v^*_{inj_\alpha} = e_{n\alpha} - v^*_{s\alpha} \\ v^*_{inj_\beta} = e_{n\beta} - v^*_{s\beta} \end{cases} \qquad (28)$$

In the above relation, $v^*_{s\alpha}$ and $v^*_{s\beta}$ are the stator reference voltages, $v^*_{inj_\alpha}$ and $v^*_{inj_\beta}$ are the reference voltages of the grid side series converter in the stationary reference frame and $e_{n\alpha}$ and $e_{n\beta}$ are the desired voltages for the grid. By the derivation of the relation Eq. (26) and replacing from Eq. (27) in terms of v_{conv_α} and v_{conv_β}, we will have:

$$\begin{bmatrix} \dot{s}_{v_\alpha} \\ \dot{s}_{v_\beta} \end{bmatrix} = \begin{bmatrix} \lambda_{v_\alpha} v^*_{inj_\alpha} \\ \lambda_{v_\beta} v^*_{inj_\beta} \end{bmatrix} - \overbrace{\begin{bmatrix} \lambda_{v_\alpha} & 0 \\ 0 & \lambda_{v_\beta} \end{bmatrix}}^{T} \begin{bmatrix} v_{conv_\alpha} \\ v_{conv_\beta} \end{bmatrix} = \mathrm{A} + \mathrm{T} \begin{bmatrix} v_{conv_\alpha} \\ v_{conv_\beta} \end{bmatrix} \qquad (29)$$

Therefore, the control rule for the switching of the series converter can be obtained as follows:

$$\begin{bmatrix} v_{conv_\alpha} \\ v_{conv_\beta} \end{bmatrix} = \frac{-1}{T} [A + \begin{bmatrix} \lambda_{v_{\alpha 1}} sign(s_{v_\alpha}) \\ \lambda_{v_{\beta 1}} sign(s_{v_\beta}) \end{bmatrix}] \qquad (30)$$

5. SIMULATION RESULTS

In this section, the simulation results for a 2 MW DFIG are presented (Appendix A). The controllers' coefficients are presented in Appendix B Simulation is performed in five tests.

Test 1: Three-phase symmetric fault which reduces the PCC voltage to 95% of the nominal grid voltage with the proposed method.

Test 2: Two-phase asymmetric fault which reduces the voltage to 63% of the nominal grid voltage at the PCC point with method used in Ref. [9].

Test 3: Two-phase asymmetric fault which reduces the voltage to 95% of the nominal grid voltage at the PCC point with the proposed method.

Test 4: DFIG with SGSC is connected to a network with voltage harmonic amplitudes: V5=6% and V7=5%.

Test 5: The robustness of the proposed control algorithm against parameter variations is cheked.

In Fig. 10, the overall scheme of the system investigated in this paper is presented. In Appendix, the system parameters are provided. Since most faults in the control systems are of the asymmetric type, the simulation results of both symmetric and asymmetric faults are presented here.

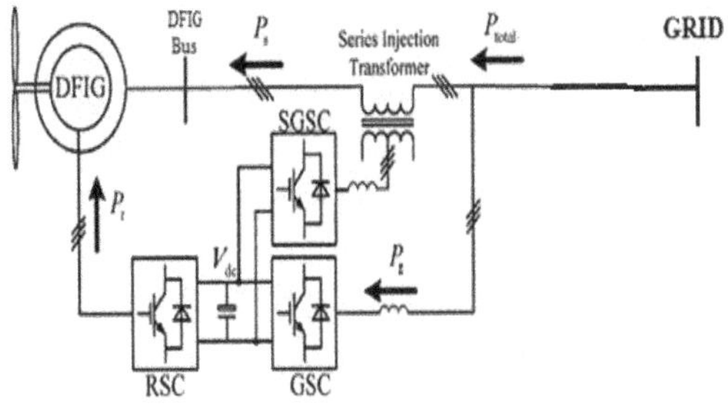

Fig. 10. the overall scheme of the proposed system

Test 1

In Fig. 11, the simulation results for the three-phase symmetric fault which reduces the voltage to 95% of the nominal grid voltage at the PCC point is presented. As it is shown in the wave shape of the PCC voltage (Fig. 11.a), the fault happens at 1.7 s which takes 150 ms time. The stator and rotor currents continue without any fluctuation at the fault time (Fig. 11.b). The DC link voltage reaches to the desired value of 1200V after a small oscillation. It is worth mentioning that while the above mentioned points are desirable in the control strategy, the active and reactive powers and the electromagnetic torque will be successfully controlled as well (Fig. 11. e, f and g, respectively). As the results indicate, the proposed strategy limits the current amplitude at a desired value even at fault condition and the DC link voltage is regulated. At the same time, the torque and power control is well achieved.

Test 2

The simulation results with the structure proposed in Ref. [9] are presented in Fig. 12. As it can be seen a two-phase fault with the grid voltage nominal value of 63% at the 1.7 s causes the voltage drop and the fault continues for 150 ms. Stator and rotor voltages have oscillations even after fault clearing to reach to the desired values (Fig 12. b and c). In this control method, the active and reactive power control has been achieved; however, as it is shown in Fig. 12. d, they have oscillations at the fault time. As well as, the DC link voltage control has not been done in this reference.

Test 3

Figure 13 is the simulation results in the two-phase fault condition which the voltage drop in two phases is 95% of the nominal voltage at the PCC point. The results for this type of faults have been attained without any change in the controller and system structure. As it is shown in Fig. 13. a, the fault is occurred at 1.7 s and takes 150 ms time. The stator and rotor currents will continue without any oscillation at the fault time (Fig. 13. b and c). The DC link voltage, with a small oscillation at the start of the fault time instant, tends to its reference value of 1200 V. Similar to the symmetric fault, the control of active and reactive powers and the electromagnetic torque is possible (Fig. 13 e, f and g).

Test 4

In this test, DFIG simulation is presented with harmonic distorted grid voltage. It is assumed that the network voltage, in addition to fundamental harmonic, has fifth and seventh order of harmonics with 8% THD. The simulation results are shown in Fig. 14. Since the resultant rotating magnetic field of the fifth order harmonic is in the opposite direction of rotating field caused by fundamental harmonic, the existence of this harmonic order creates breaking torque on machine shaft which will reduce shaft lifetime and increase T_e ripple.

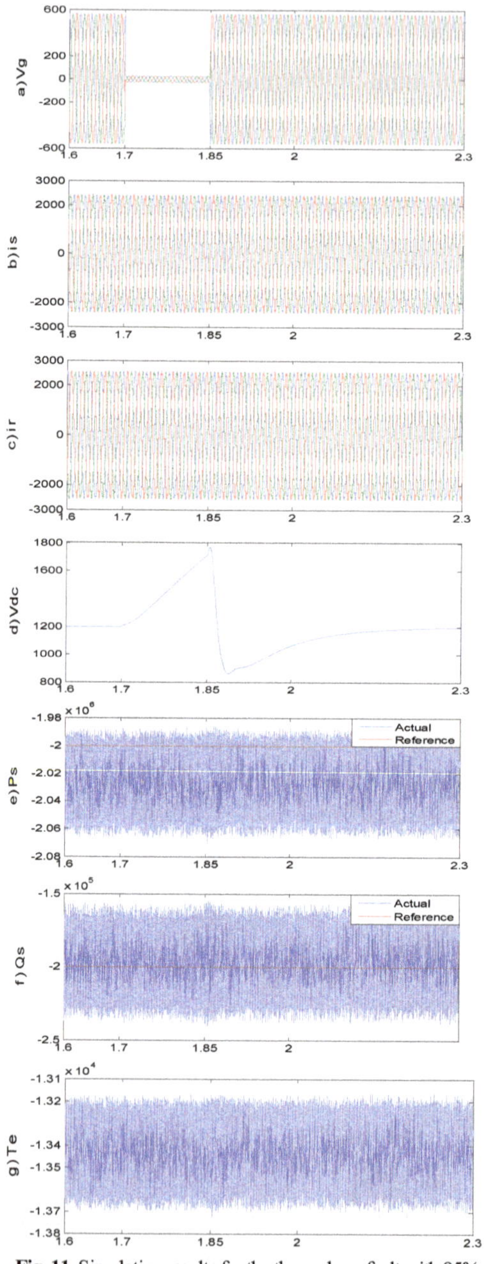

Fig. 11. Simulation results for the three-phase fault with 95% decreasing of nominal grid voltage a) grid voltage, b) stator current, c) rotor current, d) DC link voltage, e) stator's active power, f) stator's reactive power and g) electromagnetic torque.

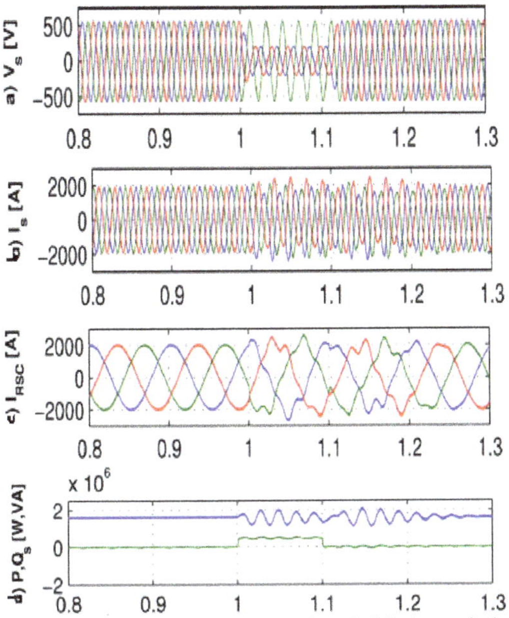

Fig. 12. Simulation results for the two-phase fault in the nominal value of 63%. a) grid voltage, b) stator current, c) rotor current and d) stator active and reactive powers [9].

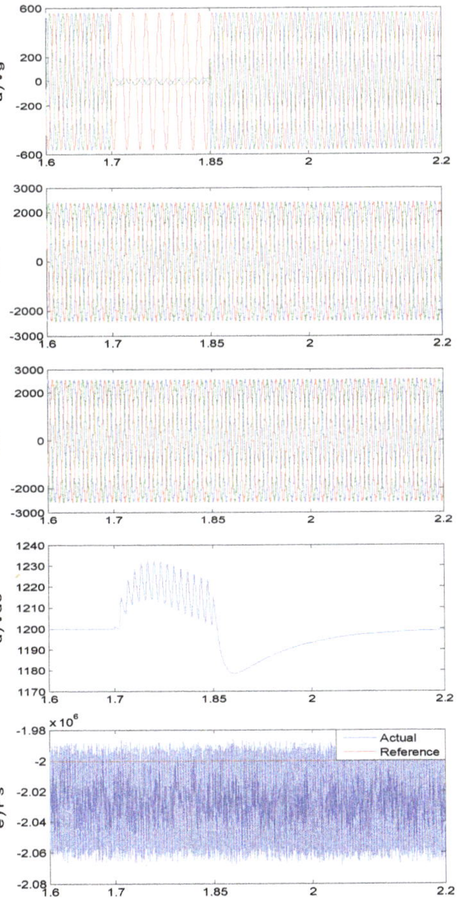

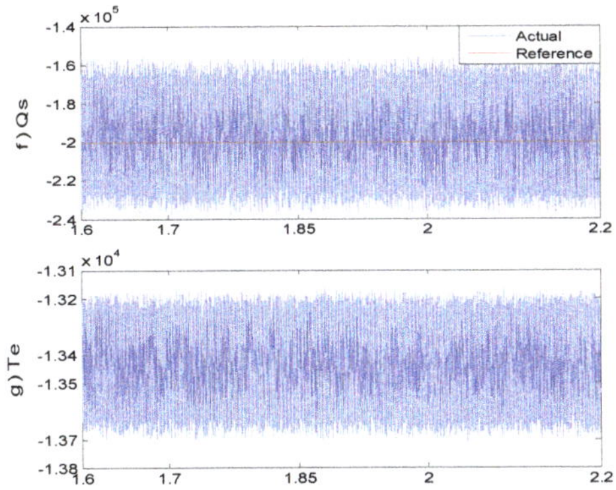

Fig. 13. Simulation results for the two-phase fault in the nominal value of 95%. a) grid voltage, b) stator current, c) rotor current, d) DC link voltage, e) stator's active power, f) stator's reactive power and g) is the electromagnetic torque.

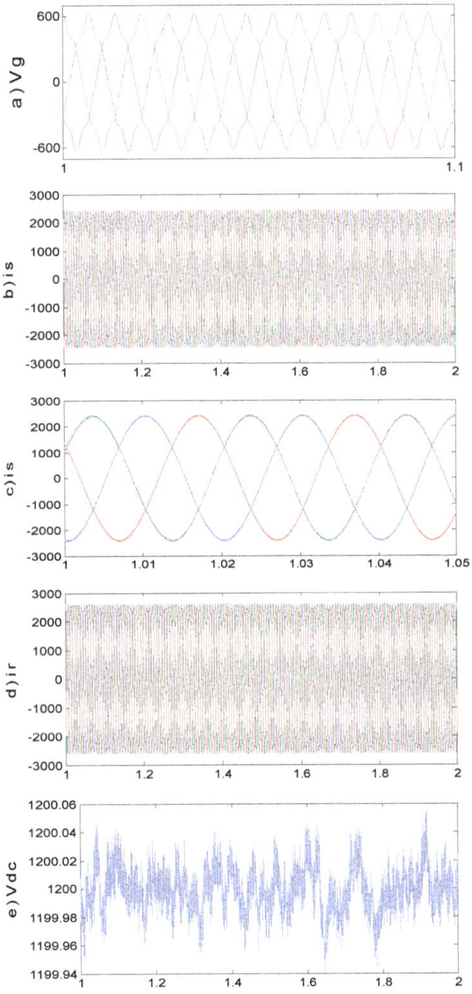

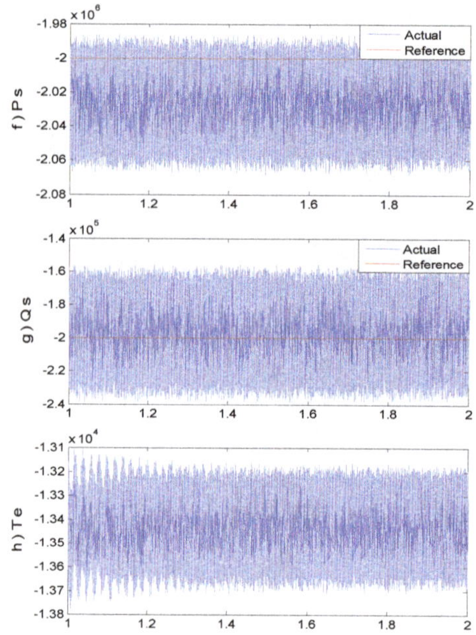

Fig. 14. Simulation results under harmonic distorted grid voltage. a) grid voltage, b) stator current, c) zooming stator current, d) rotor current,e) DC link voltage, f) stator's active power, g) stator's reactive power and h) electromagnetic torque.

Figure 14. c and h show that the problem of breaking torque and torque ripple are resolved by removing the fifth and seventh order harmonics. The stator and rotor currents in the three-phase windings are balanced while achieving the aforementioned object. (Fig. 14. b, d).

Test 5

In Fig. 15, the simulation results, similar to the two-phase are presented. The only difference is that the rotor and stator resistance is increased up to 40%. As the results show, except for the electromagnetic torque, the parameter's change does not affect the system performance.

In order to show the effectiveness of the proposed method, some control methods in different grid faults are compared and their results are reported in Appendix D. The amplitude of the stator and rotor over currents and the DC link voltage are calculated from their corresponding Refs. [9-13].

According to the results presented in Table. C, the voltage drop rate of the grid in the proposed method is considered higher than the other references, and has a better performance compared with the other published researches.

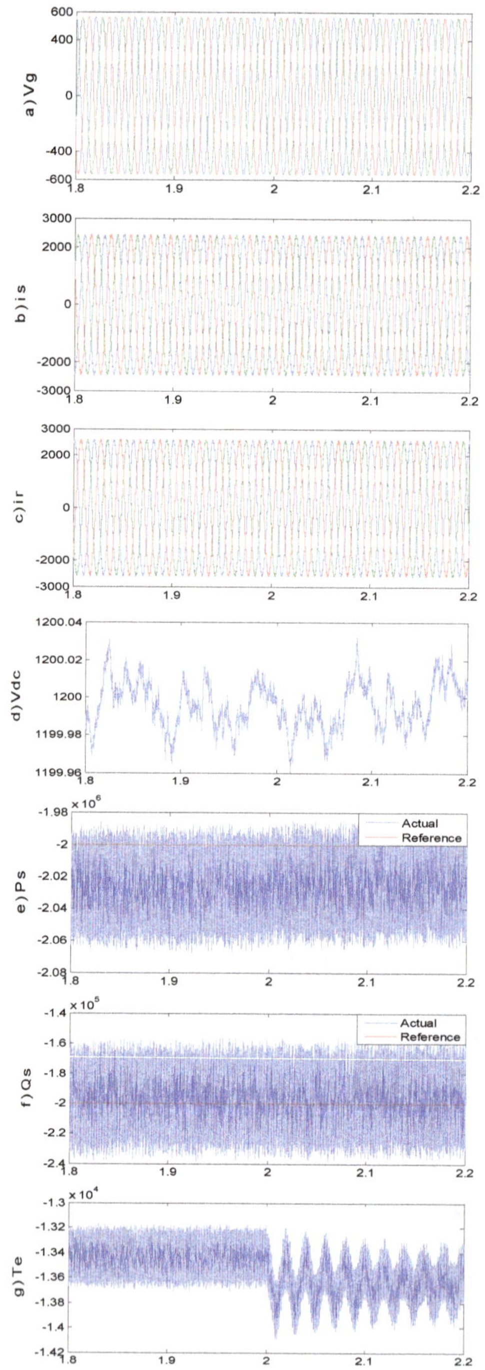

Fig. 15. Simulation results under parameter variations at t=2s. a) grid voltage, b) stator current, c) rotor current, d) DC link voltage, e) stator's active power, f) stator's reactive power and g) electromagnetic torque.

6. CONCLUSION

Since the grid regimes urge wind turbine generators to remain connected to the grid during the voltage drop conditions, a method to achieve the fault ride through condition is the application of series compensators to stator voltage correction. In this paper, sliding mode control method has been proposed to control the back-to-back RSC and GSC and the series compensator converter. In addition, the DC link voltage and the active and reactive output powers of the generator are controlled and limits the rotor and stator over currents without using the crowbar resistors at the time of fault. Furthermore, this control method has an acceptable performance in all fault conditions without any change in controller's coefficients.

AS well as, the sliding mode method is robust to generator parameter variations due to its variable structure nature. Finally an effective comparison between the proposed method and some conventional techniques is presented. Simulation results show the effectiveness of the proposed technique.

APPENDIX

Table A.1. DFIG Parameters

Parameter	Value
Rs	10 mΩ
Rr	2.9 mΩ
Lls	77.306 μH
Llr	83.369 μH
Lm	3.7 mH
Vdc*	1200 V
Rg	0 Ω
C	7.5 mF
Lg	0.25 mH
Stator voltage (rms)	690 V
Rated Power	2 MW
P (pole pairs)	3

Table A.2. Controller coefficients

Parameter	Value
$k_{P_{s1}}$	10
$k_{Q_{s1}}$	10
k_{P_s}	20000
k_{Q_s}	20000
λ_1	1
λ_2	1
λ_{v_α}	1
λ_{v_β}	1
λ_{Qg}	1
$\lambda_{v_{\alpha 1}}$	100
λ_{v_β}	100

ACKNOWLEDGMENT

The authors would like to acknowledge the partial support of Esfahan Regional Electrical Company from this research work .

Table A.3. Comparing different structures for fault ride through control

FRT Scheme	Location series converter	Fault Type	The voltage drop	Over current rotor	Over current stator	DC- Link Over voltage
vector control and adaptive fuzzy for series converter [11]	grid	Tree phase	10%	60%	85%	4%
vector control and SCESS[12]	Stator terminal	Single phase	93%	100%	100%	Not checked
		Three phase	93%	105%	105%	Not checked
Vector control and Nine switch converter [13]	Stator terminal	Single phase	33%	5%	Not checked	10%
		Two phase	33%	5%	Not checked	10%
		Three phase	80%	10%	5%	10%
vector control and resonance for DVR[9]	grid	Two phase	63%	30%	20%	Not checked
vector control for all converter[10]	grid	Three phase	90%	18%	Not checked	4%
Proposed method With sliding mode control	Stator terminal	Two phase	95%	0%	0%	11.1%
		Three phase	95%	0%	0%	48%
		Grid voltage harmonic	V5= 6% and V7=5%	0%	0%	0%

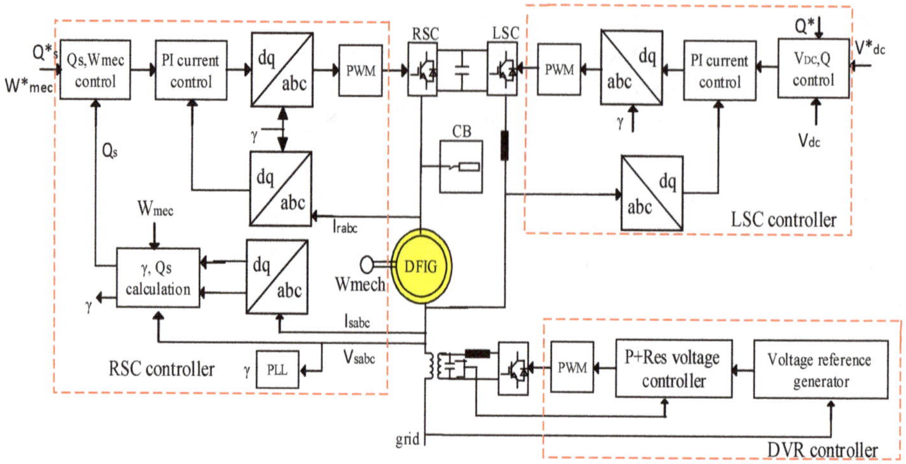

Fig. A. Fault ride through control based on vector control and dynamic voltage restore [9]

REFERENCES

[1] H.T. Jadhav and R. Roy, "A comprehensive review on the grid integration of doubly fed induction generator," *International Journal of Electrical Power and Energy Systems*, vol. 49, pp. 8-18, 2013.

[2] M. Tsili and S. Papathanassiou, "A review of grid code technical requirementsfor wind farms," *IET Proceedings on Renewable Power Generation.*, vol. 3, no. 3, pp. 308-332, 2009.

[3] A. Geniusz, S. Engelhardt and J. Kretschmann, "Optimised fault ride through performance for wind energy systems with doubly fed induction generator," in Proceedings of the *European Wind Energy Conference & Exhibition, Brussels, pp.* 1-9, 2008.

[4] M. Rahimi and M. Parniani, "Grid-fault ride-through analysis and control of wind turbines with doubly fed induction generators," *Electric Power Systems Research*, vol. 80, no. 2, pp. 184-195, 2010.

[5] A. H. Kasem, E. F. El-Saadany, H. H. El-Tamaly and M. A. A. Wahab, "An improved fault ride-through strategy for doubly fed induction generator-based windturbines," *IET Proceedings on Renewable Power Generation*, vol. 2, no. 4, pp. 201-214, 2008.

[6] J. Vidal, G. Abad, J. Arza and S. Aurtenechea, "Single-phase DC crowbar topologies for low voltage ride through fulfillmentof high-power doubly fed induction generator-based wind turbines," *IEEE Transactions on Energy Conversion*, vol. 28, no. 3, pp. 768-781, 2013.

[7] L. Peng and Y. Li, "Improved crowbar control strategy of DFIG based wind turbines for grid fault ride-through," *IEEE Transactions on Industrial Electronics*, vol. 40, no. 1, pp. 1932-1938, 2009.

[8] M. Rahimi and M. Parniani, "Efficient control scheme of wind turbines with doubly fed induction generators for low voltage ride-through capability enhancement," *IET Proceedings on Renewable Power Generation*, vol. 4, no. 3, pp. 242-252, 2010.

[9] C. Wessels, F. Gebhardt and F. Wilhelm Fuchs, "Fault ride-through of a DFIG wind turbine using a dynamic voltage restorer during symmetrical and asymmetrical grid faults," *IEEE Transactions on Power Electronics,* vol. 26, no. 3, pp. 807-815, 2011.

[10] O. Abdel, B. Nasiri and A. Nasiri, "Series voltage compensation for DFIG wind turbine low-voltage ride-through solution," *IEEE Transactions on Energy Conversion*, vol. 26, no. 1, pp. 272-281, 2011.

[11] E. El-Hawatt, M.S. Hamad, K.H. Ahmed and I.F. El Arabawy, "Low voltage ride-through capability enhancement of a DFIG wind turbine using a dynamic voltage restorer with adaptive fuzzy PI controller," *in Proceedings of the International Conference on Renewable Energy Research and Applications*, Spain, pp. 1234-1239, 2013.

[12] I. Spyros, G. kavanoudis and C. S. Demoulias, "FRT capability of a DFIG in isolated grids with dynamic voltage restorer and energy storage," *in proceedings of the IEEE 5th International Symposium on Power Electronics for Distributed Generation Systems (PEDG),* pp. 1-8, 2014.

[13] B. B. Ambati, P. Kanjiya and V. Khadkikar, "A low component count series voltage compensation scheme for DFIG WTs to enhance fault ride-through capability," *IEEE Transactions on Energy Conversion*, vol. 30, no. 1, pp. 1-10, 2015.

[14] L. Yang, Z. Xu, *J.* Østergaard, Z.Y. Dong and K. P. Wong, "Advanced control strategy of DFIG wind turbines for power system fault ride through," *IEEE Transactions on Power Systems*, vol. 27, no. 2, pp. 713-722, 2012.

[15] M. Darabian, A. Jalilvand and R. Noroozian, "Combined use of sensitivity analysis and hybrid wavelet-psoanfis to improve dynamic performance of DFIG-based wind generation," *Journal of Operation and Automation in Power Engineering,*

vol. 2, no. 1, pp. 49-59, 2014.

[16] H. Khorramdel, B. Khorramdel, M. Tayebi Khorrami and H. Rastegar, "A multi-objective economic load dispatch considering accessibility of wind power with here-and-now approach," *Journal of Operation and Automation in Power Engineering*, vol. 2, no. 1, pp. 60-73, 2014.

[17] M.I. Martinez, G. Tapia, A. Susperregui and H. Camblong, "Sliding-mode control for DFIG rotor and grid-side converters under unbalanced and harmonically distorted grid voltage," *IEEE Transactions on Energy Conversion*, vol. 27, no. 2, pp. 328-339, 2012.

[18] L. Changjin, X. Dehong, Z. Nan, F. Blaabjerg and Ch. Min, "DC-voltage fluctuation elimination through a DC-capacitor current control for DFIG converters under unbalanced grid voltage conditions," *IEEE Transactions on Power Electronics,* vol. 28, no.7, pp. 3206-3218, 2013.

[19] J. Vidal, G. Abad, J. Arza and S. Aurtenechea, "Single-phase DC crowbar topologies for low voltage ride through fulfillment of high-power doubly fed induction generator-based wind turbines," *IEEE Transactions on Energy Conversion*, vol. 28, no. 3, pp. 768-781, 2013.

[20] G. Pannell, B. Zahawi, D.J. Atkinson and P. Missailidis, "Evaluation of the performance of a DC-link brake chopper as a DFIG low-voltage fault-ride-through device," *IEEE Transactions on Energy Conversion*, vol. 28, no. 3, pp. 535-542, 2013.

[21] M. Wang, W. Xu, H. Jia and X. Yu, "A new method for DFIG fault ride through using resistance and capacity crowbar circuit," *in Proceedings of the 2013 IEEE International Conference on Industrial Technology*, pp. 2004-2009, 2013.

[22] P. Cheng and H. Nian, "An improved control strategy for DFIG system anddynamic voltage restorer under grid voltage dip," *in Proceedings of the 2012 IEEE International Symposium on Industrial Electronics,* pp.1868 - 1873, 2012.

[23] S. Zhang, K.J. Tseng, S.S. Choi, T.D. Nguyen and D. L. Yao, "Advanced control of series voltage compensation to enhance wind turbine ride through," *IEEE Transactions on Power Electronics*, vol. 27, no. 2, pp. 763-772, 2012.

[24] P.S. Flannery and G. Venkataramanan, "Evaluation of voltage sag ride-through of a doubly fed induction generator wind turbine with series grid side converter," *in Proceedings of the IEEE Power Electronics Specialists Conference,* pp. 1839-1845, 2007.

[25] V. Utkin, J. Guldner and J. Shi, "Sliding mode control in electromechanical systems," *London, U.K., Taylor and Francis,* 1999.

[26] V. Utkin, "Sliding mode control design principles and applications to electric drives," *IEEE Transaction on Industrial Electron*ics, vol. 40, no. 1, pp. 23-36, 1993.

A Multi-Objective Economic Load Dispatch Considering Accessibility of Wind Power with Here-And-Now Approach

H. Khorramdel*, B. Khorramdel, M. Tayebi Khorrami, H. Rastegar

Department of Electrical Engineering, Safashahr Branch, Islamic Azad University, Safashahr, Iran

ABSTRACT

The major problem of wind turbines is the great variability of wind power production. The dynamic change of the wind speed returns the quantity of the power injected to networks. Therefore, wind–thermal generation scheduling problem plays a key role to implement clean power producers in a competitive environment. In deregulated power systems, the scheduling problem has various objectives than in a traditional system which should be considered in economic scheduling. In this paper, a Multi-Objective Economic Load Dispatch (MOELD) model is developed for the system consisting of both thermal generators and wind turbines. Using two optimization methods, Sequential Quadratic Programming (SQP) and Particle Swarm Optimization (PSO), the system is optimally scheduled. The objective functions are total emission and total profit of units. The probability of stochastic wind power is included in the model as a constraint. This strategy, referred to as the Here-and-Now (HN) approach, avoids the probabilistic infeasibility appearing in conventional models. Based on the utilized model, the effect of stochastic wind speed on the objective functions can be readily assessed. Also a Total Index (TI) is presented to evaluate the simulation results. Also, the results show preference of PSO method to combine with HN approach.

KEYWORDS: Economics load dispatch, PSO and SQP algorithm, Wind turbine.

1. INTRODUCTION

In recent years, a growing interest in renewable energy resources has been observed. In particular, wind and solar energy are non-depletable, site-dependent, non-polluting, and constitute potential sources of alternative energy options. Due to the impeding demand of mitigating the greenhouse effect, the share of Wind Power Generation (WPG) in the total utility is daily on the increase [1]. Some European countries like Denmark and Germany are making very ambitious plans to increase the share of WPG up to 50% of the national electricity demand in the near future [2]. Electric power, generated by wind turbines, is highly erratic; therefore, the wind energy penetration in electrical power systems can lead to problems related to system operation and the planning of electrical power systems. Wind power intermittency, load mismatch, and negative impacts on grid voltage stability are some key problems which should be solved [3]. One of the major challenges associated with the generation scheduling is the way that it accommodates large amount of wind power generation. Hence, the Wind–Thermal Generation Scheduling (WTGS) problem plays an essential role to implement clean power producers in such competitive environment [4, 5]. In the literature, various approaches have been proposed to describe the impact of random parameters on electrical power systems. Numerous solutions have been proposed to solve the optimal programming problems [6,7], such as Priority List (PL), Dynamic Programming (DP), Lagrangian Relaxation (LR), Genetic Algorithm (GA), Mixed Integer Programming (MIP), Evolutionary Programming (EP), Immune Algorithm (IA), Artificial Immune System (AIS) and Particle Swarm Optimization

*Corresponding author:
H. Khorramdel (E-mail: hossein.khorramdel@gmail.com)

(PSO). In [8], Muller method was introduced to solve Economic load Dispatch (ELD) problem and Information Pre-Prepared Power Demand (IPPD) table was introduced to solve combinatorial sub problem for deregulated environment. In nodal ant colony optimization [9], to maintain the good exploitation and exploration search capabilities, the movements of the ants are represented with a search space consisting of optimal combination of binary nodes for unit on/off status. In [10], Delarue achieved the difference between the obtained profits when using perfect price forecast and without using perfect price forecast. From the literature survey, it is observed that most of the existing algorithms have some limitations to provide the qualitative solution. The first work in the minimization of emission dispatch has been done by Gent and Lamont [11]. Also, ref. [12] presented a PBUC formulation using GA which considers the softer demand constraints and allocates fixed and transitional costs to the scheduled hours. A new formulation to the Unit Commitment (UC) problems suitable for an electric power producer in deregulated markets was proposed in [13]. In addition, a hybrid LR-EP method was explored in [14] that helps Generation Companies (GENCOs) to make a decision on how much power and reserve should be sold in markets, and how to schedule generators in order to receive maximum profit by incorporating both power and reserve generation at the same time. The same problem is presented in [15] in addition to the line flow constraints to minimize the emission. Reference [16] employed an auxiliary hybrid model to solve the PBUC problem with evolutionary programming used to update the Lagrangian multiplier. The application of PSO technique to maximize the GENCOs profit is illustrated in [17]. The common ELD problem can be also presented by SQP technique by assigning weighting factors for generation and emission cost functions; the above method was proposed by [18]. Conventional ELD models need to be enhanced to characterize the stochastic behavior of wind power. In this paper, MOELD model that takes the Probability Density Function (PDF) of wind as one of the constraints is presented. One of the basic approaches to estimate the PDF has been based on Monte Carlo simulation. The convolution method was another common approach to estimate the PDF of solutions [19]. All of these approaches tried to find probabilistic characteristics of solutions of the problem under investigation. This kind of approach is called the Wait-and-See (WS) strategy in the context of Stochastic Programming (SP) [20]. In contrast, the Here-and-Now (HN) strategy introduces the probabilistic characteristics to the problem model itself, which introduces the CDF of parameters to constraints. Both WS and HN strategies are representative approaches in the discipline of SP. This paper is in line with HN approach. In the context of optimal power flow with wind power generation, there are also several representative works. The model presented in [21] is an ELD model with the objective function of the total generation cost of traditional units. The planning horizon of simulations was divided into five stages, and each stage was 30 minutes. Later this model and power flow analysis were extended in [22], where the costs of expected surplus WP and expected deficit WP were added to the objective function. A recent comprehensive review can be found in [23], where the authors described the representative models of ELD with WPG and also discussed risk management strategies in the power market. For the convenience of presentation, throughout this paper, WP means the real electric power generated by WPG units rather than the input wind power. The rest of this paper is organized as follows. In Sec. 2, an ELD model with WP is introduced. In sec. 3, we use the probability distribution of WP to the constraint. Then, Sec. 4 describes the two models of HN approach. Simulation results for a ten-generator system are reported in Sec. 5. Finally, remarks and conclusions are included in Sec. 6.

2. ELD MODEL WITH WP

In electrical power systems, the generic ELD problem takes the following form [2]:

$$Y = \sum_{i=1}^{n} (a_i + b_i p_i + c_i p_i^2) \tag{1}$$

$$P_{min,i} \leq P_i \leq P_{min,i} \qquad (i = 1,2,3,\ldots, n) \tag{2}$$

$$\sum_{i=1}^{n} P_i = P_d + P_s \tag{3}$$

The unit of P_i is megawatts (MW), then the units of a, b, and c are, respectively, \$/h, \$/MWh, and \$/MW2h. Consequently, the unit of Y is \$/h. In numerical analysis, usually per unit (P.U.) system is employed, in which the base is 100 MVA. In the present work, we introduce a new MOELD model to minimize the fuel cost and emission and maximize profit, taking the stochastic WP as a constraint. The proposed model will add a set of constraints:

$$0 \le W_j \le w_{jr} \qquad (j = 1,2,3,\ldots,m) \tag{4}$$

where, Wj and wjr are the real power and rated power generated by WPG unit jth, respectively.
Also, equation (3) can be replaced with (5).

$$\sum_{i=1}^{n} P_i + \Psi(W) = P_d + P_s \tag{5}$$

where, $\Psi(W)$ is a function of random variable (RV) W.

3. PROBABILITY OF WIND POWER

The wind speed V (m/s) is an RV. A comprehensive review for probability distributions of wind speed can be found in [24], where the authors cited more than two hundred publications and described more than ten well-known distributions. They indicated that the two-parameter Weibull distribution had become the most widely accepted model and had been included in regulatory works as well as several popular computer modeling packages. The CDF of Weibull distribution is:

$$F_V(v) = 1 - exp\left[-\left(\frac{v}{c}\right)^c\right] \qquad (v \ge 0) \tag{6}$$

where, $c > 0$ and $k > 0$ are referred to as the scale factor and shape factor, respectively. Note that there are two special cases. The cases of $k = 1$ and $k = 2$ lead to the exponential distribution and the Rayleigh distribution, respectively. In the literature, most studies adopted $k = 2$. Corresponding to its CDF, the PDF of V is:

$$f_V(v) = \frac{k}{c}\left(\frac{v}{c}\right)^{k-1} exp\left[-\left(\frac{v}{c}\right)^k\right] \tag{7}$$

$$W = \begin{cases} 0 & (V < v_{in} \text{ or } V \ge v_{out}) \\ w_r & (v_r \le V < v_{out}) \\ \frac{(V - v_{in})}{v_r - v_{in}} w_r & (v_{in} \le V < v_r) \end{cases} \tag{8}$$

The relation between the input wind power and the output electric power system relies on several factors, such as the efficiencies of generator, wind rotor, gearbox, and inverter, depending on what type of power generation unit is investigated. For a generic WPG unit, some researchers [25] used a simplified model to characterize the relation between the WP and wind speed (8). We will adopt the above model in our ELD model. According to the probability theory for function of RVs [26], in the interval $v_{in} < V < v_r$, the PDF of W is:

$$f_W(w) = \frac{khv_{in}}{w_r c}\left[\frac{\left(1 + \frac{hw}{w_r}\right)v_{in}}{c}\right]^{k-1} \times exp\left\{-\left[\frac{\left(1 + \frac{hw}{w_r}\right)v_{in}}{c}\right]^k\right\} \tag{9}$$

Where, $h = (v_r/v_{in}) - 1$. The CDF of W, however, must take into account the piecewise linear properties shown in (8). The probability of event W =0 and $W = w_r$ are:

$$Pr(W = 0) = Pr(V < v_{in}) + Pr(V \ge v_{out}) = 1 - exp\left[-\left(\frac{v_{in}}{c}\right)^k\right] + exp\left[-\left(\frac{v_{out}}{c}\right)^k\right] \tag{10}$$

$$Pr(W = w_r) = Pr(v_r \le V < v_{out}) = exp\left[-\left(\frac{v_r}{c}\right)^k\right] - exp\left[-\left(\frac{v_{out}}{c}\right)^k\right] \tag{11}$$

For the continuous part, the integration of (9) is:

$$\varphi_W(w) = 1 - exp\left\{-\left[\frac{\left(1 + \frac{hw}{w_r}\right)v_{in}}{c}\right]^k\right\} \tag{12}$$

Furthermore:

$$Pr(W > w_r) = 0 \tag{13}$$

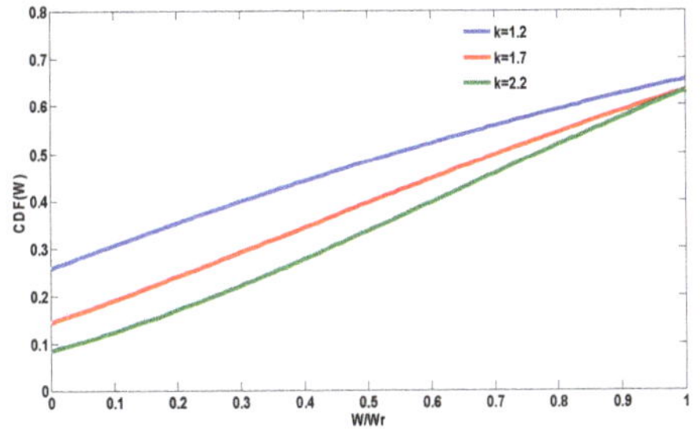

Fig. 1. Examples of cdf of WP.

According to (10-13), the CDF of W is shown in (14.1-14.3). The reader is reminded that the derivation of (14.1-14.3) has followed several axioms in the probability theory [26], including the continuity from the right. Three examples of the CDF of W are illustrated in Fig. 1, where the values of factor k are specified. Since the CDF notion includes both continuous and discrete probabilities, the overall height of CDF is affected by the probability of (14.1-14.3):

$$F_W(w) = Pr(W \le w) = 0 \qquad (w < 0) \tag{14.1}$$

$$F_W(w) = Pr(W \le w) = 1 - exp\left\{ -\left[\frac{\left(1 + \frac{hw}{w_r}\right) v_{in}}{c} \right]^k \right\} + exp\left[-\left(\frac{v_{out}}{c}\right)^k \right] \qquad (0 \le w < w_r) \tag{14.2}$$

$$F_W(w) = Pr(W \le w) = 1 \qquad (w \ge w_r) \tag{14.3}$$

4. TWO MODELS OF HN APPROACH

In this section, we describe and solve two ELD models constrained by the probabilistic metric. The first model, ELD-EQ, has a closed-form solution, which is helpful to gain some fundamental insights. The second model, ELD-INEQ, includes more constraints and has no closed-form solution [2].

4.1. ELD model with equality constraints

In this subsection, we consider the model, referred to as ELD-EQ, which consists of (1) and the following constraint:

$$Pr\left(W + \sum_{i=1}^{n} P_i \le P_d + P_s \right) = Pa \tag{15}$$

where, W represents all WP to be dispatched, and P_a is a specified threshold representing the tolerance that the total demand P_d plus power losses cannot be satisfied. For example, if P_a=0.15, then up to 15% of the chance of insufficient supply could be tolerated. Therefore, a larger P_a implies more tolerance toward insufficient supply, and vice versa. To avoid degenerated results, Pa is chosen such that $\Pr(W = 0) \le P_a < 1$. Since the total WP is characterized by a single RV here, it implies that all wind turbines are located in a coherent geographic area, represented by a small wind farm or a cluster of turbines in a large wind farm. Accordingly, constraint (15) can be rewritten as follows.

$$F_W\left(P_d + P_s - \sum_{i=1}^{n} P_i\right) = Pr\left(W \le P_d + P_s - \sum_{i=1}^{n} P_i\right) = P_a \tag{16}$$

Substituting (14) into (16), for $0 \le w < w_r$ equations 17 and 18 are obtained. The above inequality can be easily converted into expression (18), where h_p, is the penetration factor of WP and defined by (19).

$$Pr\left(W \le P_d + P_s - \sum_{i=1}^{n} P_i\right) = 1 + exp\left[-\left(\frac{v_{out}}{c}\right)^k \right] - exp\left\{ -\frac{1}{w_r^k c^k}\left[v_{in} w_r + (v_r - v_{in})\left(P_d + p_s - \sum_{i=1}^{n} P_i\right)\right]^k \right\} = p_a \tag{17}$$

$$\sum_{i=1}^{n} P_i = P_d + P_s + \frac{v_{in} w_r}{v_r - v_{in}} - \frac{w_r c}{v_r - v_{in}} \left\{ -ln\left[1 + exp\left(-\frac{v_{out}^k}{c^k} \right) - p_a \right] \right\}^{1/k} = P_d + P_s + \frac{v_{in} w_r}{v_r - v_{in}} - \frac{w_r c}{v_r - v_{in}} \left| ln\left[1 + exp\left(-\frac{v_{out}^k}{c^k} \right) - p_a \right] \right|^{1/k} = P_d + P_s - w_r h_p \tag{18}$$

$$h_p = \frac{c}{v_r - v_{in}} \left| ln\left[1 + exp\left(-\frac{v_{out}^k}{c^k} \right) - p_a \right] \right|^{1/k} - \frac{v_{in}}{v_r - v_{in}} \tag{19}$$

In (19), note that

$$1+exp\left(-\frac{v_{out}^k}{c^k}\right)-p_a<1 \quad (20)$$

$$p_a \geq Pr\,(W=0)=1-exp\left(-\frac{v_{in}^k}{c^k}\right)+exp\left(-\frac{v_{out}^k}{c^k}\right) \quad (21)$$

As a result, constraint (15) in model ELD-EQ becomes:

$$\sum_{i=1}^{n} P_i = P_d + P_s - w_r h_P \quad (22)$$

Finally, the solution of ELD_EQ is as follows [2]:

$$P_{optm,\,j} = \frac{2(P_d + P_s - w_r h_P) + \sum_{i=1}^{n}\left(\frac{b_i}{c_i}\right)}{2c_j \sum_{i=1}^{n}\left(\frac{1}{c_i}\right)} - \frac{b_j}{2c_j}, \quad (j = 1, 2, 3, \ldots, n) \quad (23)$$

For a system consisting of ten thermal generators and one wind farm, C=15, V_{in} =5, V_{out} =45, V_r=15, w_r = 1(p.u), P_s = 0.5 (P.u) are chosen.

4.2. ELD model with inequality constrains

In this subsection, we consider the model, referred to as ELD_INEQ, which consists of (1), (2), and the following constraints [2]:

$$Pr\left(W + \sum_{i=1}^{n} P_i \leq P_d + P_s\right) \leq p_a \quad (24)$$

Similar to Model ELD_EQ, constraint (24) can be converted into the following expression:

$$\sum_{i=1}^{n} P_i \geq P_d + P_s - w_r h_P \quad (25)$$

Where h_P was defined in (19). Note that model ELD_INEQ involves two sets of inequality constraints. Therefore, the classic lagrange multiplier method cannot be directly applied [2]. Therefore, a numerical optimization procedure is needed. Thus, we have developed a computer program to solve Model ELD_INEQ and implemented it in MATLAB. The minimum and maximum value of produced active power for units are 0.03 p.u. and 1.5 p.u., respectively.

4.2.1. Minimization of total emission (stage 1)

Minimization of emission is one important issue with regard to economic and optimal operation of electrical power system. Consequently, using the probability of wind turbine output, the cost function of stage 1 is considered as minimization of emission. It is expressed in the following formula.

$$Min\ EC_i = \alpha_i + \beta_i(P_{it}) + \gamma_i(P_{it})^2 \quad (26)$$

Where α_i , β_i and γ_i are the emission co-efficient of ith unit. The objective function is subjected to the following constraints. Where P_{it} is the output power of ith unit at hour *t*.

4.2.2. Maximization of total profit (stage 2)

Maximization of profit is very important issue with regard to ELD and optimal operation of electrical power systems. As ELD model plays key role in electrical power systems in terms of cost and revenue, its effects should be considered in many electrical power system scheduling. Therefore, the cost function of stage 2 is considered as the maximization of total profit. It is expressed in the following formula with equations (27) to (30) [3].

$$Max\ PF = (RV - TC) \quad (27)$$

$$RV = \sum_{t=1}^{T}\sum_{i=1}^{N} P_{it}.SP_i \quad (28)$$

$$TC = \sum_{t=1}^{T}\sum_{i=1}^{N} FC_i(P_{it}) \quad (29)$$

$$FC_i(P_{it}) = a_i + b_i(P_{it}) + c_i(P_{it})^2 \quad (30)$$

4.2.3. Multi-objective optimization problem (stage 3)

In the third stage, the optimization algorithm minimizes multi-objective cost function (MCF) using the results of two previous stages. In [27, 28] presented a multi-objective mathematical programming to find the best reactive power control strategy in a microgrid with uncertainty of wind farms. Based on the concept of this algorithm, in this paper an MCF is proposed to minimize total emission and maximize the total profit, which can be written by (31).

$$Min.\,MCF = \sqrt{\alpha\left(\frac{\text{Emission}}{\text{Emission}^*}\right)^2 + \beta\left(\frac{\text{Profit}^*}{\text{Profit}}\right)^2} \quad (31)$$

In stage 1, the total emission is minimized with setting $\alpha = 1, \beta = 0$; and in the stage 2, the total profit is maximized when $\alpha = 0, \beta = 1$; Then, a compromise programming is employed in the third stage with $\alpha = 1, \beta = 1$; which is designed to minimize emission and maximize profit. Emission* and profit* are the minimum and maximum amount of emission and profit respectively. Fig. 2 shows the flowchart of SQP based optimization algorithm.

PSO is an evolutionary computational algorithm derived from a natural system. On a given iteration, a set of particles or solutions move around the search space in consecutive iterations. The movement rules of particles are expressed in [29]. Because of abilities of PSO algorithm, in this method, the three stages are combined and solved simultaneously which may result in a better global optimum.

To apply the PSO algorithm in ELD, the following steps should be taken:

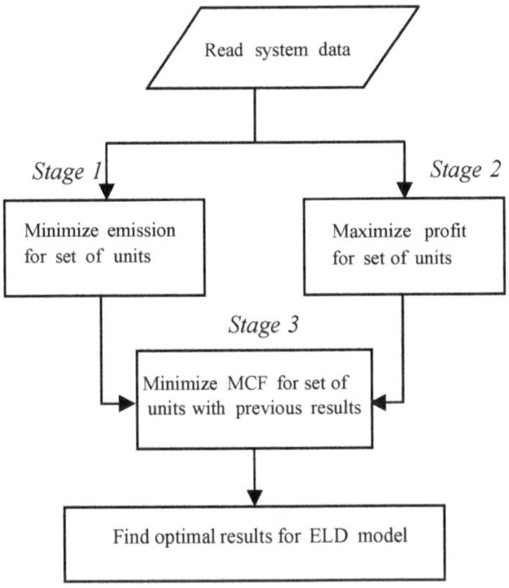

Fig. 2. The flowchart of the proposed SQP based optimization algorithm.

Step 1: The input data should be specified.

Step 2: The initial population and initial velocity for each particle should randomly be produced.

Step 3: The objective functions should be calculated for each individual.

Step 4: The value of objective functions should be normalized in accordance with related fuzzy membership function.

Step 5: The Minimum value of normalized Objective Functions (MOF) should be chosen for each individual as ith row of MOF matrix.

Step 6: The individual that has the maximum value of MOF should be selected as global position (Gbest).

Step 7: The ith individual is selected.

Step 8: The best local position (Pbest) for the ith individual is the individual with the minimum value for the ith row of MOF matrix.

Step 9: The modified velocity and position for each individual should be calculated based on expressed movement rules in [29].

Step 10: If all individuals are chosen, go to next step, otherwise i =i+1 and go to the ***Step*** *7*.

Step 11: If the current iteration is the maximum iteration number, PSO is stopped, otherwise go to ***Step*** *3*.

The last Gbest is selected as optimal solution. The proposed PSO algorithm optimizes the generated active power of units. Finally, a Total Index (TI) is computed in accordance with (32) to find the better solution. This index is defined to distinguish between two methods of ELD. As there should be low emission and high profit (low cost) in a power system, the smallest value for TI expresses that optimal situation is dominated in the grid.

$$TI = \sqrt{\left(\frac{\text{Emission}}{\text{Emission}_{\min}}\right)^2 + \left(\frac{\text{Cost}}{\text{Cost}_{\min}}\right)^2} \tag{32}$$

5. SIMULATION RESULTS

In order to demonstrate the accuracy and effectiveness of the used algorithm, it is applied to IEEE 39-bus test system [3].

The PSO and SQP formulation and solution methodology has been implemented using MATLAB 7.10 and executed on a corei5 (2.53 GHz) personal computer with 4 GB RAM, and average computing time is around 4 minutes. The control parameters of PSO algorithm are simply adjusted as following:

$c_1=c_2=2,\ w=0.9-((0.5)/iter_{max})*iter$

This work analyzed the impact of wind power on generation scheduling problem with the test system consists of ten thermal units and one wind farm to solve a multi objective problem. If there is *N* number of units in the system, some of them have high fuel cost and other generating units have low fuel cost. Therefore, the GENCOs decide to save production cost by starting up the units with low fuel cost over a period of scheduling.

Before economic load dispatch, the GENCOs want to get an accurate hourly demand and price forecast for the period of scheduling horizon.

Developing the forecasted data is an important matter, but it is beyond the scope of this paper. For the results existing in this section, the forecasted load and price are taken as shown in Figs. 3 and 4, respectively. The amount of base load and peak load of the system is 700MW at 01:00 am and 1300MW at 11:00 am, respectively. In addition to the forecasted hourly price and demand, which are shown in table 1 and the generator parameters listed in table 2, ELD program needs the parameters of each generating unit.

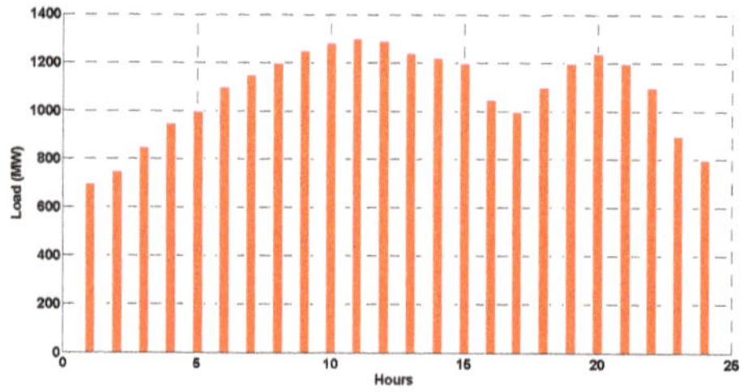

Fig. 3. Base load and peak load unit operating cycles [3].

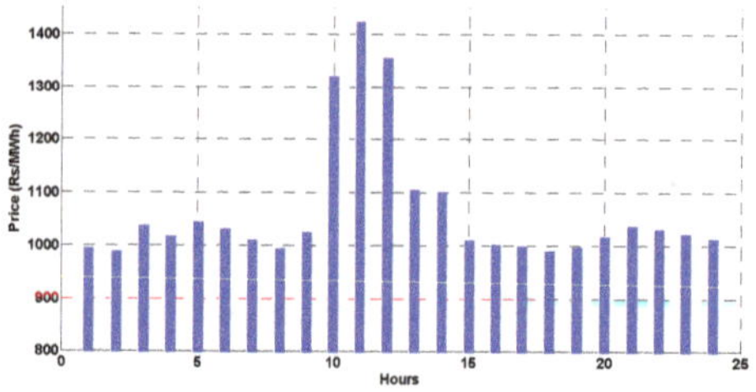

Fig. 4. Forecasted prices for 10 generator units [3].

Table 1. Forecasted demand and prices (10 units) [3].

Hour	Load	Price	Hour	Load	Price
h	MW	Rs/MWh	h	MW	Rs/MWh
1	700	996.75	13	1240	1107.00
2	750	990.00	14	1220	1102.50
3	850	1039.50	15	1200	1012.50
4	950	1019.25	16	1050	100.35
5	1000	1046.25	17	1000	1001.25
6	1100	1032.75	18	1100	992.25
7	1150	1012.50	19	1200	999.00
8	1200	996.75	20	1240	1019.25
9	1250	1026.00	21	1200	1039.50
10	1280	1320.75	22	1100	1032.75
11	1300	1424.25	23	900	1023.75
12	1290	1356.75	24	800	1014.75

Emissions co-efficient of coal-fired, petroleum and natural gas power plants are quite different. It is assumed that conventional thermal units are coal-fired because of low operating cost. The operating data for 10- unit case is shown in Tables 2 and 3.

Table 2. Operating parameters of units [3].

Units	P_i(Max)	P_i(Min)	a_i	b_i	c_i
U-1	455	150	1000	16.19	0.00048
U-2	455	150	970	17.26	0.00031
U-3	130	20	700	16.60	0.00200
U-4	130	20	680	16.50	0.00211
U-5	162	25	450	19.70	0.00398
U-6	80	20	370	22.26	0.00712
U-7	85	25	480	27.74	0.00079
U-8	55	10	660	25.92	0.00413
U-9	55	10	665	27.27	0.00222
U-10	55	10	670	27.79	0.00173

Table 3. Generator emission coefficients [3].

Units	$\alpha_i(ton/h)$	$\beta_i(ton/MWh)$	$\gamma_i(ton/MW^2h)$
U-1	10.33908	-0.024444	0.00312
U-2	10.33908	-0.024444	0.00312
U-3	30.03910	-0.406950	0.00509
U-4	30.03910	-0.406950	0.00509
U-5	32.00006	-0.381320	0.00344
U-6	32.00006	-0.381320	0.00344
U-7	33.00056	-0.390230	0.00465
U-8	33.00056	-0.390230	0.00465
U-9	33.00056	-0.395240	0.00465
U-10	36.00012	-0.398640	0.00470

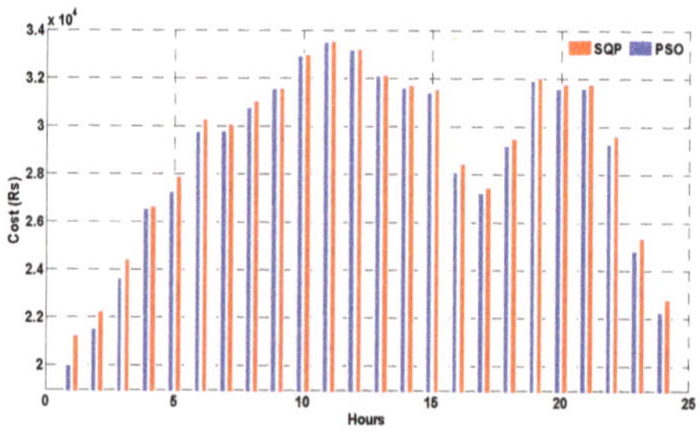

Fig. 5. Comparison of fuel cost by PSO and SQP algorithms over 24 h for 10 units with wind turbine.

Having calculation of the cost of such a scheduling, the algorithm ensures that the profit is based on a valid scheduling by considering reserved units. Figures 5 and 6 show the total cost and emission of 10-unit system, for each hour of optimization. Although the value of emission in PSO algorithm is a little, more than SQP one, TI shows that profit will outweigh the emission.

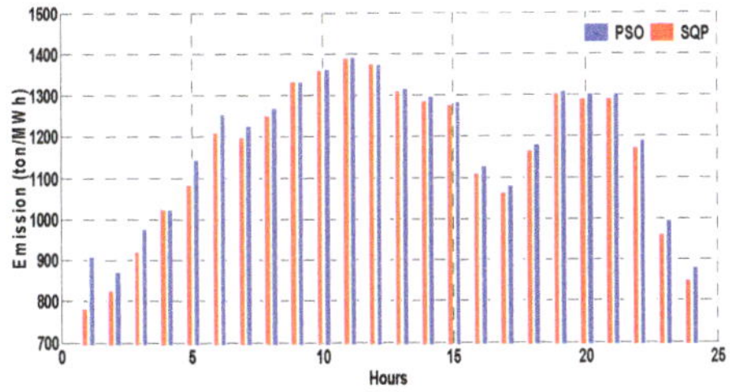

Fig. 6. Comparison of Emission by PSO and SQP algorithms over 24 h for 10 units with wind turbine.

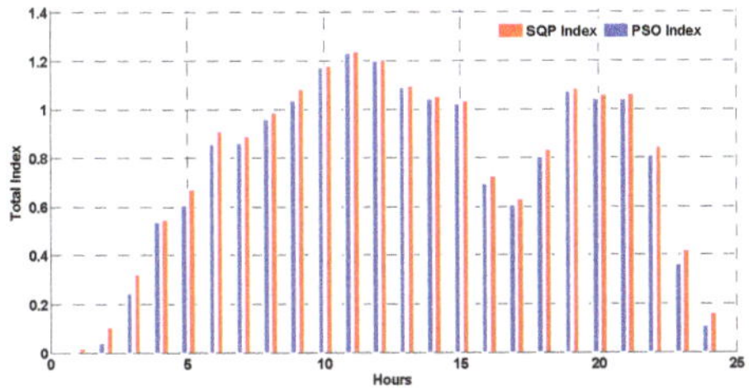

Fig. 7. Comparison of Total Index (TI) over 24 h for 10 units with wind turbine.

Figure 7 shows TI that is computed in accordance with (32). It is obvious that there has to be a trade-off between maximization of total profit and minimization of total emission over 24-hour for 10 units.

TI is improved by 3.68 % when PSO algorithm is used. According to (27), the produced total active power is constant for all units therefore for increasing total profit the fuel cost should be decreased. Therefore, the total index is defined based on emission and cost. Tables 4 and 5 indicate the optimal generated active power of units based on per-unit system and the value of objective functions over 24-houre period of time. Fig.8 shows the convergence process of PSO algorithm for the best solution. The value of the objective functions settles at the minimum value after 500 iterations, and would be constant after that.

The two-dimensional Pareto front with its surface which contains optimal and non-optimal solutions for the objective functions is shown in Fig. 9. The best solution for objectives Emission and Cost are 6920969 (ton) and 28442 (Rs) over 24-hour period of time which is shown in Fig. 9 with cursor.

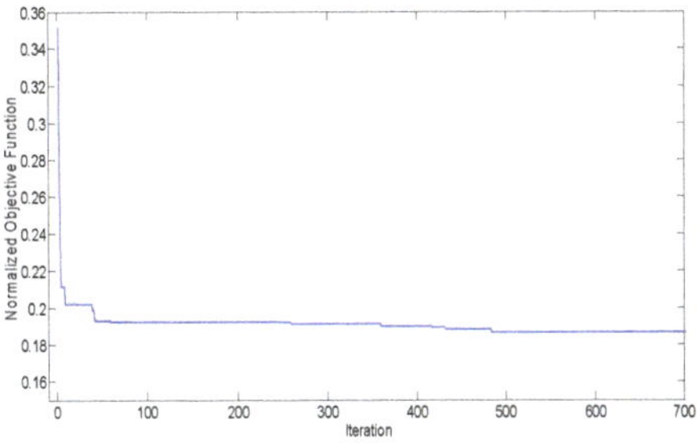

Fig. 8. Convergence process of the best solution obtained by PSO algorithm.

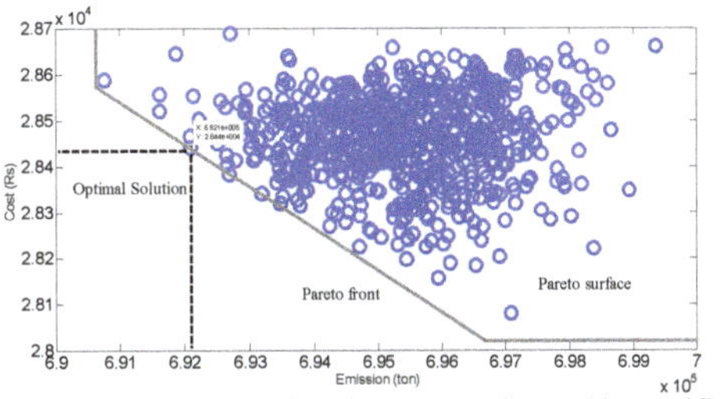

Fig. 9. Two-dimensional Pareto surface with specified Pareto front for Emission and Cost.

5. CONCLUSIONS

In this paper, Here-and-Now approach is used for solving the generation scheduling problem by considering thermal and wind energy systems. In this paper, the probability of stochastic wind power is included in the constraints set. This approach avoids the probabilistic infeasibility caused by using the average of RVs. In particular, we used a threshold parameter Pa into the constraints to characterize the tolerance that the total load demand cannot be satisfied. In addition, it has proposed a multi objective problem for optimizing profit and Emission over 24-hour period of time based on the algorithms for ten units in the presence of WPG units.

PSO algorithm with HN approach decreases TI about 3.68% than SQP algorithm. Furthermore, it provides better solution particularly for systems containing larger number of generating units. PSO algorithm with HN approach can provide a fast solution and the GENCOs can maximize their profit and minimize their emission.

Table 4. The objective functions values for PSO based algorithm.

Hour	U-1	U-2	U-3	U-4	U-5	U-6	U-7	U-8	U-9	U-10	Emission	Cost	Profit
1	1.50	1.50	1.50	1.50	1.50	0.75	0.75	0.75	0.75	0.75	911.388	20050.54	717584.3
2	1.48	1.48	1.05	1.10	1.06	0.80	0.21	0.21	0.21	0.21	872.730	21554.55	755585.6
3	1.48	1.48	1.19	1.29	1.30	0.82	0.30	0.30	0.30	0.30	977.778	23656.55	891103.5
4	1.76	1.64	0.94	0.82	1.06	1.09	0.63	0.67	0.59	0.53	1023.864	26582.11	967176.5
5	1.46	1.46	1.46	1.46	1.46	0.59	0.60	0.60	0.59	0.59	1146.266	27265.70	1047756
6	1.50	1.50	1.50	1.50	1.50	0.75	0.75	0.75	0.75	0.75	1255.853	29789.12	1135669
7	1.41	1.41	1.41	1.41	1.41	1.16	0.74	0.74	0.74	0.74	1227.064	29820.29	1104159
8	1.50	1.50	1.50	1.25	1.44	1.11	0.81	0.81	0.81	0.83	1269.930	30797.60	1125412
9	1.50	1.50	1.49	1.50	1.45	1.34	0.80	0.80	0.80	0.80	1333.193	31571.42	1199618
10	1.50	1.50	1.38	1.29	1.50	1.41	0.95	0.98	0.94	0.91	1363.286	32948.75	1604768
11	1.50	1.50	1.34	1.37	1.47	1.44	0.97	1.02	0.97	1.00	1393.817	33520.45	1763897
12	1.50	1.50	1.34	1.36	1.50	1.38	1.00	1.00	0.95	0.95	1377.215	33208.42	1662716
13	1.50	1.50	1.24	1.45	1.37	1.26	0.92	0.92	0.92	0.92	1318.940	32101.86	1301833
14	1.50	1.50	1.21	1.44	1.50	1.20	0.87	0.91	0.87	0.85	1297.559	31614.34	1278211
15	1.50	1.50	1.36	1.27	1.46	1.23	0.86	0.86	0.86	0.86	1285.412	31423.05	1163327
16	1.50	1.50	1.27	1.26	1.30	1.06	0.65	0.65	0.65	0.65	1130.608	28128.68	1025556
17	1.50	1.50	1.33	1.07	1.12	1.13	0.65	0.60	0.60	0.59	1081.431	27259.37	984013.1
18	1.49	1.49	1.30	1.21	1.49	1.09	0.70	0.71	0.70	0.70	1182.468	29238.83	1057275
19	1.50	1.50	1.24	1.35	1.50	1.42	0.84	0.87	0.84	0.90	1310.222	31923.13	1166877
20	1.49	1.49	1.25	1.49	1.44	1.23	0.87	0.87	0.87	0.87	1304.520	31617.58	1181331
21	1.46	1.46	1.34	1.40	1.46	1.32	0.84	0.85	0.87	0.84	1304.601	31612.45	1205372
22	1.50	1.50	1.33	1.30	1.38	1.20	0.68	0.71	0.68	0.70	1192.861	29292.81	1106763
23	1.50	1.50	1.17	1.18	1.31	0.79	0.43	0.43	0.43	0.43	997.792	24837.51	917022.7
24	1.49	1.49	1.04	0.96	1.18	0.77	0.27	0.30	0.29	0.26	883.275	22281.82	799655.5

Table 5. The objective functions values for SQP based algorithm.

Hour	U-1	U-2	U-3	U-4	U-5	U-6	U-7	U-8	U-9	U-10	Emission	Cost	Profit
1	1.47	1.41	0.71	0.72	0.91	0.74	0.33	0.40	0.34	0.32	784.190	21263.75	716331.2
2	1.50	1.47	0.76	0.76	0.96	0.79	0.37	0.44	0.39	0.36	827.356	22253.75	754906.2
3	1.50	1.50	0.88	0.88	1.10	0.93	0.48	0.54	0.49	0.46	921.751	24433.12	890316.5
4	1.50	1.50	0.98	0.99	1.24	1.07	0.60	0.66	0.60	0.57	1023.943	26670.96	967097.8
5	1.50	1.50	1.50	1.50	1.50	1.50	0.03	1.18	0.03	0.03	1084.926	27904.56	1047117
6	1.50	1.50	1.15	1.16	1.47	1.30	0.78	0.84	0.79	0.75	1210.193	30296.02	1135162
7	1.50	1.50	1.14	1.15	1.46	1.28	0.77	0.83	0.78	0.74	1199.198	30095.47	1103915
8	1.50	1.50	1.19	1.20	1.50	1.33	0.83	0.88	0.83	0.80	1251.051	31057.95	1125172
9	1.50	1.50	1.24	1.25	1.50	1.41	0.88	0.94	0.89	0.85	1333.193	31571.42	1199618
10	1.50	1.50	1.30	1.30	1.50	1.47	0.94	0.99	0.94	0.91	1360.756	32993.73	1604736
11	1.50	1.50	1.32	1.33	1.50	1.50	0.98	1.03	0.98	0.95	1391.688	33545.34	1763872
12	1.50	1.50	1.31	1.32	1.50	1.49	0.96	1.01	0.96	0.92	1375.150	33233.86	1662704
13	1.50	1.50	1.25	1.25	1.50	1.42	0.89	0.94	0.89	0.86	1311.638	32148.57	1301797
14	1.50	1.50	1.23	1.23	1.50	1.39	0.87	0.92	0.87	0.84	1288.337	31738.13	1278032
15	1.50	1.50	1.21	1.22	1.50	1.38	0.86	0.91	0.86	0.83	1277.512	31545.67	1163214
16	1.50	1.50	1.07	1.07	1.35	1.18	0.69	0.75	0.69	0.66	1111.947	28434.78	1025240
17	1.50	1.50	1.02	1.03	1.29	1.12	0.64	0.70	0.64	0.61	1064.253	27493.81	983779
18	1.50	1.50	1.12	1.12	1.42	1.25	0.74	0.80	0.75	0.71	1167.590	29497.89	1057016
19	1.50	1.50	1.24	1.25	1.50	1.41	0.88	0.94	0.89	0.85	1304.747	32027.97	1166782
20	1.50	1.50	1.23	1.23	1.50	1.40	0.87	0.92	0.87	0.84	1291.056	31786.40	1181121
21	1.50	1.50	1.23	1.23	1.50	1.40	0.87	0.92	0.87	0.84	1291.069	31786.67	1205229
22	1.50	1.50	1.13	1.13	1.43	1.25	0.75	0.80	0.75	0.72	1173.920	29615.88	1106409
23	1.50	1.50	0.92	0.92	1.17	0.99	0.53	0.59	0.54	0.51	963.175	25383.96	916466
24	1.50	1.50	0.78	0.79	1.00	0.82	0.40	0.47	0.41	0.38	851.272	22825.38	799122.1

NOMENCLATUER

c	Scale factor of the Weibull distribution
$F_W(w)$	Cumulative distribution function (CDF) of random variable W
$f_W(w)$	Probability density function (PDF) of random Variable W
k	Shape factor of the Weibull distribution
m	Number of wind power generation (WPG) units
N	Number of generators
P_d	Total load demand
$Pr\ (E)$	Probability of event E
P_s	Total transmission losses
P_a	Upper bound of probability that the sum of real power not greater than $P_d + P_s$
$a_i,\ b_i, c_i$	Cost coefficients of generator i
W_j	Real power generated by WPG unit j
w_{jr}	Rated power of WPG unit j
w_r	Rated power of WPG units if all the same
p_i	Real power generated by generator j
$P_{optm,i}$	Optimal value of P_i
Y	Cost index in the economic load dispatch (ELD) model
RV	Revenue
TC	Total Cost
PF	Profit
EC_i	Emission cost function of unit i
$GENCO$	Generation Company
WPG	Wind Power Generation
v_r, v_{in}, v_{out}	Rated, cut-in, and cut-out wind speeds

ACKNOWLEDGMENT

The authors gratefully acknowledge the technical support given by Mr. Abdollah Khorramdel and Mr. Ali Khorramdel who are electrical engineers with more than 35 and 15 years background in Fars Electrical Company, respectively. Also, authors would like to thank Safashahr Branch, Islamic Azad University for financial support.

REFERENCES

[1] A. Rabiee, H. Khorramdel and J. Aghaei, "A review of energy storage systems in microgrids with wind turbines," *Renewable and Sustainable Energy Reviews*, vol. 18, pp. 316–326, 2013.

[2] X. Liu and X. Wilsun, "Economic load dispatch constrained by wind power availability: A Here-and-Now Approach," *IEEE Transaction on sustainable energy*, vol. 1, no. 1, pp. 2-9, 2010.

[3] T. Venkatesan and M.Y. Sanavullah, "SFLA approach to solve PBUC problem with emission limitation," *International Journal of Electrical Power and Energy Systems*, vol. 46, pp. 1-9, 2013.

[4] H. Siahkali and M. Vakilian, "Stochastic unit commitment of wind farms integratein power system,"*Electric Power System Research*, vol. 80, pp. 1006-17, 2010.

[5] H. Moghimi, A. Ahmadi, J. Aghaei and A. Rabiee. "Stochastic techno-economic operation of power systems in the presence of distributed energy resources," *International Journal of Electrical Power and Energy Systems*, vol. 45, no. 1, pp. 477-488, 2013.

[6] N. Padhy. "Unit commitment-a bibliographical survey," *IEEE Transaction on Power Systems*, vol. 19, no. 2, pp. 1196-1205, 2004.

[7] N. Padhy. "Unit commitment problem under deregulated environment a review," *IEEE Conference on Power engineering society general meeting*, vol. 2, pp. 1088–1094, 2003.

[8] K. Chandram and N. Subrahmanyam, "New approach with Muller method for solving profit based unit commitment," *Proc. of the 3th International conference on Electric utility: Deregulation and Restructuring and power technologies*, pp. 1–8, 2008.

[9] C. C. Christopher, K. Chandrasekaran and S. Sishaj "Nodal ant colony optimization for solving profit based unit commitment problem for GENCOs," *Applied Soft Computing*, vol. 12, no. 1, pp. 145–160, 2012.

[10] E. Delarue, P. Van den Bosch and W. D'haeseleer. "Effect of the accuracy of price forecasting on profit in a price based unit commitment," *International journal of Electrical Power and Energy Systems*, vol. 80, no. 10, pp. 1306-1313, 2010.

[11] M. Gent and W. Lamont John, "Minimum-emission dispatch," *IEEE Transaction on Power Apparatus and Systems*, vol. 90, no. 6, pp. 2650-2660, 1971.

[12] C. Richter and G. Sheble, "Profit based unit

commitment GA for competitive environment," *IEEE Transaction on Power Systems*, vol. 15, no. 2, pp.715–21, 2000.

[13] J. Valenzuela and M. Mazumdar. "Making unit commitment decisions when electricity is traded at spat market prices," *Proc. of the IEEE conference on Power Engineering Society Winter Meeting*, vol. 3, pp. 1509- 1512, 2001 .

[14] P. Attaviriyanupap, H. Kita, E. Tanka and J. Hasegawa. "A hybrid LR-EP for solving newprofit-based UC problem under competitive environment," *IEEE Transaction on Power Systems,* vol. 18, no. 1, pp. 229-237, 2003.

[15] L. Nanda, J. Hari and ML. Kothari, "Economic emission load dispatch with line flow constraints using a classical technique," *IET Proceedings on Generation, Transmission and Distribution*, vol. 141, no. 1, pp. 1–10, 1994.

[16] H. Yamin and SM. Shahidehpour. "A new approach for GENCOs profit based unit commitment in day-ahead competitive electricity markets considering reserve uncertainty," *International Journal of Electrical Power and Energy Systems*, vol. 29, no. 8, pp. 609-16, 2007.

[17] C. Jacob Reglend, G. Raghuveer, R. Avinash, N. Padhy and D. Kothari, "Solution to profit based unit commitment problem using particle swarm optimization," *Applied Soft Computing*, vol. 4, pp. 247-256, 2010.

[18] P. Hota, R. Chakrabarti and PK. Chattopadhyay, "Economic emission load dispatch with line flow constraints using sequential quadratic programming technique," *Proc. of the 12th Conference on Power Technology*, vol. 81, pp. 21-5, 2000.

[19] R.N. Allan, A.M.L. Da Silva and R.C. Burchett, "Evaluation methods and accuracy in probabilistic load flow solutions," *IEEE Transaction on Power Systems*, vol. 100, no. 5, pp. 2539–2546, 1981.

[20] J.R. Birge and F. Louveaux, *"Introduction to Stochastic Programming,"* Springer, New York, 1997.

[21] H. Chen, J. Chen and X. Duan, "Multi-stage dynamic optimal power flow in wind power integrated system," *Proc. of the IEEE International Conference on Transmission and Distribution and Exhibition Asia and Pacific*, pp. 1-5, 2005.

[22] R. A. Jabr and B. C. Pal, "Intermittent wind generation in optimal power flow dispatching," *IET Proceedings on Generation Transmission Distribution*, vol. 3, no. 1, pp. 66-74, 2009.

[23] B. Ren and C. Jiang, "A review on the economic dispatch and risk management considering wind power in the power market," *Renewable and Sustainable Energy Reviews*, vol. 13, no. 8, pp. 2169-2174, 2009.

[24] J. A. Carta, P. Ramirez and S. Velázquez, "A review of wind speed probability distributions used in wind energy analysis: Case studies in the canary islands," *Renewable and Sustainable Energy Reviews*, vol. 13, no. 5, pp. 933-955, 2009.

[25] S. Roy, "Market constrained optimal planning for wind energy conversion systems over multiple installation sites," *IEEE Transaction on Energy Conversion*, vol. 17, no. 1, pp. 124-129, 2002.

[26] A. Leon-Garcia, "*Probability, Statistics, and Random Processes for Electrical Engineering*," Englewood Cliffs, Prentice-Hall, 2008.

[27] B. Khorramdel and M. Raoofat. "Optimal stochastic reactive power scheduling in a microgrid considering voltage droop scheme of DGs and uncertainty of wind farms," *Energy*, vol. 45, no. 1, pp. 994-1006, 2012.

[28] B. Khorramdel, H. Khorramdel and H. Marzooghi, "Multi-objective optimal operation of microgrid with an efficient stochastic algorithm considering uncertainty of wind power", *International Review on Modeling and Simulation*, vol. 4, no. 6, pp. 3079-3089, 2011.

[29] T. Niknam, B. Bahmani Firouzi and A. Ostadi. "A new fuzzy adaptive particle swarm optimization for daily Volt/Var control in distribution networks considering distributed generators," *Applied Energy*, vol. 87, no. 6, pp.1919-1928, 2010.

Novel Hybrid Fuzzy-Intelligent Water Drops Approach for Optimal Feeder Multi Objective Reconfiguration by Considering Multiple-Distributed Generation

H. Bagheri Tolabi*[1], M. H. Ali[2], and M. Rizwan[3]

[1]Faculty of Engineering, Islamic Azad University, Khorramabad Branch, Khorramabad, Iran
[2]Department of Electrical and Computer Engineering, University of Memphis, Tennessee, United States
[3]Department of Electrical Engineering, Delhi Technological University, Delhi-110042, India

ABSTRACT

This paper presents a new hybrid method for optimal multi-objective reconfiguration in a distribution feeder in addition to determining the optimal size and location of multiple-Distributed Generation (DG). The purposes of this research are mitigation of losses, improving the voltage profile and equalizing the feeder load balancing in distribution systems. To reduce the search space, the improved analytical method has been employed to select the optimum candidate locations for multiple-DGs, and the intelligent water drops approach as a novel swarm intelligence based algorithm is used to simultaneously reconfigure and identify the optimal capacity for installation of DG units in the distribution network. In order to facilitate the algorithm for multi-objective search ability, the optimization problem is formulated for minimizing fuzzy performance indices. The proposed method is validated using the Tai-Power 11.4-kV distribution system as a real distribution network. The obtained results proved that this combined technique is more accurate and has the lowest fitness value as compared with other intelligent search algorithms. Also, the obtained results leadto the conclusion that multi-objective simultaneous placement of DGs along with reconfiguration can be more beneficial than separate single-objective optimization.

KEYWORDS: Multi objective reconfiguration, Intelligentwater drops algorithm, Distribution system, Power loss, Load balancing, Voltage profile.

1. INTRODUCTION

Distribution system is an interface between consumers and transmission network. Due to the advantages such as lower short circuit current and easier protection coordination, they are generally utilized with radial configuration. On the other hand, this radial structure may lead to reduce the reliabilityofconsumers feeding, increase thepower losses and voltage drop at the load points. Electrical power distribution systems have two types of tie and sectionalizing switches, whose statuses determine the configuration of distribution network. Bychanging the switches states andtransition of sectionsbetweenfeeders duringoperation, the construction of distribution network will change [1]. Since network reconfiguration is a complex combinatorial, non-differentiable constrained optimization problem, many algorithms were proposed in the past. Manyresearchesin the literature have presented several methods for the optimal reconfiguration ofthe distribution networks with different objectives.

Reconfiguration of distribution network for loss reduction was first proposed by Merlin and Back [2] in 1975. They have used a branch and bound optimization method to determine the configuration that has the minimum total loss. After that, many algorithms have been developed for reconfiguration of distribution system with different aims. Goswami and Basu [3] presented a heuristic algorithm for

*Corresponding author:
H. B. Tolabi, (E-mail: hajar.bagheri@hotmail.com)

reconfiguration, which is determined using a power flow program. The main advantages of this research are using a very fast power flow method and the independence of the final configurations upon the initial configuration of the feeders. Gomes *et al.* [4] reported a heuristic algorithm for the large distribution systems that begins in a meshed configuration with all switches closed. Several tests are performed using the procedure described in [4], and more efficient configurations are obtained when compared with the methods proposed in three classical papers. The obtained results proved that the proposed procedure [4] presents a very good compromise, as it tends to find a near-optimum or even the optimum solution without the risk of combinatorial explosion. A new path to node based modeling and its application to reconfiguration of distribution system has been proposed by Ramos *et al.* [5] in 2005. In this work, the authors suggested to employ a power flow method-based heuristic algorithm for determining the minimum loss configuration of radial distribution networks. Also, two different optimization algorithms-one resorting to a genetic algorithm and the other solving a conventional mixed-integer linear problem-are fully developed. Schmidt *et al.* [6] have introduced a method for loss minimization based on the standard Newton technique. Zhou *et al.* [7] have presented two reconfiguration algorithms for service restoration and load balancing in distribution systems. They have suggested the operation cost reduction and it is based on the long term operation of the power system. An optimization technique to determine the network structure with minimum energy losses for a given period has proposed by Taleski and Rajicic [8]. In this research, a new method for checking system radiality which is based on upward-node expression is developed for solving the problem of restorative planning of power system. Kavousi-Fard and Niknam [9] solved the multi-objective distribution feeder reconfiguration problem from the reliability point of view. The investigated objective functions are: System Average Interruption Frequency Index (SAIFI), Average Energy Not Supplied (AENS), total active power losses and the total network cost. The obtained results show that neglecting the uncertainty effect and so studying in a deterministic environment can deprive the operator from real optimal and dependable final solutions. In [10] multi objective reconfiguration of distribution network has solved using NSGA-II algorithm. It was shown that in addition to reduction of network losses, voltage regulation the load balancing on the system branches were also optimally improved.

Deregulation of electricity markets in many countries world-wide brings new perspectives for Distributed Generation (DG) of electrical energy using renewable energy sources with small capacity. Since the selection of optimal locations and sizes of DG units in distribution system, is also a complex combinatorial optimization problem, many methods have been proposed in this area in the recent past. Among recent works in this area, Ishak *et al.* [11] present a method to identify the optimal location and size of DGs based on the power stability index and Particle Swarm Optimization (PSO) algorithm. In this paper, the Maximum Power Stability Index (MPSI) is utilized as an objective function to determine the optimal DG locations. Next, a PSO-based model with randomized load is developed to optimize DG sizing in view of the system's real power losses. Doagu-mojarrad *et al.* propose an interactive fuzzy satisfying method, which is based on hybrid modified shuffled frog leaping algorithm to solve the problem of the Multi-objective optimal placement and sizing of DG units in the distribution network [12]. One of the advantages of this work is to account the technical, economical and environmental protection considerations.

Solving simultaneous reconfiguration and allocation of DGs problemstogether,despitethe complexity has more advantages rather than separatesolutions of them, andhas been discussedrecently inseveralstudies. Tolabi *et al.* [13] used a method based on the combination of fuzzy sets and Bees Algorithm (BA) for simultaneous reconfiguration and optimal allocation of multiple-DG units in a distribution network. The proposed approach is tested on Taiwan power company system with three DGs. The obtained results are compared with GA, PSO and Harmony Search Algorithm (HSA) at nominal load and found better result than the above mentioned approaches because of the lowest optimal fitness and more reliable convergence behavior.

In this paper, a novel Intelligent Water Drops (IWD) approach is used for both multi objective reconfiguration and optimal allocation of multiple-DG units in a distribution network. Also, a fuzzy logic technique is used to achieve a compromise between the objective functions. Along with the combination of fuzzy-IWD techniques, an effective approach is used in order to reduce the search space and simplify the selection of candidate buses for installation of DG units using IA method.

The main contribution of the paper is to solve the multi-objective problem using the combination of IWD algorithm and fuzzy approach in order toreduction of losses, improve the voltage profile and equalize the feeder load balancing in power distribution system.

The remainder of this paper is organized in the following manner: DG types are presented in section 2. Sec. 3 gives the problem formulation. Sec. 4 gives the idea about multi-objective function and constraints of the problem. Section 5 explains the IA method. Sec. 6 presents the optimization in fuzzy environment. Intelligent water drop approach is presented in Sec. 7. In Sec. 8, Fuzzy-IWD method is discussed. Results are presented in Sec. 9. A conclusion followed by references is presented in Sec. 10.

2. DGTYPES

Four different types of DGs are introduced as follows:

Type 1 DG: This type only injects the real power.

Type 2 DG: This type only injects the reactive power.

Type 3 DG: This typeiscapable ofinjecting both real powerandreactive power.

Type 4 DG: This type is capable of injecting real power,but consuming reactive power [14].

3. PROBLEM FORMULATION

3.1. Power flow equations

The problem is formulated using the power flow equations. Power flows in a distribution system are computed by the following set of simplified recursive equations [15]:

$$\begin{aligned} P_{k+1} &= P_k - P_{loss,k} - P_{Lk+1} \\ &= P_k - \frac{R_k}{|V_k|^2}\left\{P_k^2 + (Q_k + Y_k|V_k|^2)^2\right\} - P_{Lk+1}, \\ Q_{k+1} &= Q_k - Q_{loss,k} - Q_{Lk+1} \\ &= Q_k - \frac{X_k}{|V_k|^2}\left\{P_k^2 + (Q_k + Y_{k1}|V_k|^2)^2\right\} - Y_{k1}|V_k|^2 - Y_{k2}|V_{k+1}|^2 - Q_{Lk+1} \end{aligned} \tag{1}$$

$$\begin{aligned} |V_{k+1}|^2 &= |V_k|^2 + \frac{R_k^2 + X_k^2}{|V_k|^2}(P_k^2 + Q_k^2) - 2(R_k P_k + X_k Q_k) \\ &= |V_k|^2 + \frac{R_k^2 + X_k^2}{|V_k|^2}(P_k^2 + (Q_k^2 + Y_k|V_k|^2)^2) \\ &\quad - 2(R_k P_k + X_k(Q_k + Y_k|V_k|^2)). \end{aligned} \tag{2}$$

where, P_k, Q_k are real and reactive power flowing out of bus k; $P_{loss,k}$, $Q_{loss,k}$ are real and reactive power loss at bus k; P_{Lk+1}, Q_{Lk+1} are real and reactive load power at bus $k+1$; R_k, X_k are resistance and reactance of the line section between buses k and $k+1$; Y_k, V_k are Shunt admittance and voltage amplitude at bus k.

The power loss when a DG is installed at an arbitrary is given by:

$$\begin{aligned} P_{DG,loss} &= \frac{R_k}{V_k^2}(P_k^2 + Q_k^2) \\ &+ \frac{R_k}{V_k^2}(P_G^2 + Q_G^2 - 2P_k P_G - 2Q_k Q_G)\left(\frac{G}{L}\right) \end{aligned} \tag{3}$$

where, P_G, Q_G are real and reactive power supplied by DG; G and L are distance and length of the feeder from source to bus in Km.

4. MULTI OBJECTIVE FUNCTION AND CONSTRAINTS OF THE PROBLEM

The objective function $f(x)$ is a constrained optimization problem to find an optimal configuration of the distribution system and DG allocation. $f(x)$ is a multi objective function that consists of three goals: reducing the loss, increasing the load balancing, and improving the voltage that is formulated as a follows:

$$MinF(X) = \min[P_{loss}, LBI, VPI] \tag{4}$$

The constraintsof the problem are:

$1: V_{k\min} \le V_k' \le V_{k\max}$

$2: |I'_{k,k+1}| \le |I_{k,k+1\max}|$

$3: \sum_{k=1}^{nf} P_{Gk} \le \sum_{k=1}^{nf} (P_k + P_{loss,k})$

4 : Radial structure of network should be maintained

5 : All available nodes of considered distribution system should be fed.

Where,
V'_k : Voltage at bus k after reconfiguration.
$V_{k\max}$: Maximum bus voltage.
$V_{k\min}$: Minimum bus voltage.
$I'_{k,k+1}$: Current in line section between buses k and $k+1$ after reconfiguration.
$I_{k,k+1\max}$: Maximum current limit of line section between buses k and $k+1$.
nf : Total number of lines sections in the system.

The first term of the objective function reflects real power losses that are defined by (5):

$$P_{loss} = \sum_{k=1}^{n_f} R_k \frac{P_k^2 + Q_k^2}{V_k^2} \tag{5}$$

The second term of the objective function is considered for the Load Balancing Index (LBI) of the lines in the feeder, which is given by:

$$LBI = \sum_{F_j} \left(\frac{I_{Fj}}{I_{Favg}}\right)^2 \tag{6}$$

where, I_{Fj} is the current passing through line j and I_{Favg} is defined by (7):

$$I_{Favg} = \frac{1}{n_f} \sum_{j=1}^{n_f} I_{Fj} \tag{7}$$

The decrease inthisindeximpliesincreaseofload balancing of lines in the distribution feeder.
The third term of the objective function reflects the improvement of the voltage profile, which is shown by Voltage Profile Index (VPI) in (8):

$$VPI = \sum_{k \in LB} \left| V_k - V_{ref,k} \right| \tag{8}$$

where LB is the collection of the load buses and $V_{ref,k}$ is the nominalvoltage at load bus k.
The decrease inthisindeximpliesimprovementthe profile of voltages in the distribution feeder buses.

5. REDUCE THE NUMBER OF SOLUTION SPACE

An effective method is used in order to simplify the selection of candidate buses for installation of DG units using Improved Analytical (IA) method. This method is chosen because it is effective as corroborated by Exhaustive Load Flow (ELF) and Loss Sensitivity Factor (LSF) solutions in terms of loss reduction and computational time [16]. Itis based on IA expressions to calculate the optimal size of different DG types and a methodology to identify the best location for DG allocation, which helps reduce the number of solution space. To reduce the search space in this paper, IA has been employed to select the candidate locations for multiple-DG [16] and the sizes of DG unit at candidate buses are calculated using fuzzy-IWD method. Because detailed description about multiple-DG placement using the AI method is presented in [16], only anoverall view to this method is presented in this paper as follows:

First, a single DG is addedto the system. After that, the load data are updated with the first DG placed and then another DG is added. Similarly, the algorithm continues to allocate other DG units until it does not satisfy at least one of the following constraints:

a) The voltage at a particular bus is over the upper limit;
b) The total size of DG units is over the total load plus loss;
c) The maximum number of DG units is unavailable;
d) The new iteration loss is greater than the previous iteration loss.

6. OPTIMIZATION IN FUZZY ENVIRONMENT

Since the different terms of the multi objective function are in various ranges, a fuzzy system [17] is used in order to compare these terms during reconfiguration and DG placement. In this plan each variable has a membership function (μ) that determines the rank and effectiveness of its variable. The membership values for each variable are between zero and unity in the fuzzy domain and may be different for each element. The membership function are presented by (9) and Fig. 1.

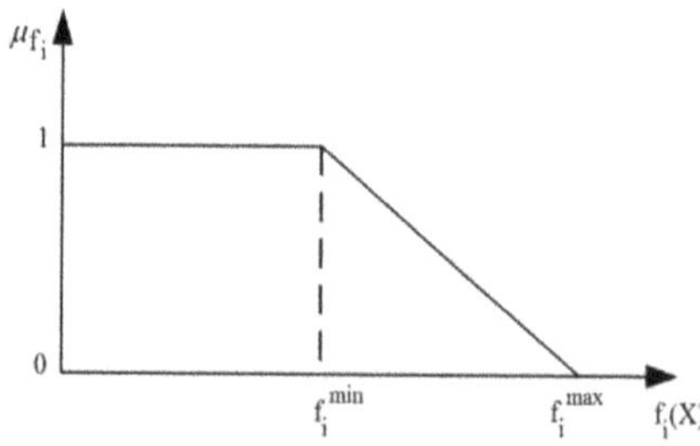

Fig. 1. Membership functions for three different terms of the objective function

$$\mu_{fi}(X)=\begin{cases}1, & f_i(X)\le f_i^{\min}\\ \dfrac{f_i^{\max}-f_i(X)}{f_i^{\max}-f_i^{\min}}, & f_i^{main}<f_i(X)<f_i^{\max}\\ 0, & f_i^{\max}\le f_i(X)\end{cases} \quad (9)$$

where, f_i represents the *i*th term of the objective function (*i=1, 2, 3*), f_i^{min}, and f_i^{max} are the best and worst answers that observed in the single-objective optimization area for the *i*th term in the objective function, respectively.

By using of anotheradvantageoffuzzy sets, three different objective functions are combined with each other in the form of a Fuzzy Interface System (FIS), so the multi-objective optimization problem will be converted into an optimized fuzzy single-objective function. To achieve this purpose, the value of each objective function, which is considered as an input in the FIS, is divided into several regions using fuzzy membership functions and the final objective function that wants to optimize it, is made through the appropriate rules [18].

Table 1 shows the fuzzy rules that were employed for reconfiguration process simultaneously allocation the optimum size for DGs. In these Table, B, A, G, VB, EB, VG, EG, and EX stand for bad, average, good, very bad, extremely bad, very good, extremely good, and excellent, respectively. In this system, the Mamdani's inference mechanism and the center of the area defuzzification method is used.

Table 1. Fuzzy rules when μ_{LBI} is **a.** bad, **b.** average, **c.** good.

a

		μ_{VPI}		
		G	A	B
$\mu_{P_{loss}}$	G	G	B	EB
	A	A	VB	EB
	B	B	EB	EB

b

		μ_{VPI}		
		G	A	B
$\mu_{P_{loss}}$	G	EG	A	VB
	A	VB	B	EB
	B	G	EB	EB

c

		μ_{VPI}		
		G	A	B
$\mu_{P_{loss}}$	G	EX	A	VB
	A	EG	A	VB
	B	VG	VB	EB

7. INTELLIGENT WATER DROPS APPROACH

IWD algorithm is inspired by the observation of natural water flow in the rivers formed by a swarm of water drops. The swarms of water drops find their own way to the lakes or oceans, even though it has to overcome a number of obstacles in its path. Without the presence of these obstacles, the water drops tend to be pulled straight towards the destination by the gravitational force. However, being blocked by different kinds of obstacles and constraints, there exist lots of twists and turns in the real path of the river. The interesting point is that the path of the river, constructed by the flow of water drops, seems to be optimized in terms of distance from the source to the destination under the constraints of the environment. By mimicking the features of water drops and obstacles of the environment, the IWD algorithm uses a population of water drops to construct paths and obtain the optimal or near-optimal path among all these paths over time. The environment represents the optimization problem needed to be solved. A river of IWDs looks for an optimal route for the given problem [19]. Hosseini [20] presented the basics of the IWD algorithm, then applied it to solve different optimization problems. As described in [20], an IWD model is proposed with two important parameters:

- The amount of soil it carries or its soil load, "$soil^{IWD}$".
- The velocity at which it is moving, "vel^{IWD}".

The values of these two parameters may change as the IWD flows in its environment from the source to a destination. An IWD moves in discrete finite-length steps and updates its velocity by an amount Δvel^{IWD} when it changes the position from point *i* to point *j* as follows:

$$\Delta vel^{IWD}=\frac{a_v}{b_v+c_v[soil(i,j)]^2} \quad (10)$$

where, $soil(i,j)$ is the soil on the bed of the edge between two points *i* and *j*; *av*, *bv* and *cv* are pre-defined positive parameters for the IWD algorithm. The relationship between velocity and the amount of soil of the edge is decided by *av* and *cv*, meanwhile *bv* is a small number used to prevent the singularity problem. Equation (10) indicates that the rate of changing the velocity, Δvel^{IWD} is dependent on the soil of the edge, *i.e.*, edge with more soil provides more resistance to the water flow that results in a smaller increment in velocity and vice versa. Thus, the velocity at $time(t+1)$, vel_{t+1}^{IWD} is given by:

$$vel_{t+1}^{IWD} = vel_t^{IWD} + \Delta vel^{IWD} \quad (11)$$

where, vel_t^{IWD} is the velocity of the IWD at time *t*.

The amount of soil removed from the bed of $edg(i,j)$ is inversely proportional in a non-linear manner to the time needed for the IWD to move from point i to point j and can be calculated by using (12):

$$\Delta soil(i,j) = \frac{a_s}{b_s + c_s[time(i,j;vel^{IWD})]^2} \quad (12)$$

where, *as*, *bs* and *cs* are pre-defined positive parameters for the IWD algorithm. *as* and *cs* define the relationship between the amount of soil and the period of time IWD takes to move through the $edg(i,j)$, and *bs* is a small number used to avoid the singularity problem. Meanwhile, the duration of time is calculated by the simple laws of physics for linear motion. The time spent by the IWD to move from point *i* to *j* with velocity vel^{IWD} is given by:

$$time(i,j;vel^{IWD}) = \frac{HUD(i,j)}{\max(\varepsilon_v; vel^{IWD})} \quad (13)$$

wherea local heuristic function $HUD(i,j)$ has to be defined for a given problem to measure the undesirability of anIWD to move from point *i* to point *j*, 1v is the threshold of velocity to avoid the negative value ofvel^{IWD}. Equations (12) and (13) represent the assumption that the water drop which moves faster or spends less time to pass from point *i* to point *j* can gather more soil than the one which has a slower velocity. Once the IWD moves from point *i* to point *j*, the following formulae are used to calculate the updated soil of the edge and the soil load of the IWD, respectively.

$$soil(i,j)_{(t+1)} = (1-\rho_n)(soil(i,j)_{(t)} - \rho_n \Delta soil(i,j) \quad (14)$$

$$soil^{IWD}{}_{(t+1)} = soil^{IWD}{}_{(t)} + \Delta soil(i,j) \quad (15)$$

where ρ_n is the local soil updating parameter, which is chosen from [0, 1], and $\Delta soil(i,j)$ is calculated in (12).

To present the behavior of an IWD that prefers the easier edge or the edge with less soil on their beds, the edge selection of an IWD is based on the probability, $P(i,j;IWD)$ defined as follows which is inversely proportional to the amount of soil on the available edges.

$$p(i,j;IWD) = \frac{f(soil(i,j))}{\sum_{k \notin v_c(IWD)} f(soil(i,k))} \quad (16)$$

where, $f(soil(i,k)) = 1 + \varepsilon_S + g(soil(i,j))$.

The constant ε_s is a small positive number to prevent singularity. The set $v_C(IWD)$ denotes the group of nodes that the IWD should not visit to satisfy the constraints of the problem. The function $g(soil(i,j))$ is used to shift $soil(i,j)$ of the edge connecting point *i* and point *j* towards a positive value and is described below:

$$g(soil(i,j)) = \begin{cases} soil(i,j) & if: \min_{l \notin v_C(IWD)} (soil(i,l)) \geq 0 \\ soil(i,j) - \min_{l \notin v_C(IWD)} (soil(i,l)) & otherwise \end{cases} \quad (17)$$

The function $\min(.)$ returns the minimum value of its arguments. A uniform random distribution is used to generate a random number which can be compared with this probability in order to decide which is the next location that the IWD will move to.

For a given problem, an objective or quality function is needed to evaluate the fitness value of the solutions. A set of IWDs can be utilized and work together to find the optimal solution. The function $q(.)$ is denoted as the quality function and T^{IWD} is a solution founded by an IWD. When all the IWDs have constructed their solutions, one iteration can be considered complete. At the end of the iteration, the current iteration best solution T^{IB} is calculated by:

$$T^{IB} = \arg\max_{\forall IWDs} q(T^{IWD}) \quad (18)$$

Therefore, the iteration-best solution T^{IB} is the solution that has the highest quality over all solutions T^{IWD}.

Equation (14) updates the soil of each edge whenever an IWD traverses through a particular path based on the current amount of soil of the edge and the current velocity of the IWD. The soil is updated in (14) by using local information at each edge of the tree, and thus it may result in a local optimum. In order to increase the opportunities of finding the global optimum, the amount of soil on the edges of the current iteration best solution T^{IB} is updated according to the goodness of the solution after the iteration is complete and the overall knowledge of the solution is acquired. Equation (19) can be used to update the $soil(i,j)$ belonging to the current iteration best solution T^{IB}.

$$soil(i,j) = (1+\rho_{IWD})soil(i,j) - \rho_{IWD}\frac{1}{N_{IB}-1}soil_{IB}^{IWD}, \quad \forall(i,j)\in T^{IB} \tag{19}$$

where, $soil_{IB}^{IWD}$ represents the soil of the current iteration best IWD when it reaches the destination, $N_{IB}$is the number of nodes in the solution T^{IB} and ρ_{IWD} is the global soil updating parameter which is chosen from [0, 1]. The first term on the right-hand side of (19) is the amount of soil that remains from the previous iteration. Meanwhile, the second term of the right-hand side of (19) represents the quality of the current solution, obtained by the IWD. This way of updating the soil assists the reinforcement of the best-iteration solutions gradually, and thus, the IWDs are guided to search near good solutions with the expectation of finding the global optimum.

At the end of each iteration of the algorithm, the total best solution $T^{TB}$is updated by the current iteration-best solution T^{IB} as follows:

$$T^{TB} = \begin{cases} T^{TB} & if \quad q(T^{TB}) \geq q(T^{IB}) \\ T^{IB} & otherwise \end{cases} \tag{20}$$

By doing this, it is guaranteed that T^{IB} holds the best solution obtained so far by the IWD algorithm.The algorithm implementation details are specified in the following steps:

Step 1: Initialize soil updating parameters (*as, bs* and *cs*) and velocity updating parameters (*av, bv, cv*), the quality of total best solution ($q(T^{IWD})$), the maximum number of iterations (MaxIter), the iteration count (Itercount), the local soil updating parameter (ρ_n), the global soil updating parameter (ρ_{IWD}), the initial soil on each path (Initsoil) and the initial velocity (Initvel).

Step 2: Every IWD has visited node of list $v_c(IWD)$, which is initially empty. The IWDs velocity is set to Initvel and the entire IWDs are set to have zero amount of soil.

Step 3: Spread the IWDs on the nodes of the graph and then update the visited nodes.

Step 4: Repeat steps 5 to 8 for those IWDs with the partial solutions.

Step 5: For the IWD in node *i*, select the next node *j* by using the probability $P(i,j;IWD)$ presented in (16) such that doesn't violate any constraints of the problem and make certain it is not in the visited node list $v_c(IWD)$ and then add the recently visited node *j* to the list $v_c(IWD)$.

Step 6: For every IWD from node *i* to node *j*, updating its velocity $vel(t)$ to $vel(t+1)$) by (11).

Step 7: For the IWD moving on the path from node *i* to *j* calculate the $\Delta soil(i,j)$ by using the (12) and (13).

Step 8: Update $soil(i,j)$ of the path from node *i* to *j* traversed by that IWD, and also update the soil that IWD carries $soil^{IWD}$ by (14) and (15).

Step 9: Find the iteration based best solution T^{IB} from all the solutions T^{IWD} found by the IWDs using (18).

Step 10: Update the soils on the paths that form the current iteration based best solution T^{IB} by (19).

Step 11: Update the total best solution T^{TB} by using (20).

Step 12: Increment the iteration number by one. Itercount = Itercount +1 and then, go to step 2 if Itercount<Itermax .

Step 13: The algorithm stops with the total-best solution T^{TB}.

8. EXPLANATION OF THE PROPOSED FUZZY-IWD METHOD

This section describes the application of proposed fuzzy-IWD in optimal network reconfiguration and multiple-DG allocation problems. Since both reconfiguration and DG(s) allocation problems are complex combinatorial optimization problems, to reduce the search space, first IA method has been employed to select the best candidate locations for DGs, then optimal configuration and optimal sizes of DG units at candidate buses are discovered using hybrid fuzzy-IWD technique with the objectives of mitigating power loss, improving voltage profile and equal load balancing of the lines.

To reconfigure and DG allocation in the distribution feeder using proposed method, optimal buses candidate for DGs installation are suggested using AI method (The sizes of DG units will vary in discrete steps at suggested locations during optimization process). Thus, assuming to know theoptimal location for DG(s)installation, in order to represent an optimal feeder topology, it is enough to know the positions of open (tie) switches and DG(s) sizes in the network. Accordingly, first solution vector using reconfiguration and DG installation without violating the constraints of problem is formed as follows:

$$IWD^1 = [proposedtieswitches^1, proposedDG(s)size^1]$$

wherelength of the first part of solution vector for reconfiguration problem (*proposedtieswitches*), is equal to the number of tie switches, and the length of second part for DGs size, (*proposedDGs size*)is equal to the number of DG units.

By updating IWD parameters, second, third, and …, *i*th solution vector is generated with new proposed tie and new DG sizes at the same locations as follows:

$$IWD^i = \{proposedtieswitches^i, proposedDG(s)size^i\}$$

For each solution *i*, power flow program is carried out, the membership value for each objective and the fuzzified objective function are evaluated and compared with the previous solution, and the better solution will be selected and replaced. This procedure is repeated until a termination criterion is satisfied.

The proposed method is described as following steps that is summarized as a flowchart in Fig. 2.

Step 1) read data of distribution system (bus, load, branch, sectionalizing and tie switches numbers, DG types and numbers) and initialize the IWD parameters.

Step 2) give the best buses location for DG(s) using IA proposal.

Step 3) run the power flow program [15] based on equations (1-3), generate the solution vector as IWD for reconfiguration and determine DG sizes in the network without violating of five constraints that are presented in section 3.2.

Step 4) run the power flow program, calculate three terms of the objective function (P_{loss}, LBI, VPI) using (5-8), evaluate the membership value for each objective, compute $\mu_{P_{loss}}$, μ_{LBI} , and μ_{VPI} using (9). Compute fuzzified objective function value according to linguistic variable. Store the solution results.

Step 5) update the IWD algorithm parameters using (10-20). Go to step3 to generate a new solution using updated IWD parameters.

Step 6) if the fuzzified objective function value of the new solution is better than stored solution, update the IWD vector by storing solution=new solution.

Step 7) if *Itercount*<*Itermax*, *Itercount* = *Itercount* +*1* and go to step 5.

Step 8) Best solution=stored solution.

Stop 9) defuzzification of best solution and print the result.

Step 10) stop.

9. SIMULATION AND NUMERICAL RESULTS

Based on the proposed methodology, an analytical software tool has been developed in MATLAB environment. In order to investigate the effectiveness of the proposed method, the prepared program is appliedon a test system. Although the tool can handle four different DG types, only the results of applying three numbers of type 1 DG and three number of type 3 DG at the nominal load are presented.

In the simulation of network, six scenarios are considered to analyze the superiority of the proposed method for both type 1 and type 3 DGs as follow:

Scenario I: the base system without reconfiguration and DG;

Scenario II: the base system only with reconfiguration;

Scenario III: the base system only with DG type 1 allocation;

Scenario IV: the base system only with DG type 3 allocation;

Scenario V: the base system with simultaneous reconfiguration and DG type 1 allocation.

Scenario VI: the base system with simultaneous reconfiguration and DG type 3 allocation.

Using IA method the candidate bus locations to install the DGs are determined for scenarios III, IV, V, and VI. The limits of total DG unit sizes chosen for installation at candidate bus locations are 0 to 6 MVA.

The selected IWD parameters for simulation are: *as*=1, *bs*=0.01, *cs*=1, *av*=1, *bv*=0.01, *cv*=1, $q(T^{IWD}) = -\infty$, *MaxIter*=300, *Itercount*=1, $\rho_n = 0.88$,

$\rho_{IWD} = -0.85$, *Initsoil* = 1200, *Initvel*= 4, and ε_s =0.001.

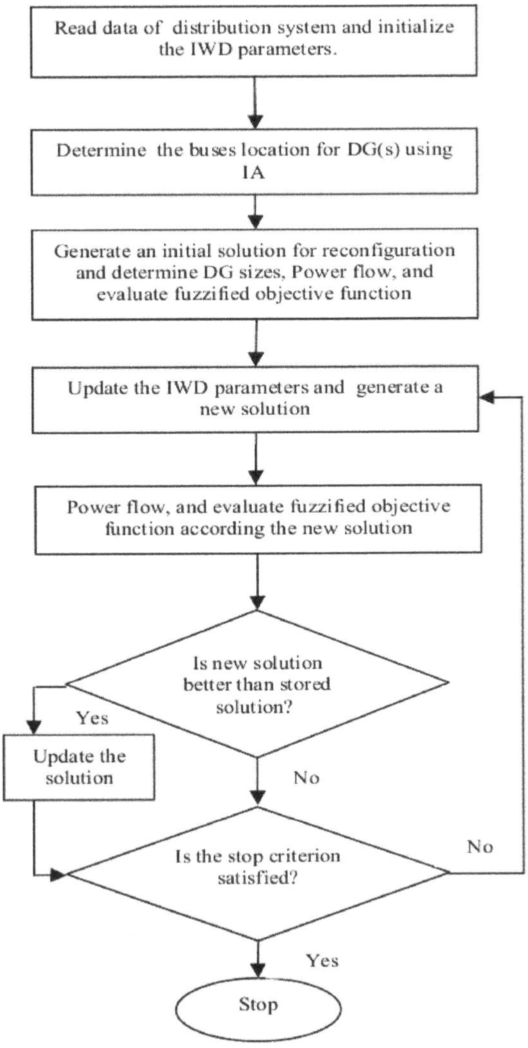

Fig. 2. Flow chart of the proposed method.

9.1 Test system

The test system is a real distribution network of the Taiwan power company. This practical 11.4-kV system is equipped with 83 sectionalizing switches and 13 tie switches.The total system load, which is considered as balanced and constant, is 28.35 kW and 20.7 kVAr. Other information can be obtained from [21]. The power flow calculation is performed based on S_{base}=100 MVA and V_{base}= 11.4 kv. The single line diagram of the Tai-Power 11.4-kV distribution systemis shown in Fig. 3.

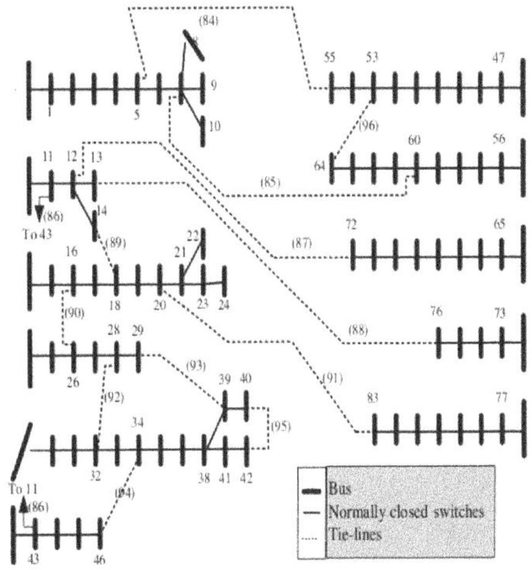

Fig. 3. Single line diagram of the Taiwan power company system.

9.2 Test result

The results of applying the proposed method on the test system are shown in Table 2 for all scenarios. It is observed from this Table that base case power loss in the system is 531.5 kW, which is reduced to 406.91, 326.43, 298.79, 210.62, and 197.23 kW using scenarios II, III, IV, V, and VI , respectively. VPI index is obtained 2.5, 2.35, 1.96, 1.83, 1.66, and 1.47 and LBI index is calculated 140.4, 117.01, 112.54, 114.89, 104.11, and 110.86 for scenarios I to VI, respectively. Also, Table 2 included the optimal locations and sizes for DG units. The total size of DG units is equal to 4.81, 4.83, 5.88, and 5.97 MVA for scenarios III to VI, respectively.

The percentage improvement in P_{loss}, VPI, and, LBI as compared with the base system (scenario I) are presented in Table 3 for scenarios II to VI. As can be seen in this Table, the most improvements in loss reduction, and the voltage profile are 62.89% and 41.2%, respectively for scenario V(simultaneous reconfiguration and DG type 1 allocation). The maximum improvement in equal load balancing(LBI index) is 25.84% for scenario VI(simultaneous reconfiguration and DG type 3 allocation). These results prove that the superiority of the scenarios V and VI (proposed hybrid method) in comparison with others. Also, among all scenarios which DG is presented, itisseen that the presence ofDGtype 1 lead to more improvement in three-indexes of P_{loss}, and VPI in compared to DG type3, while DGtype 3 has led to more improvement in equal load balancing (LBI index)than DGtype 1.

By investigating various scenarios involving reconfiguration, DG allocation, and hybrid of them, it is found that simultaneous multi-objective reconfiguration and placement of DG units is more beneficial than separate single-objective optimization. Scenario V (at nominal load) are simulated using GA [22], PSO [23], and HSA [24], Fuzzy-BA [13], and Honey Bee Mating Optimization (HBMO) and Shuffled Frog Leaping Algorithm (SFLA) (HBMO-SFLA) [25] to be compared with the results obtained by IWD and fuzzy-IWD (proposed method).

Table 2. Results of Taiwan power company

Scenario	Tie switches	Total DG sizes (MVA) @ buses	$P_{loss}(KW)$	VPI	LBI
Scenario I	84, 85, 86, 87, 88, 89, 90, 91, 92, 93, 94, 95, 96	-	531.50	2.5	140.4
Scenario II	7, 13, 34, 39, 41, 61, 84, 86, 87, 89, 90, 91, 92	-	406.91	2.35	117.01
Scenario III	7, 13, 34, 39, 41, 61, 84, 86, 87, 89, 90, 91, 92	4.81 @ 8, 42, 95	326.43	1.96	112.54
Scenario IV	7, 13, 34, 39, 42, 55, 72, 86, 89, 90, 91, 92, 96	4.83 @ 26, 31, 80	298.79	1.83	114.89
Scenario V	7, 13, 34, 39, 42, 55, 72, 86, 89, 90, 91, 92, 96	5.88 @ 17, 36, 50	210.62	1.66	104.11
Scenario VI	7, 13, 34, 39, 42, 55, 72, 86, 89, 90, 91, 92, 96	5.97 @ 33, 59, 66	197.23	1.47	110.68

Table 3. Comparison of results for all tested scenarios

Improvements	Scenario II	Scenario III	Scenario IV	Scenario V	Scenario VI
P_{loss} (%)	23.44	38.58	43.78	60.37	62.89
VPI (%)	6.00	21.6	22.8	33.6	41.2
LBI (%)	16.65	19.84	18.16	25.84	21.16

Table 4. Comparison of simulation results for different methods

Method	Case	Scenario V
GA	Tie-switches	7, 13, 34, 39, 42, 55, 72, 86, 89, 90, 91, 92, 96
	P_{loss}	370.09
	VPI	2.03
	LBI	129.8
	Total DG size (MVA) @ buses	4.76 @ 14, 43, 95
PSO	Tie-switches	7, 13, 34, 39, 41, 61, 84, 86, 87, 89, 90, 91, 92
	P_{loss}	323.98
	VPI	1.89
	LBI	112.41
	DG size (MVA) @ buses	4.93 @ 22, 43, 69
HSA	Tie-switches	7, 13, 34, 39, 41, 61, 84, 86, 87, 89, 90, 91, 92
	P_{loss}	341.60
	VPI	1.94
	LBI	118.23
	Total DG size (MVA) @ buses	4.79 @ 14, 43, 95
Fuzzy-BA	Tie-switches	7, 13, 34, 39, 42, 55, 72, 86, 89, 90, 91, 92, 96
	P_{loss}	232.18
	VPI	1.71
	LBI	107.53
	Total DG size (MVA) @ buses	5.82 @ 10, 73, 84
HBMO-SFLA	Tie-switches	7, 13, 34, 39, 41, 61, 84, 86, 87, 89, 90, 91, 92
	P_{loss}	287.31
	VPI	1.73
	LBI	109.04
	Total DG size (MVA) @ buses	5.46 @ 2, 44, 78
IWD	Tie-switches	7, 13, 34, 39, 42, 55, 72, 86, 89, 90, 91, 92, 96
	P_{loss}	294.55
	VPI	1.86
	LBI	110.57
	Total DG size (MVA) @ buses	5.12 @ 49, 52, 73
Fuzzy-IWD	Tie-switches	7, 13, 34, 39, 42, 55, 72, 86, 89, 90, 91, 92, 96
	P_{loss}	210.62
	VPI	1.66
	LBI	104.11
	Total DG size (MVA) @ buses	5.88 @ 17, 36, 50

From Table 4, it is observed that the performance of the fuzzy-IWD is better than GA, PSO, HSA in all terms of the loss reduction [26], voltage profile, and equal load balancing improvement for this scenario. In Figure 4, the obtained values for the optimal fitness of the different algorithms are compared with together on Taiwan power company test system.

As it is shown in this figure, the fuzzy-IWD method has better performance than the others because of the lowest optimal fitness equal to 0.1306 in comparison with the fuzzy-BA (0.1352), HBMO-SFLA (0.1483), IWD (0.1527), PSO (0.1595), HSA (0.1874), and GA (0.2118) methods, which confirms the ability of the proposed method.

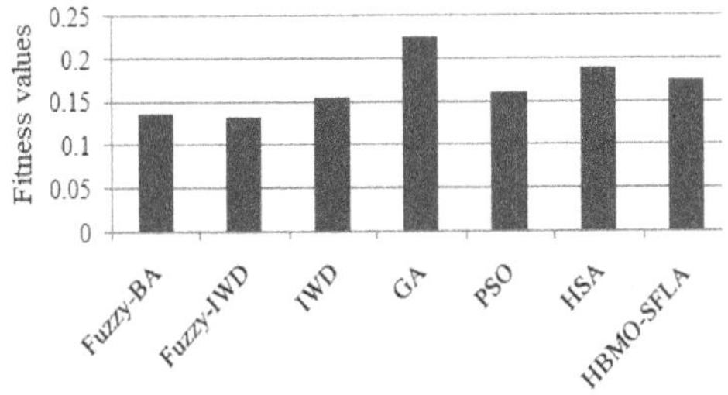

Fig. 4. Optimal fitness for different methods.

10. CONCLUSION

In this paper, a new hybrid method based on fuzzy-intelligent water drops approach has been proposed to simultaneous multi objective reconfiguration and installation of multiple DG units in order to loss reduction, improving the voltage profile, and equalizing the feeder load balancing in distribution system. To reduce the search space, the improved analytical method is employed to select the optimal candidate locations for multiple-DG. Six different scenarios have been tested on a Taiwan power company test system by considering three numbers of DG type 1 and DG type 3 to demonstrate the effectiveness of the proposed technique. From the simulation and analysis of the results, among the six scenarios, the scenario that includes simultaneous reconfiguration and DG type 1 allocation generated the best result in loss reduction (62.89%), and improving the voltage profile (41.2%) as compared with the base test system. The best result in equalizing the feeder load balancing has been obtained by the scenario that proposes simultaneous reconfiguration and DG type 3 allocation. This scenario led to load balancing improvement about 25.84 % as compared with the base case. By investigating all obtained results, it is proved that simultaneous reconfiguration and placement of multiple DG units is more beneficial than separate single objective optimization.

The obtained results by applying the proposed hybrid method are compared with the obtained results based on another intelligent methods i.e. IWD, GA, PSO, HSA, fuzzy-BA, and HBMO-SFLA at nominal load for scenario V. The results of this comparison showed that performance of the proposed technique is better than others.

REFERENCES

[1] H.B. Tolabi, M. Gandomkar, and M.B. Borujeni, "Reconfiguration and load balancing by software simulation in a real distribution network for loss reduction", *Canadian Journal on Electrical and Electronics Engineering*, vol. 2, no. 8, pp. 386-391, 2011.

[2] A. Merlin, and H. Back, "Search for a minimal-loss operating spanning tree configuration in an urban power distribution system", *Proceedings of the5th Power System Computation Conference*, pp.1-18, 1975.

[3] S.K. Goswami, and S.K. Basu, "A new algorithm for the reconfiguration of distribution feeders for loss minimization", *IEEE Transactionson Power Delivery*, vol. 7, no. 3, pp.1484-1491,1992.

[4] F.V. Gomes, S. Carneiro, and J.L. R. Pereira, M. P.V.P.A. N. Garcia, and L. RamosAraujo, "A new heuristic reconfiguration algorithm for large distribution systems", *IEEE Transactions on Power Systems*, vol. 20 no.3, pp.1373-1378, 2005.

[5] E. Ramos, A.G. Exposito, J.R Santos, and F.L. Iborra, "Path-based distribution network modeling: application to reconfiguration for loss reduction", *IEEE Transaction Power Systems*, vol. 20, no. 2, pp. 556-564, 2005.

[6] H.P. Schmidt, and N. Kagan, "Fast reconfiguration of distribution systems considering loss minimization", *IEEE Transaction Power Systems*, vol. 20, no. 3, pp.1311-1319, 2005.

[7] Q. Zhou, D. Shirmohammadi, and W.H. E. Liu, "Distribution feeder reconfiguration for service restoration and load balancing", *IEEE Transactions on Power Systems*, vol. 12, no. 2, pp. 724-729, 1997.

[8] R. Taleski, and D. Rajicic, "Distribution network reconfiguration for energy loss reduction", *IEEE Transaction on Power Systems*, vol. 12, no. 1, pp. 398-406, 1997.

[9] A. Kavousi-Fard, and T. Niknam, "Multi-objective

stochastic distribution feeder reconfiguration from the reliability point of view", *Energy*, vol. 64, no. 1, pp. 342-354, 2014.

[10] J. Moshtagh and S. Ghasemi, "Optimal distribution system reconfiguration using non-dominated sorting genetic algorithm (NSGA-II)", *Journal of Operation and Automation in Power Engineering*, vol. 1, no. 1, pp. 12-21, 2013.

[11] R. Ishak, A. Mohamed, A.N. Abdalla, and M. Z.C. Wanik, "Optimal placement and sizing of distributed generators based on a novel MPSI index", *International Journal of Electrical Power & Energy Systems*, vol. 60, pp. 389-398, 2014.

[12] H. Doagou-Mojarrad, G.B. Gharehpetian, and H. Rastegar, JavadOlamaei, "Optimal placement and sizing of DG (distributed generation) units in distribution networks by novel hybrid evolutionary algorithm", *Energy*, vol. 54, pp. 129-138, 2013.

[13] H.B. Tolabi, M.H. Ali, S.B. M. Ayob, and M. Rizwan, "Novel hybrid fuzzy-bees algorithm for optimal feeder multi-objective reconfiguration by considering multiple-distributed generation", *Energy*, vol. 71, pp. 507-515, 2014.

[14] D. Quoc Hung, N. Mithulananthan, and R. C. Bansal, "Analytical expressions for DG allocation in primary distribution networks", *IEEE Transactions on Energy Conversion*, vol. 25, no. 3, pp. 814-820, 2010.

[15] S. Ghosh, and K.S. Sherpa, "An efficient method for load flow solution of radial distribution networks," *International Journal of Electrical Power & Energy Systems* ,vol. 1, no. 2, pp. 108-115, 2008.

[16] D. Quoc, and N. Mithulananthan, "Multiple distributed generator placement in primary distribution networks for loss reduction", *IEEE Transactions on Industrial Electronics*, vol. 60, no. 4, pp. 1700-1708, 2013.

[17] L.A. Zadeh, "Fuzzy sets", Information and Control, vol. 8, pp. 338-353, 1965.

[18] E. Seyedi, M.M. Farsangi, M. Barati, and H. Nezamabadipour, "SVC multi-objective VAR planning using SFL", *International Journal on Technical and Physical Problems of Engineering*, vol. 3, pp. 76-80, 2010.

[19] H. Shah-Hosseini, "Problem solving by intelligent water drops". *Proceedings of the IEEE Congress on Evolutionary Computation*, pp. 3226-3231, 2007.

[20] S.H. Hosseini, "The intelligent water drops algorithm: a nature-inspired swarm-based optimization algorithm", *International Journal of Bio-Inspired Computing*, vol. 1, no.1/2, pp. 71-79, 2009.

[21] L.W. Oliveira, S. Carneiro, E.J. Oliveira, J.L.R. Pereira, I.C. Silva, and J.S. Costa, "Optimal reconfiguration and capacitor allocation in radial distribution systems for energy losses minimization", *Electrical Power and Energy Systems*, vol. 32, pp. 840-848, 2010.

[22] K. Nara, A. Shiose, M. Kitagawa, and T. Ishihara, "Implementation of genetic algorithm for distribution system loss minimum reconfiguration", *IEEE Transactions on Power Delivery*, vol. 7, no.3, pp.1044-1051, 1992.

[23] J. Olamaei, T. Niknam, and G. Gharehpetian, "Application of particle swarm optimization for distribution feeder reconfiguration considering distributed generators", *Applied Mathematics and Computation*, vol. 201, no.1-2, pp. 575-586, 2008.

[24] R.S. Rao, S.V.L. Narasimham, M.R. Raju, and A.S. Rao, "Optimal network reconfiguration of large-scale distribution system using harmony search algorithm", *IEEE Transactions on Power Systems*, vol. 26, no.3, pp.1080-1088, 2011.

[25] B. Bahmanifirouzi , E. Farjah , T. Niknam , and E. Azad Farsani, "A new hybrid HBMO-SFLA algorithm for multi-objective distribution feeder reconfiguration problem considering distributed generator units", *Iranian Journal of Science and Technology, Transactions of Electrical Engineering*, vol. 36, no. 1, pp. 51-66, 2012.

[26] R. Baghipour, and S.M. Hosseini, "A hybrid algorithm for optimal location and sizing of capacitors in the presence of different load models in distribution network", *Journal of Operation and Automation in Power Engineering*, vol. 2, no. 1, pp. 10-21, 2014.

18

Performance Scrutiny of Two Control Schemes Based on DSM and HB in Active Power Filter

R. Kazemzadeh[1,*], E. Najafi Aghdam[1], M. Fallah[1], Y. Hashemi[2]

[1]Renewable Energy Research Center, Faculty of Electrical Engineering, Sahand University of Technology, Tabriz, Iran

[2]Department of Electrical Engineering, University of Mohaghegh Ardabili, Ardabil, Iran

ABSTRACT

This paper presents a comparative analysis between two current control strategies, constant source power and generalized Fryze current, used in Active Power Filter (APF) applications having three different modulation methods. The Hysteresis Band (HB) and first-order Delta-Sigma Modulation (DSM) as well as the second-order DSM is applied. The power section of the active power filter is viewed as an Analogue to Digital Converter (ADC), then as a result a three-phase shunt active filter modulator controller which, uses Delta-Sigma analogue to digital converter is presented to improve modulator performance. As a result, using second-order Delta-Sigma modulator makes low switching rate compared with first-order Delta-Sigma and hysteresis modulators under same sampling frequency. So, applying this modulator increases system efficiency and reduces cost of switches. In addition, simulation results on MATLAB software show that by using the Delta-Sigma modulator, Total Harmonic Distortion (THD) can be significantly decreased. Moreover, active filter based on the second-order DSM with constant source power has high efficiency and provides lower source current THD.

KEYWORDS: Active power filter, Constant source power, Delta-Sigma modulation, Generalized Fryze current, Hysteresis band.

1. INTRODUCTION

For many years, shunt active power filter has been developed to suppress the harmonics caused by nonlinear loads as well as to compensate the reactive power. The shunt type active power filter eliminates the reactive power and harmonic currents from the grid current by injecting compensation currents intended to result in sinusoidal grid current [1].

The control strategy for a shunt active filter (as shown in Fig. 1) generates the reference current, i_{cref}, that must be provided by the power filter to compensate reactive power and harmonic currents demanded by the load. This involves a set of currents in the phase domain, which should track switching signal applied to the electronic converter by means of the appropriate closed-loop switching control technique such as hysteresis control or dead-beat control [2, 3]. The performance of an active filter mainly depends on the reference current generation strategy. Several papers have studied and compared the performances of different reference current generation strategies under balanced sinusoidal, unbalanced or distorted Alternating Currents (AC) and voltage conditions [4-7].

In general, the shunt active filter consists of two distinct main blocks: (1) active filter controller, and (2) power converter. The controller is responsible for determining the instantaneous compensating reference current, which is continually passed to the power converter. The power converter is responsible for synthesizing the compensating current that should be drawn from the power system by utilizing an adequate modulation method. The power converter should have a high switching frequency

*Corresponding author:
R. Kazemzadeh (E-mail: r.kazemzadeh@sut.ac.ir)

(f_s) in order to reproduce accurately the compensating current. Usually, we have $f_s > 10 f_{hmax}$, where f_{hmax} represents the frequency of the highest order of harmonic current that must be compensated. Hysteresis control is a well-known current control technique used in voltage-fed PWM converters that forms a non-linear feedback loop. Its advantages are simplicity, outstanding robustness, lack of tracking error, independence of load parameter changes and good dynamics. The hysteresis control drawbacks are needed to extra and variable switching rate, uncontrolled frequency and higher noise compared with other techniques [8, 9]. High switching frequency leads to increasing switching losses and imperfections. Moreover, higher noise causes incorrect performance. Therefore, in this paper, to solve these drawbacks and reducing the THD amount of compensated current (increasing precision of compensation), Delta-Sigma modulation is suggested. Since DSM is an example of pulse-density modulation, the switching pulse waveform can be configured without calculation the on/off time duration. This means that the output switching frequency can be randomly varied under the constant sampling frequency. In addition, DSM has the advantage of harmonic-spreading affects, and pushes off the low-frequency noise [10].

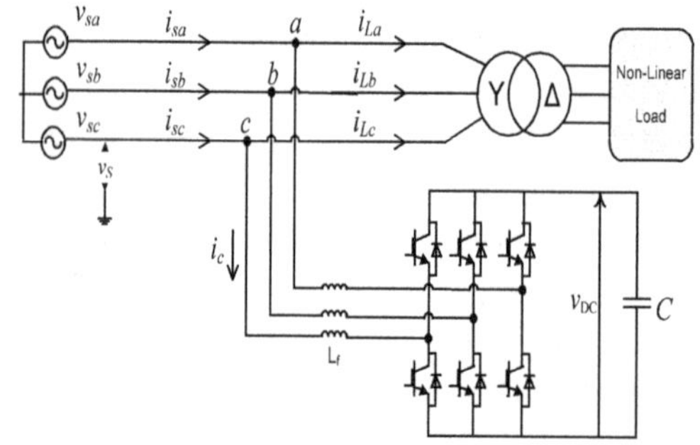

Fig. 1. Shunt active power filter

For this reason, DSM recently has been used for converter in [11-13]. In sections II and III, a review of two control strategies (constant source power and generalized Fryze current) for the extraction of the reference currents is presented. Section IV describes Delta-Sigma modulation and its performance by applying the first and second order loop filters in an active filter. Then, the power section of the active power filter is viewed as an analogue to digital converter, and as a result a three-phase shunt active filter modulator controller which uses Delta-Sigma analogue to digital converter is presented. The remainder of the paper presents comparison of these control strategies using simulation in MATLAB software. The results show that using the first and second order DSM based strategies decreases THD. Moreover, constant source power based on the second-order DSM provides the lowest THD.

2. CONSTANT SOURCE POWER CONTROL STRATEGY

Several methods have been proposed to identify harmonic current such as the methods based on the FFT in frequency domain and the methods based on instantaneous power calculation in time domain. The instantaneous power theory, known as p-q theory, has been developed for three-phase three-wire systems with balanced and sinusoidal source voltage. The compensation target is assumed to get a constant instantaneous source power [14]. This process is shown in Fig. 2. It is based on coordinate transformation from the phase reference system (*abc*) to the *αβ*. The transformation matrix is associated as follows:

$$\begin{bmatrix} v_\alpha \\ v_\beta \end{bmatrix} = \sqrt{\frac{2}{3}} \begin{bmatrix} 1 & -\frac{1}{2} & -\frac{1}{2} \\ 0 & \frac{\sqrt{3}}{2} & -\frac{\sqrt{3}}{2} \end{bmatrix} \begin{bmatrix} v_a \\ v_b \\ v_c \end{bmatrix} \quad (1)$$

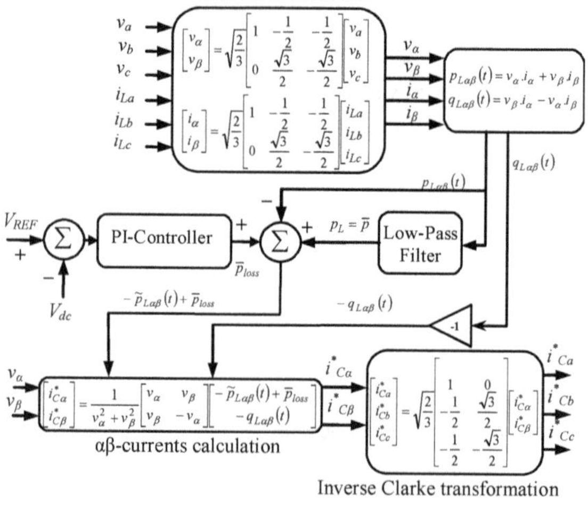

Fig. 2. Block diagram of the constant source power control strategy

$$\begin{bmatrix} i_\alpha \\ i_\beta \end{bmatrix} = \sqrt{\frac{2}{3}} \begin{bmatrix} 1 & -\frac{1}{2} & -\frac{1}{2} \\ 0 & \frac{\sqrt{3}}{2} & -\frac{\sqrt{3}}{2} \end{bmatrix} \begin{bmatrix} i_{La} \\ i_{Lb} \\ i_{Lc} \end{bmatrix} \quad (2)$$

The different power terms are defined as follows:

$$\begin{bmatrix} p_{L\alpha\beta}(t) \\ q_{L\alpha\beta}(t) \end{bmatrix} = \begin{bmatrix} v_\alpha & v_\beta \\ -v_\beta & v_\alpha \end{bmatrix} \begin{bmatrix} i_\alpha \\ i_\beta \end{bmatrix} = [T] \begin{bmatrix} i_\alpha \\ i_\beta \end{bmatrix} \quad (3)$$

Considering inverse matrix [T], it is possible to calculate the current components by means of the different power terms. The expression is given in the next equation:

$$\begin{bmatrix} i_\alpha \\ i_\beta \end{bmatrix} = \frac{1}{v_{\alpha\beta}^2} \begin{bmatrix} v_\alpha & -v_\beta \\ v_\beta & v_\alpha \end{bmatrix} \begin{bmatrix} p \\ q \end{bmatrix} \quad (4)$$

where, $v_{\alpha\beta}^2 = v_\alpha^2 + v_\beta^2$, p is the $\alpha\beta$ real instantaneous power, and q is the imaginary instantaneous power.

The strategy assumed in this theory has been obtained using a constant instantaneous power as the source side with the only restriction of getting a null average instantaneous power exchanged by the compensator *pc*.

In order to calculate the compensator current, it is verified that:

$$p_c(t) = p_L(t) - p_s(t) = p_L(t) - p_L \quad (5)$$

wherep_L is the total active power incoming to the load. Equation (5) can be expressed in the following way:

$$p_{c\alpha\beta}(t) = p_{L\alpha\beta}(t) - p_{\alpha\beta} = \tilde{p}_{L\alpha\beta}(t) \quad (6)$$

Which, effectively, fulfils the average value becomes zero as $\langle p_c(t) \rangle = 0$, where $\tilde{p}_L(t)$ represents the AC part of the $p_L(t)$. On the other hand, the instantaneous imaginary power exchanged by the compensator must be the same as the instantaneous imaginary power required by the load: $q_c(t) = q_{L\alpha\beta}(t)$.

3. GENERALIZED FRYZE CURRENT CONTROL STRATEGY

The fundamentals of the *pq* theory are exploited to develop a control strategy, the generalized Fryze current control based on the minimization method of equation [15, 16]. The number of equations is reduced since it does not utilize any reference frame transformation. The control strategy for a shunt active filter that has denominated as the generalized Fryze currents is shown in Fig. 3. The product between the phase voltages v_a, v_b, v_c and load currents i_{La}, i_{Lb} and i_{Lc} will determine the instantaneous active three-phase power of the load. At the same time, the square sum of phase voltages (v_a, v_b, v_c) is determined.

The conductance G_e is determined by the division between the active three-phase power and squared instantaneous aggregate voltage (v_a, v_b, v_c). This conductance comprises all current components that produce active power with voltages v_a, v_b and v_c.

A low-pass filter will extract the average value of the conductance $\overline{G}_e$. The DC voltage regulator control circuit determines the signal $\overline{G}_{loss}$. Control signal, $\overline{G}_e + \overline{G}_{loss}$ with the control voltages v_a, v_b and v_c which are used to determine the active current are pure sinusoidal waves in phase with v_a, v_b and v_c include only the portion (proportional $\overline{G}_e$) of load current that produces active power, and the active current (proportional to G_{loss}) that is necessary for losses compensation in the shunt active filter. Since the shunt active filter compensates the difference between the calculated active current and measured load current, it is possible to guarantee that the source currents i_{as}, i_{bs} and i_{cs} drawn from the network are always sinusoidal, balanced and in phase with positive-sequence voltages. The DC voltage regulator is used to generate control signal $\overline{G}_{loss}$ as shown in Fig. 3. This signal forces the shunt active filter to drawn an additional active current from the network to compensate losses in the power circuit of the shunt active filter. Additionally, it corrects DC voltage variations caused by abnormal operation and transient compensation errors. It consists only of a closed loop PI-controller $[G(s) = k_p + \frac{k_i}{s}]$.

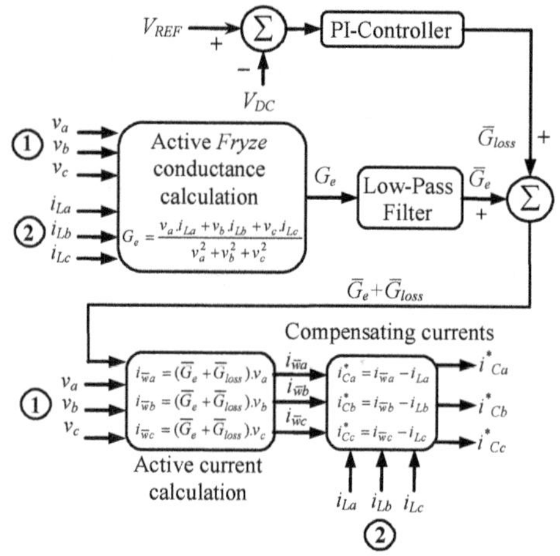

Fig. 3. Block diagram of the generalized Fryze current control strategy

4. DELTA-SIGMA MODULATION IN POWER ELECTRIC APPLICATIONS

In the proposed switching power converter, an analogue to digital (A/D) converter is used to properly prepare the state of switches. In other words, the analogue input is the compensating current reference signal to be synthesized, and the quantized digital output is the state of the circuit switches.

A/D conversion is an inherent non-linear operation and introduces errors to the conversion (known as quantization noise) [17]. In both communications and power electronics implementations, the purpose is to design the system so that the input signal is passed through the system with minimal distortion from noise.

Oversampling Delta-Sigma technique is one of the best quantization noise reduction methods [18]. This modulator does not reduce the magnitude of the quantization noise, but instead, shapes the noise. It shapes power density spectrum of noise by moving the energy toward higher frequencies and reducing quantization noise within the signal band as depicted in Fig. 4 [10, 18]. Increased noise in the high frequency band can be removed by using a simple low pass filter. In fact, Delta-Sigma modulators achieve better conversion performance by using a low-resolution quantizer in a feedback loop with linear filtering [19]. The quantizer can be modeled using an input-independent additive white noise [20], which is very simple and useful for many practical purposes. Fig. 5 shows the structure of Delta-Sigma modulator [10].

Since power switching converters typically switch at frequencies well in excess of the input Nyquist rate, the Delta-Sigma modulation technique can be employed in power electronics application as the cases [21].

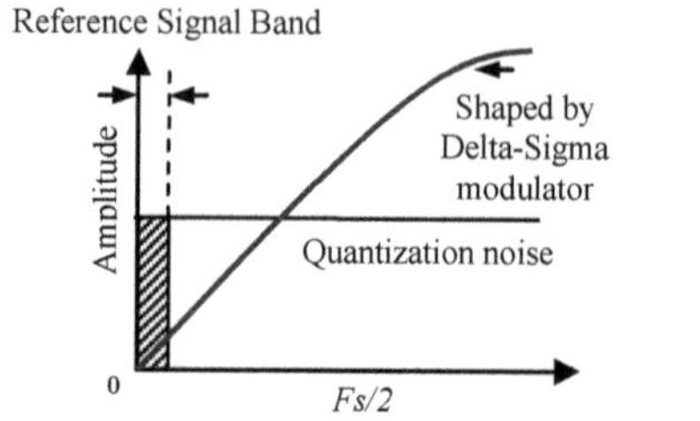

Fig. 4. The Delta-Sigma modulator noise-shaping characteristic (first-order)

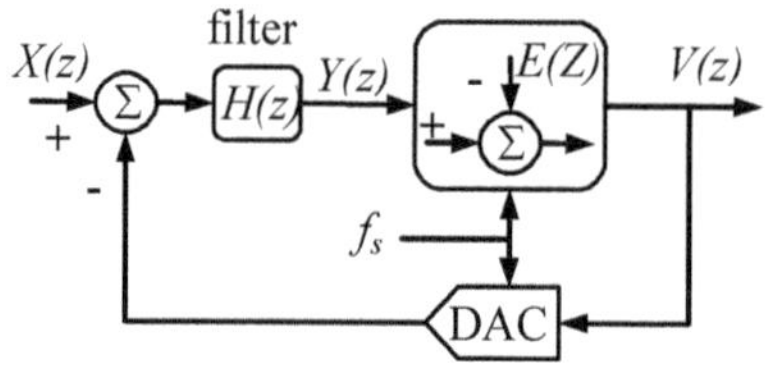

Fig. 5. Block diagram of Delta-Sigma modulator

To illustrate how a power electronic circuit can be embedded in a Delta-Sigma modulator, consider the modulator for the half-bridge converter shown in Fig. 6. In this arrangement, the gating circuitry and half-bridge are embedded into the loop following the latch. The comparator and latch set the switch state for each sampling period according to the sign of the comparator input at the sampling instant. The switch state impresses the voltage at the output. Thus, taking the input signal to be the desired output voltage, the actual output voltages can follow the desired state. Output passive low pass filter exists in the APF's inverter serves as a converter on the feedback path which transforms two level output voltage of the inverter to continuous current that follows reference signal. In addition, the nature of the low pass transfer function in the APF's inverter acts the role of a decimation filter. In other words, the passive filter in active shunt filter structure, serves as a digital to analogue (D/A) converter mentioned in Fig. 4, as well as filtering out of band quantization noise shaped towards high frequencies.

A typical input-output transfer characteristic of a monobit quantizer is a stair function. Obviously, an infinite number of input amplitudes can be mapped to a unique output level resulting in loss of exact information of the input magnitude and hence impose the quantization error (*E(z)*). Based on the Bennette's method for approximating quantization function [22], the quantizer can be linearized using an input-independent additive white noise model in order to make the analysis tractable. Under such an approximation, the nonlinear modulator becomes a linear system with a stochastic input, and the performance can be easily derived in *Z* domain.

$$\mathrm{V}(z) = STF(z)X(z) + NTF(z)E(z) \tag{7}$$

$$\mathrm{STF}(z) = \frac{V(z)}{X(z)} = \frac{\mathrm{H}(z)}{1+\mathrm{H}(z)} = z^{-L} \tag{8}$$

$$NTF\ (z)\frac{V(z)}{E(z)} = \frac{1}{1+H(z)} = (1-z^{-1})^{L} \tag{9}$$

The first order Delta-Sigma modulator is the simplest structure; where $H(z)= z\text{-}1/1\text{-}z^{-1}$ is discrete mode or equivalent $1/s$ in continuous-time type of modulator [22].

A way to further noise reduction is to use a higher order Delta-Sigma modulator. A second-order continuous time Delta-Sigma modulator is shown in Fig. 6 (b), where, a first-order modulator is embedded in the second loop with an integrator in the feed-forward path. Compared with the first-order modulator, the second-order modulator has one more design parameter available; it is the ratio of the gains of two paths; this makes it capable to determine low frequency properties of the circuit by outer path and improve system stabilizing while determining high frequency properties by inner path. Moreover, by a deliberate increase in the inner loop, delay in the outer loop can be compensated [18].

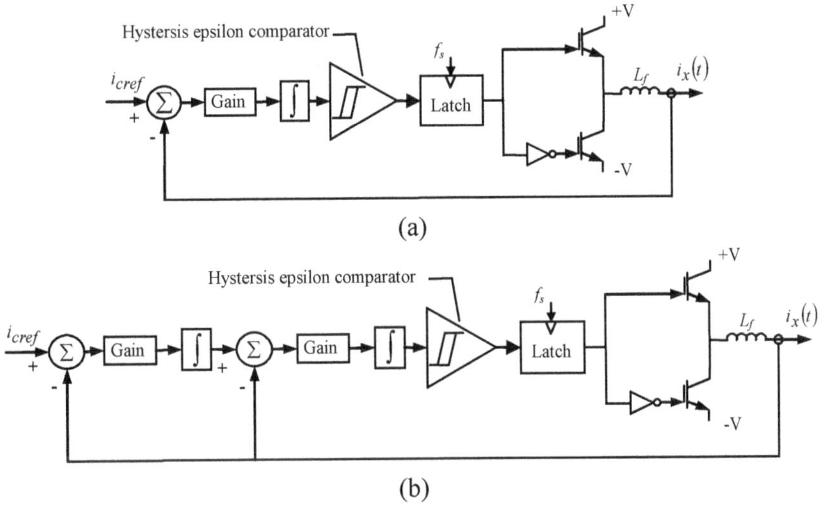

Fig. 6. Half-bridge embedded in Delta-Sigma modulator loop; a) first-order, b) second-order

Higher order modulators also are possible to construct, but they cannot simply be made by adding further stages as above. The reason is that the phase shift caused by more than two integrators will make the system unstable [18].

The first statement of (7) presents Signal Transfer Function (STF), and the second statement shows Noise Transfer Function (NTF). STF is generally flat in the band for the first and second order modulators, however the difference between first, second and third order modulator can affect differently on the modulator's performance.

The NTF of systems is shown in Fig. 7. Regarding to comparison of STF and NTF, and having stability problem, system complexity and implementation problems of the modulators of orders higher than 2 in view, it becomes manifest that a second-order Delta-Sigma modulator is a better compromise between circuit complexity and signal to noise ratio. Therefore, a second order Delta-Sigma modulator is chosen for emerging in the proposed active power filter.

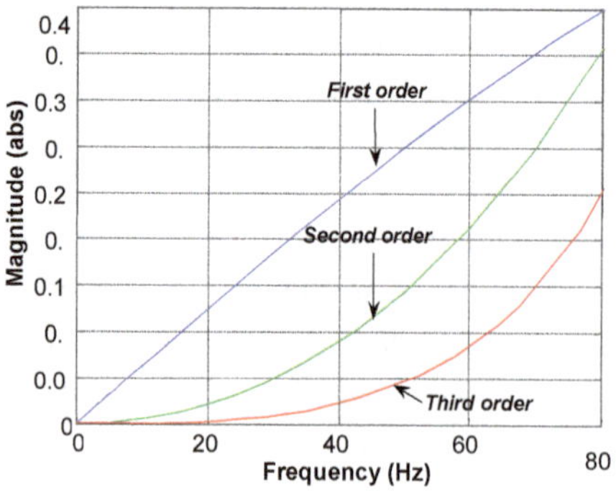

Fig. 7. Magnitude spectra for NTFs

Since, the proposed modulator will be implemented using continuous time integrator, once the order of modulator is chosen (second order here as depicted in Fig. 6(b)), it is usually designed in discrete or *Z* domain then, transferred to continuous time domain using the following equation as impulse-invariant transform method [10]:

$$Z^{-1}\{H(z)\}=L^{-1}\{R_D(s).H(s)\}_{t=nTs} \tag{10}$$

where*H(s)* is equivalent Laplace transform of *H(z)* together with *RD(s)* which is the impulse response of the DAC on the feedback path. In the simplest second order modulator case, discrete to continuous time conversion results in the structure are shown in Fig.8, where all zeros of the selected NTF are non-optimized and laid at real zero.

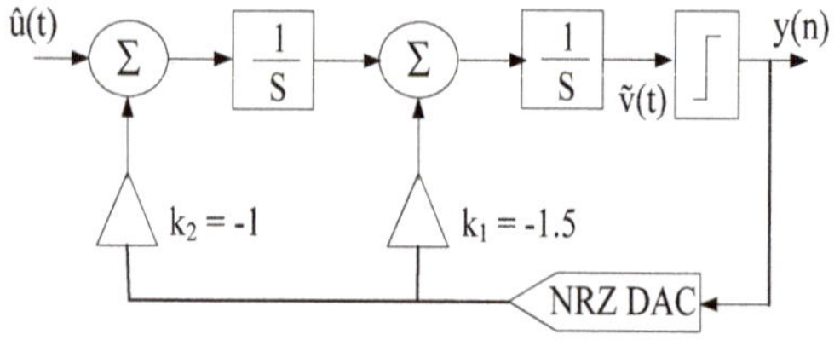

Fig. 8. Discrete to continuous time conversion of second order modulator

5. SIMULATION RESULTS

The harmonic currents and reactive power compensation by APF is implemented in a three-phase power system which the power system voltage of 380^V and thyristor rectifier with three types of resistive-inductive, resistive and inductive (are intended as A, B and C in Table 1, respectively) loads as the harmonic current compensation objects are considered. The design specifications and the circuit parameters used in the simulation are indicated in Table 1.

In the second loop control strategy, a structure of the second-order Delta-Sigma modulator, as mentioned above in the Fig. 6, is used. For the simulation of the active filter operation, it is used a balanced and sinusoidal three-phase voltage system where two different methods to control the active filter are applied. Judging from these results, it is clear that the Delta-Sigma modulator system compared to the HB modulated system has better performance in two different methods to control the active filter in terms of lower THD and better-compensated currents wave shape. Table 2-3 show the general characteristics of both systems in terms of harmonic components. Main AC voltage is illustrated in Fig. 8. In the case of resistive-inductive load, line current is shown in Fig. 9.

Table 1. Parameter values of simulation

Parameter	Parameter value
System voltage	380 V
Load	3-phase thyristor by 20° firing angle rectifier with A) 10 mH- 20 ohm, B) 10 ohm and C)2 H loads, fed by a YΔ transformer
Filter Inductance	0.5 mH
Dc link capacitors	1200 μF
DC link reference voltage	800 V
Sampling frequency	30 kHz

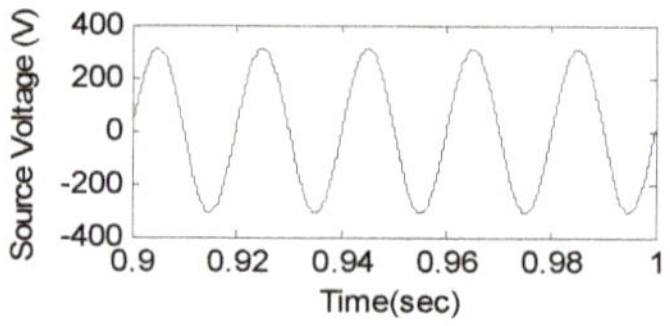

Fig. 9. Balanced and sinusoidal voltage

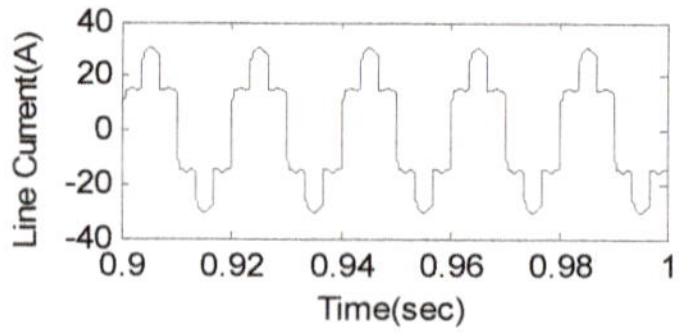

Fig. 10. Line current waveform of resistive-inductive load

Compensated line current and filter current in case of resistive-inductive load by the constant source power and the generalized Fryze current control strategies for HB, first-order Delta-Sigma and second-order Delta-Sigma modulation systems are illustrated in Figs. 10-13.

In the cases of pure resistive and inductive loads, line current is depicted in Fig. 14. Also, compensated line current of pure resistive and inductive loads by mentioned control strategies with different modulations are shown in Figs. 15-16 and Figs. 17-18, respectively. Note that simulations are made for a three-phase system, but only one phase (phase A) is shown in figures to make images clear to present avoidance of visual error.

Finally, general characteristics of load harmonics and compensated current with three types of modulation systems for pure resistive and inductive loads are presented in Tables 4-5 and 6-7, respectively.

Another advantage of Delta-Sigma modulation in comparison with hysteresis modulation is the lower switching rate under the same sampling frequency. The benefits of low switching of Delta-Sigma modulation technique are low switching losses, imperfection and cost. Thus, reducing the switching losses will increase the system efficiency. A comparison of relative switching losses in the power circuit of the shunt active power filter by the constant source power control strategy with three types of modulation is summarized in Table 8.

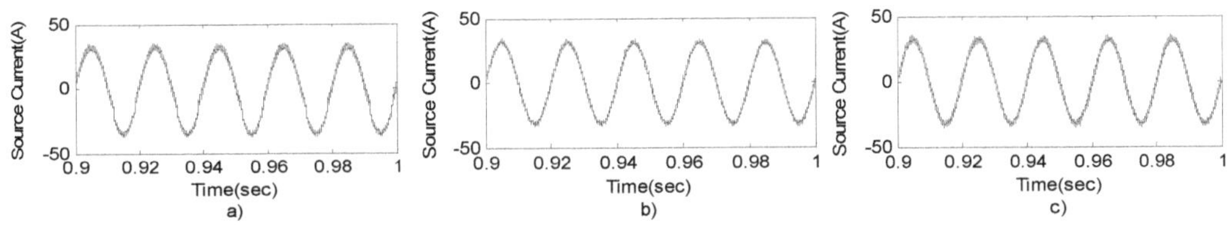

Fig. 11. Compensated currents by the constant source power control strategy using (a) HB, (b) First-order Delta-Sigma (c) Second-order Delta-Sigma modulation

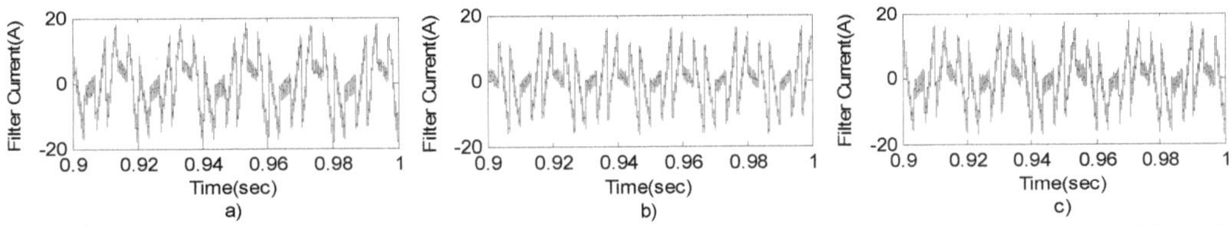

Fig. 12. APF currents by the constant source power control strategy using (a) HB (b) First-order Delta-Sigma (c) Second-order Delta-Sigma modulation

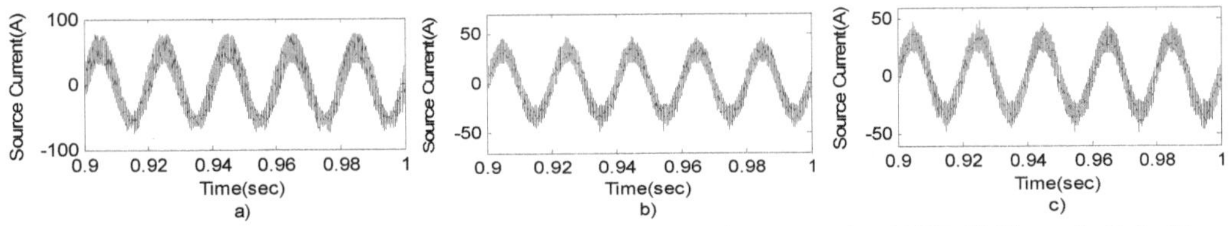

Fig. 13. Compensated currents by the generalized Fryze current control strategy using (a) HB (b) First-order Delta-Sigma (c) Second-order Delta-Sigma modulation

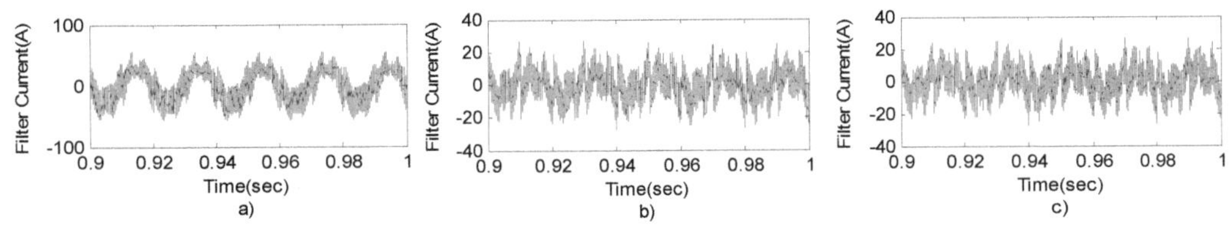

Fig. 14. APF currents by the generalized Fryze current control strategy using (a) HB (b) First-order Delta-Sigma (c) Second-order Delta-Sigma modulation

Table 2. Harmonic componets demanded to AC mains by resistive-inductive load wihout APF and with mentiond APFs by the constant source power control strategy

	Without APF	HB controlled APF	First delta-sigma controlled APF	Second delta-sigma controlled APF
I_3	0.00%	0.01%	0.01%	0.01%
I_5	21.81%	1.71%	0.34%	0.11%
I_7	11.61%	1.26%	0.48%	0.22%
I_9	0.00%	0.02%	0.03%	0.04%
I_{11}	8.37%	0.73%	0.10%	0.09%
I_{13}	6.22%	0.78%	0.09%	0.13%
I_{15}	0.00%	0.03%	0.02%	0.01%
I_{17}	4.81%	0.46%	0.04%	0.11%
I_{19}	3.92%	0.48%	0.18%	0.12%
THD	28.06%	7.74%	1.94%	1.68%

Table 3. Harmonic componets demanded to AC mains by resistive-inductive load wihout APF and with mentiond APFs by the generalized Fryze current control strategy

	Without APF	HB controlled APF	First delta-sigma controlled APF	Second delta-sigma controlled APF
I_3	0.00%	0.14%	0.16%	0.08%
I_5	21.81%	3.24%	2.95%	2.72%
I_7	11.61%	0.83%	0.68%	0.63%
I_9	0.00%	0.12%	0.10%	0.13%
I_{11}	8.37%	1.31%	0.16%	0.22%
I_{13}	6.22%	1.16%	0.63%	0.68%
I_{15}	0.00%	0.31%	0.02%	0.07%
I_{17}	4.81%	0.41%	0.62%	0.25%
I_{19}	3.92%	0.47%	0.79%	0.48%
THD	28.06%	12.58%	7.43%	6.83%

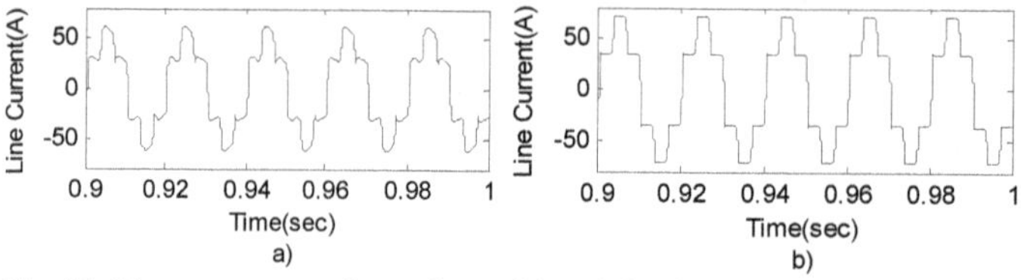

Fig. 15. Line current waveform of pure (a) resistive load, (b) inductive load

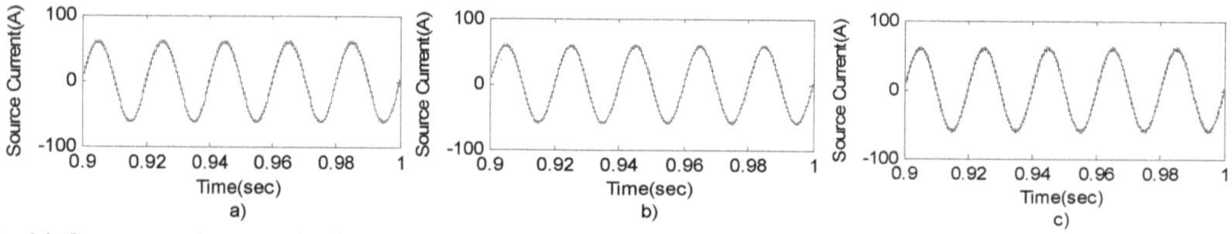

Fig. 16. Compensated currents by the constant source power control strategy using (a) HB, (b) First-order Delta-Sigma (c) Second-order Delta-Sigma modulation

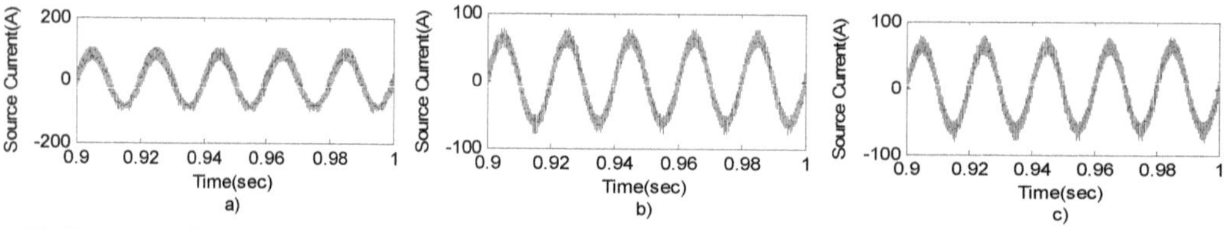

Fig. 17. Compensated currents by the generalized Fryze current control strategy using (a) HB (b) First-order Delta-Sigma (c) Second-order Delta-Sigma modulation

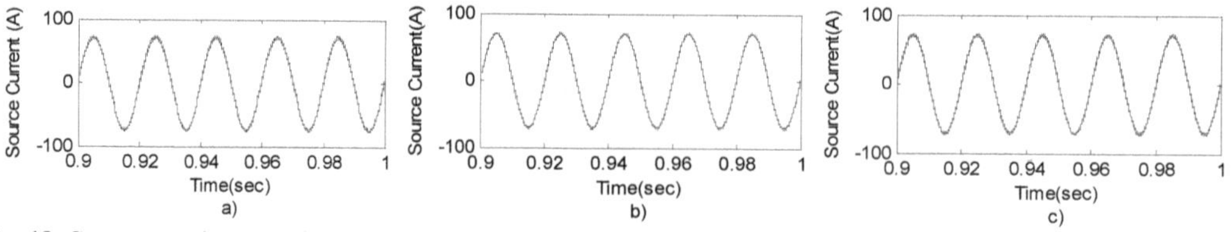

Fig. 18. Compensated currents by the constant source power control strategy using (a) HB, (b) First-order Delta-Sigma (c) Second-order Delta-Sigma modulation

6. CONCLUSION

This paper has provided a comparative analysis of two control strategies and three modulation techniques for shunt APF. For instance, in constant source power and generalized Fryze current control strategies, using a first-order DSM results in lower line current distortion when compared to HB modulation. Furthermore, for the same conditions, using the second-order DSM guarantees better performance than that of the first-order DSM. Based on theoretical discussion and simulation results, it can be concluded that the use of DSM instead of HB increases the compensation accuracy and reduces switching losses. In addition, under same sampling frequency, switching frequency of Delta-Sigma modulation and therefore its switching losses is lower than HB modulation. Also, among mentioned methods, the performance of the constant source power control strategy compared with the generalized Fryze current control strategy is better. Thus, based on our comparative analysis, constant source power control strategy with second DSM can be proposed as the best control structure for shunt active filter.

Table 4. Harmonic componets demanded to AC mains by pure resistive load wihout APF and with mentiond APFs by the constant source power control strategy

	Without APF	HB controlled APF	First delta-sigma controlled APF	Second delta-sigma controlled APF
I_3	0.00%	0.03%	0.01%	0.04%
I_5	22.59%	0.94%	0.15%	0.08%
I_7	10.46%	0.68%	0.20%	0.20%
I_9	0.00%	0.04%	0.03%	0.04%
I_{11}	8.25%	0.48%	0.20%	0.18%
I_{13}	5.08%	0.48%	0.07%	0.04%
I_{15}	0.00%	0.02%	0.03%	0.02%
I_{17}	4.30%	0.24%	0.18%	0.16%
I_{19}	2.78%	0.20%	0.05%	0.09%
THD	27.44%	4.31%	1.08%	0.97%

Table 5. Harmonic componets demanded to AC mains by pure resistive load wihout APF and with mentiond APFs by the generalized Fryze current control strategy

	Without APF	HB controlled APF	First delta-sigma controlled APF	Second delta-sigma controlled APF
I_3	0.00%	0.12%	0.03%	0.07%
I_5	22.59%	1.87%	2.31%	2.15%
I_7	10.46%	0.48%	0.71%	0.71%
I_9	0.00%	0.02%	0.05%	0.07%
I_{11}	8.25%	0.05%	0.02%	0.03%
I_{13}	5.08%	0.65%	0.38%	0.35%
I_{15}	0.00%	0.04%	0.05%	0.04%
I_{17}	4.30%	0.40%	0.38%	0.31%
I_{19}	2.78%	0.16%	0.15%	0.14%
THD	27.44%	6.66%	5.04%	4.68%

Table 6. Harmonic componets demanded to AC mains by pure inductive load wihout APF and with mentiond APFs by the constant source power control strategy

	Without APF	HB controlled APF	First delta-sigma controlled APF	Second delta-sigma controlled APF
I_3	0.00%	0.03%	0.02%	0.02%
I_5	19.47%	0.93%	0.08%	0.06%
I_7	13.53%	0.71%	0.25%	0.20%
I_9	0.00%	0.02%	0.02%	0.02%
I_{11}	7.92%	0.46%	0.07%	0.10%
I_{13}	6.34%	0.50%	0.12%	0.11%
I_{15}	0.00%	0.00%	0.02%	0.01%
I_{17}	4.21%	0.24%	0.08%	0.09%
I_{19}	3.45%	0.18%	0.09%	0.08%
THD	26.63%	4.37%	1.08%	0.94%

Table 7. Harmonic componets demanded to AC mains by pure inductive load wihout APF and with mentiond APFs by the generalized Fryze current control strategy

	Without APF	HB controlled APF	First delta-sigma controlled APF	Second delta-sigma controlled APF
I_3	0.00%	0.09%	0.04%	0.14%
I_5	19.47%	1.89%	2.27%	1.55%
I_7	13.53%	0.39%	0.75%	0.17%
I_9	0.00%	0.02%	0.08%	0.08%
I_{11}	7.92%	0.10%	0.10%	0.14%
I_{13}	6.34%	0.58%	0.37%	0.35%
I_{15}	0.00%	0.10%	0.01%	0.15%
I_{17}	4.21%	0.35%	0.39%	0.20%
I_{19}	3.45%	0.17%	0.22%	0.53%
THD	26.63%	6.32%	4.66%	4.29%

Table 8. Comparison of relative switching losses

APF Control Strategy	Constant source power		
Modulation	Hysteresis	First-Order DS	Second-Order DS
Relative Losses	1	0.97	0.93

REFERENCES

[1] H.R. Imanijajarmi, A. Mohamed, H. Shareef, , "Active power filter design by a novel approach of multi-objective optimization", *Journal of Operation and Automation in Power Engineering*, vol. 1, no. 1, pp. 54-62, 2013.

[2] Y. Han, L. Xu, G. Yao, L. Zhou, M. Khan and C. Chen, "A robust deadbeat control scheme for active power filter with LCL input filter," *Electrical Review,* vol. 86, no. 2, pp. 14-19, 2010.

[3] Y. He, J. Liu, J. Tang, Z. Wang and Y. Zou, "Deadbeat control with a repetitive predictor for three-level active power filters," *Journal of Power Electronics,* vol. 11, no. 4, pp. 583-590, 2011.

[4] G.W. Chang and T.C. Shee, "A novel reference

compensation current strategy for shunt active power filter control" *IEEE Transactions on Power Delivery*, vol. 19, no. 4, pp. 1751-1758, 2004.

[5] G.W. Chang and C.M. Yeh, "Optimization-based strategy for shunt active power filter control under non-ideal supply voltages," *IEE Proceedings on Electric Power Applications,* vol. 152, no. 2, pp. 182-190, 2005.

[6] K.S. Hong and C.S. Kim, "A DSP based optimal algorithm for shunt active filter under non-sinusoidal supply and unbalanced load conditions," *IEEE Transactions on Power Electronics*, vol. 22, no. 2, pp. 593-601, 2007.

[7] M.I.M. Montero, E.R. Cadaval and F.B. Gonzalez, "Comparison of control strategies for shunt active power filters in three-phase four-wire systems," *IEEE Transactions on Power Electronics*, vol. 22, no. 1, pp. 229-236, 2007.

[8] A. M. Razali, M. A. Rahman and N. A. Rahim, "An analysis of current control method for grid connected front-end three phase AC-DC converter," *Proceedings of the IEEE ECCE Asia Downunder Conference,* pp. 45-51, 2013.

[9] C. Attaianese, M. Di Monaco and G. Tomasso, "High performance digital hysteresis control for single source cascaded inverter," *IEEE Transactions on Industrial Informatics,* vol. 9, no. 2, pp. 620-629, 2013.

[10] P.M. Aziz, H.V. Sorensen and J. Vn der Spiegel, "An overview of sigma-delta converters," *IEEE Signal Processing Magazine,* vol. 13, no. 1, pp. 61-84, 1996.

[11] Y. Wang, and D. Ma, "Design of Integrated Dual-Loop Delta–Sigma Modulated Switching Power Converter for Adaptive Wireless Powering in Biomedical Implants" *IEEE Transactions on Industrial Electronics,* vol. 58, no. 9, pp. 4241-4249, 2011.

[12] J. Paramesh and A. Von iouanne, "Use of sigma-delta modulation to control EMI from switch-mode power supplies," *IEEE Transactions on Industrial Electronics*, vol. 48, no. 1, pp. 111-117, 2001.

[13] J. Amini, R. Kazemzahed, H. Madadi Kojabadi, "Performance enhancement of indirect matrix converter based variable speed Doubly-Fed induction generator", *Proceedings of the 1st Conferenceon Power Electronic & Drive Systems & Technologies*, pp. 450-455, 2010.

[14] M. Popescu, A. Bitoleanu and V. Suru, "A DSP-Based implementation of the p-q theory in active power filtering under nonideal voltage conditions," *IEEE Transactions on Industrial Informatics*, vol. 9, no. 2, pp. 880-889, 2013.

[15] M. Aredes, L.F.C. Monteiro and J. Mourente, "Control strategies for series and shunt active filters," *Proceedings of the IEEE Power Technology Conference*, vol. 2, 2003.

[16] A.E. Leon, S.J. Amodeo, J.A. Solsona and M. I. Valla, "Non-linear optimal controller for unified power quality conditioners," *IET Power Electronics,* vol. 4, no. 4, pp. 435-446, 2011.

[17] L.F.C. Monteiro and M. Aredes, "A compare-ative analysis among different control strategies for shunt active filters," *Proceedings of the Industrials Applications Conference,* pp. 345-350, 2002.

[18] S.R. Norsworthy, R. Schreier and G.C. Temes, "Delta-Sigma data converters-theory, design, and simulation," *IEEE Press*, 1997.

[19] E.N. Aghdam, "Nouvelles techniques d'appariement dynamique dans un CNA multibit pour les convertisseurs sigma-delta," Doctoral Dissertation," *University of Paris*, 2006.

[20] W. R. Bennett, "Spectra of quantized signals," *Bell System Technical Journal,* vol. 27, no. 3, pp. 446-472, 1948.

[21] E. Monmasson, "Power Electronic Converters," *France, Wily*, 2009.

[22] L. Breems and J.H. Huijsing, "Continuous- time sigma-delta modulation for A/D conversion in radio receivers," *Springer*, 2001.

19

Multi-Stage DC-AC Converter Based on New DC-DC Converter for Energy Conversion

E. Salary, M. R. Banaei*, A. Ajami

Department of Electrical Engineering, Azarbaijan Shahid Madani University, Tabriz, Iran

ABSTRACT

This paper proposes a multi-stage power generation system suitable for renewable energy sources, which is composed of a DC-DC power converter and a three-phase inverter. The DC-DC power converter is a boost converter to convert the output voltage of the DC source into two voltage sources. The DC-DC converter has two switches operates like a continuous conduction mode. The input current of DC-DC converter has low ripple and voltage of semiconductors is lower than the output voltage. The three-phase inverter is a T-type inverter. This inverter requires two balance DC sources. The inverter part converts the two output voltage sources of DC-DC power converter into a five-level line to line AC voltage. Simulation results are given to show the overall system performance, including AC voltage generation. A prototype is developed and tested to verify the performance of the converter.

KEYWORDS: Renewable energy, Multi-stage inverter, DC-DC converter, Multi-level inverter.

1. INTRODUCTION

Distributed generation (DG) systems as local power sources have great potential to contribute toward energy sustainability, energy efficiency and supply reliability. In the electrical power system, the most important driving forces for the proliferation of DGs are [1-3]:

- Liberalized electricity market
- environmental concerns with greenhouse gas emissions
- energy efficiency
- diversified energy sources constitute
- peak shaving capability

The power conversion interface is important to load or grid-connected DG power generation systems. Some DGs such as Fuel cell (FC) and photovoltaic (PV) generate DC voltage so an inverter is necessary to convert the DC power to AC power [4-6]. Since the output voltage of a PV or FC array is low, a DC-DC power converter is used in a power generation system to boost the output voltage. Variable voltage is one of the most important problems of the fuel cells. PVs are small power resources and it is essential to choose a suitable method of maximum power point tracking (MPPT). Output voltages of the PV modules are variable because of changes of temperature and sunlight irradiation. Different type of MPPT algorithms have suggested in different papers. Among the MPPT techniques, the perturbation and observation (P&O) method and incremental conductance are the most popular because of the simplicity of its control structure [3]. The PV system requires DC-DC converters with a flexible control range to obtain maximum power and increasing voltage. Input current ripple is an important feature for DC-DC converters, used in fuel cell applications. The low current ripple increases the efficiency and lifetime of the fuel cells. In multi-stage converter, inverter part generates AC voltage and injects power to the grid.

The DC-DC converter with high step-up voltage gain is widely used for energy conversion systems. Conventionally, the classic DC-DC boost converter is used for voltage step-up applications. In high step-up voltage gain this converter will be operated at a high duty ratio [7]. Some literatures have researched

*Corresponding author:
M. R. Banaei (m.banaei@azaruniv.edu)

the high step-up DC-DC converters that do not incur an extremely high duty ratio [8-12].

Multilevel inverter technology has appeared recently as a very important alternative in the area of the power system. In theory, multilevel inverters should be designed with higher voltage levels in order to improve the conversion efficiency and to reduce harmonic content and electromagnetic interference. This technology can be used in energy conversion. They can generate output voltages with extremely low distortion and lower dv/dt [13,14]. The different topologies presented in the literature as multilevel converters. The most popular multilevel converters are the diode-clamped, flying capacitor and cascaded H-bridge structures. In recent years, novel topologies of multilevel inverters using a reduced number of switches and gate driver circuits have presented [15-17]. Unfortunately, multilevel inverters have some disadvantages. One particular disadvantage is the great number of power semiconductor switches needed. Although low voltage rate switches can be utilized in a multilevel inverter, each switch requires a related gate driver circuit. This may cause the overall system to be more expensive and complex. So, in practical implementation, reducing the number of switches and gate driver circuits is very important.

It seems that using of multilevel inverter with high number of voltage levels and switches is not commodious in low voltage application. The multilevel inverter with a low number of levels needs a less number of switches and can be suitable for low voltage application. In Ref. [18], a five-level inverter is developed and applied for injecting the real power of the renewable power into the grid to reduce the switching power loss, harmonic distortion, and electromagnetic interference caused by the switching operation of power electronic devices. This topology uses six switches and two diodes in one phase. In Ref. [19] the hybrid seven-level cascaded active neutral-point-clamped (ANPC) based multilevel converter is presented. The converter topology is the cascaded connection of a three-level ANPC converter and an H-bridge per phase. This topology uses ten switches in one phase. In Ref. [20], a seven-level inverter topology, configured by a level generation part and a polarity generation part, is proposed. There, only power electronic switches of the level generation part switch in high frequency, but ten power electronic switches and three DC capacitors are used. The T-type multilevel converter is one type of multilevel inverters that is introduced in Ref. [21]. This inverter is suitable for using grid distribution voltage (400 or 380 V). This converter uses two DC-link capacitor and nine switches.

In this paper, one DC-AC converter based T-type multilevel inverter is presented. The proposed converter has two parts: DC-DC converters and T-type multilevel inverter. In T-type, the conventional six switches inverter is converted to a multilevel inverter topology that requires only nine active switches. A new type of DC-DC converter is used to provide DC-link voltages for multilevel inverter. The main advantages of DC-DC converter are low input current ripple and high voltage gain. In this letter, the operating principle of the developed system is described, and a prototype of inverter is constructed for verifying the effectiveness of the topology. A case study about fuel cell power generation is studied.

2. MULTI-STAGE DC-AC CONVERTER

Figure 1 shows the configuration of the proposed multi-stage system. The proposed system is composed of a DC voltage source array, a DC-DC power converter and T-type multilevel inverter. The DC voltage source is connected to the DC-DC converter that is a boost converter. The boost DC-DC converter converts the output power of the DC voltage into two voltage sources, which supply the inverter.

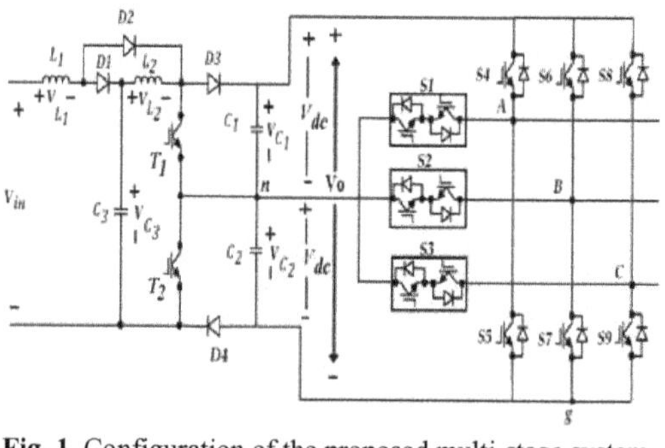

Fig. 1. Configuration of the proposed multi-stage system.

The T-type multilevel inverter is composed of

three bidirectional switches and a conventional six switches power converter, connected in cascade. As can be seen, T-type multilevel inverter contains only nine switches, so, the power circuit is simplified.

2.1. DC-DC converter

In a classic DC-DC boost converter, the voltage stresses on the switch and diode, which are equal to the output voltage, are high. In the proposed multi-stage converter, a step-up DC-DC converter is used, as shown in Fig. 1. Inside DC-DC converter, the semiconductor device voltage rating is only half of the output voltage. Special modulation technique, offers lower input current ripple and output voltage ripple. The peak inverse voltage of switches and diodes is half of output voltage.

In order to simplify the circuit analysis of the converter, all components are assumed ideal. The voltage of capacitors is equal. To operation analysis of DC-DC converter, it is assumed that converter has one resistive load.

$$V_{C1} = V_{C2} = V_{dc} \quad (1)$$

The output voltage is equal to sum of voltage of output capacitors.

$$V_O = V_{C1} + V_{C2} = 2V_{dc} \quad (2)$$

The proposed converter operates in continuous conduction mode (CCM) and discontinuous conduction mode (DCM). Here, CCM is analyzed and discussed.

Based on the aforementioned assumptions, there are three operating modes discussed in one switching period under CCM operation.

Figure 2 shows the topology stages of the proposed converter. The operating modes are described as follows.

Mode 1: Fig. 2(a) shows mode 1 equivalent circuit. During this mode T_1 and T_2 are turned on. The DC-source energy is transferred to L_1 and L_2 is charged by C_3, so currents of inductors are increased. In this mode D_2 is turned on and D_1, D_3 and D_4 are turned off. Energy of output capacitors are given to load.

$$V_{L1} = V_{in} \quad (3)$$

$$V_{L2} = V_{C3} \quad (4)$$

Duration of mode 1 is equal as:

$$t_1 = DT = \frac{D}{f_D} \quad (5)$$

where, D, T and f_D are the duty cycle switching period and switching frequency, respectively.

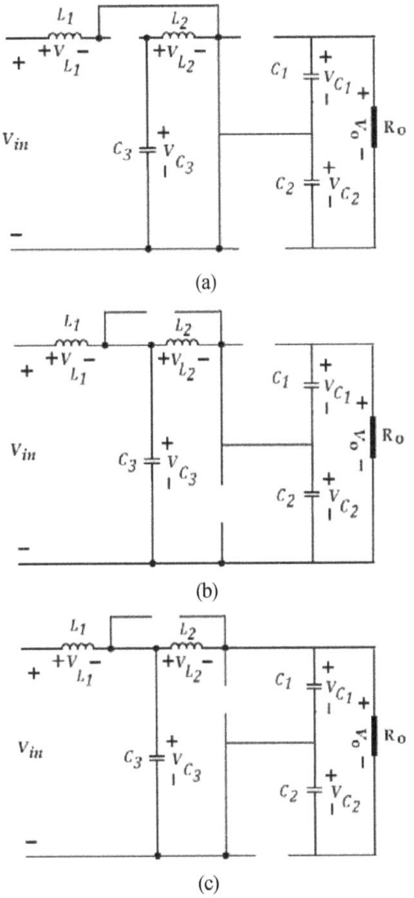

Fig. 2. Topological stages of the proposed converter (a) mode 1 (b) mode 2 and (c) mode 3.

Mode 2: T_1 is turned on and T_2 is turned off. The energy is pumped to C_2 and C_3 while the energy of C_1 is given to load. The currents of inductors decrease. Fig. 2(b) shows mode 2 equivalent circuit. During mode 2, the voltage across the inductors is:

$$V_{L1} = V_{in} - V_{C3} \quad (6)$$

$$V_{L2} = V_{C3} - V_{C2} = V_{C3} - \frac{V_O}{2} \quad (7)$$

Duration of mode 2 is equal as:

$$t_2 = (\frac{1-D}{2})T \quad (8)$$

Mode 3: T_1 is turned off and T_2 is turned on. The energy is pumped to C_1 through T_2, and D_3, so, the currents of inductors are decreased. Fig. 2(c) shows mode 3 equivalent circuit. During mode 3, the voltage across the inductor is:

$$V_{L1} = V_{in} - V_{C3} \tag{9}$$

$$V_{L2} = V_{C3} - V_{C1} = V_{C3} - V_O/2 \tag{10}$$

Duration of mode 3 is equal as:

$$t_3 = ((1-D)/2)T \tag{11}$$

The inductor average voltage and capacitor average current over one cycle is zero [11].

$$\overline{V_{L1}} = 0 = D.V_{in} + \frac{2(1-D)}{2}(V_{in} - V_{C3}) \\ \Rightarrow V_{C3} = \frac{V_{in}}{(1-D)} \tag{12}$$

$$\overline{V_{L2}} = 0 = D.V_{C3} + \frac{2(1-D)}{2}(V_{C3} - \frac{V_O}{2}) \\ \Rightarrow V_O = \frac{2V_{C3}}{(1-D)} \tag{13}$$

$$\overline{i_{C3}} = 0 = -D.I_{L2} + (1-D)(I_{L1} - I_{L2}) \\ \Rightarrow I_{L2} = \frac{I_{L1}}{(1-D)} \tag{14}$$

Substituting Eq. (12) into Eq. (13) yields the voltage conversion ratio of the proposed converter,

$$V_O = \frac{2V_{in}}{(1-D)^2} \tag{15}$$

Figure 3 shows modulation waveforms and gate signals in one period. The modulation technique is expressed as follows:

If (-D ≤ triangular ≤ D) then T_1: on, T_2: on

If (triangular > D) then T_1: on, T_2: off

If (triangular < -D) then T_1: off, T_2: on

Figure 4 shows the signal gates, voltage and current of input inductor (L_1). The charge and discharge of L_2 are the same as L_1. The frequency of inductor current is doubling of switching frequency. Based on Eq. (15), the input current I_{L1} can be expressed as:

$$I_{L1} = \frac{2V_O}{R(1-D)^2} = \frac{2I_O}{(1-D)^2} \tag{16}$$

Where, I_O is the output current. In addition, the current ripple of i_{L1} and i_{L1} denoted by Δi_{L1} and Δi_{L2}, respectively. The current ripple of inductors can be expressed to be

$$\Delta i_{L1} = \frac{DT}{2L_1}V_{in} = \frac{D(1-D)^2T}{4L_1}V_O \tag{17}$$

$$\Delta i_{L2} = \frac{(1-D)DT}{4L_2}V_O \tag{18}$$

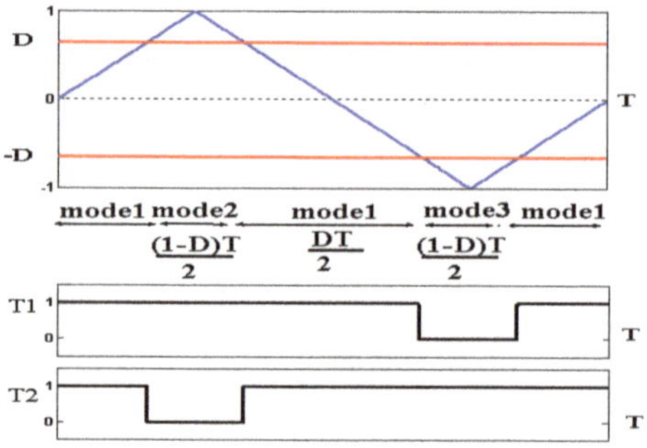

Fig. 3. Modulation waveforms and gate signals in one period.

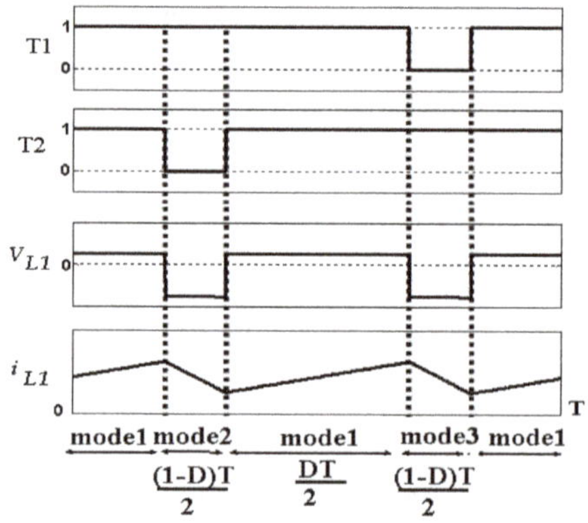

Fig. 4. The signal gates, voltage and current of input inductor (L_1).

The ripple of inductor current is half of ripple of inductor current in classic boost DC-DC converter. This is one advantage of proposed converter.

From Eqs. (15) and (16), for steady-state analysis of the boundary condition mode (BCM) we have:

$$I_{L1} = \frac{\Delta i_{L1}}{2} = \frac{D.T}{4L_1}V_{in} = \frac{2I_O}{(1-D)^2} \tag{19}$$

$$L_1 = \frac{D(1-D)^4TV_O}{16I_O} = \frac{D(1-D)^4TR_O}{16} \tag{20}$$

Where

$$R_O = V_O / I_O \tag{21}$$

The normalized magnetizing-inductor time constant is defined as Eqs. (22) and (25). BCM condition for L_1 and L_2 is shown in Fig. 5.

$$\tau_{L1} = \frac{L_1}{T.R_O} = \frac{D(1-D)^4}{16} \tag{22}$$

$$I_{L2} = \frac{2I_O}{(1-D)} = \frac{\Delta i_{L2}}{2} = \frac{DT}{4L_2}\frac{V_{in}}{(1-D)} \quad (23)$$

$$L_2 = \frac{D(1-D)^2 TR_O}{16} \quad (24)$$

$$\tau_{L2} = \frac{L_2}{T.R_O} = \frac{D(1-D)^2}{16} \quad (25)$$

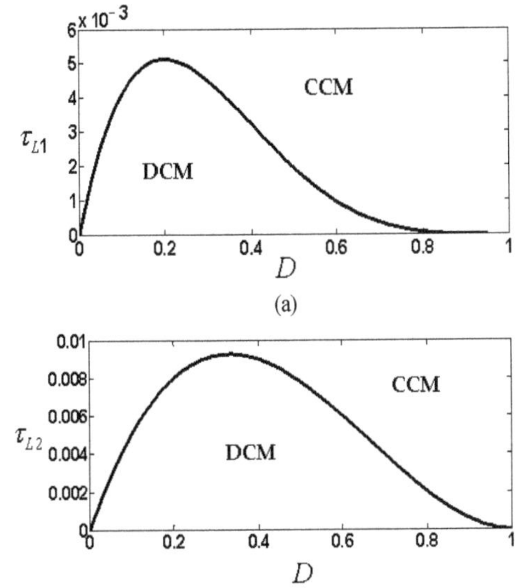

(b)

Fig. 5. BCM condition for L_1 and L_2.

During mode 1, L_2 discharges the C_3 capacitor therefore; the voltage ripple across the C_3 capacitors can be expressed as:

$$\Delta V_{C3} = \frac{DT}{2C_3}I_{L2} = \frac{DT}{(1-D)C_3}I_O \quad (26)$$

If C_1 is equal to C_2 then the voltage ripple of output capacitors is shown as Eq. (27).

$$\Delta V_{C1} = \frac{(1+D)T}{2C_1}I_O \quad (27)$$

Another important problem in power electronic converters is the ratings of switches. In other word, voltage and current ratings of the switches in a converter play important roles on the cost and realization.

The PIV on semiconductors are given as:

$$V_{T1} = V_{T2} = V_{D3} = V_{D4} = \frac{V_O}{2} \quad (28)$$

$$V_{D1} = \frac{(1-D)V_O}{2} \quad (29)$$

$$V_{D2} = DV_O / 2 \quad (30)$$

If inductors are large enough, the current of inductors can be calculated as Eqs. (15) and (23) and they are constant in one period. The current of semiconductor is given as bellow based on current of inductors.

$$i_{D1} = \begin{cases} 0 & DT \\ I_{L1} & (1-D)T \end{cases} \quad (31)$$

$$I_{D1} = \sqrt{(1-D)}I_{L1} \quad (32)$$

$$I_{D1av} = (1-D)I_{L1} \quad (33)$$

$$i_{D2} = \begin{cases} I_{L1} & DT \\ 0 & (1-D)T \end{cases} \quad (34)$$

$$I_{D2} = \sqrt{D}I_{L1} \quad (35)$$

$$I_{D2av} = DI_{l1} \quad (36)$$

$$i_{T1} = i_{T2} = \begin{cases} I_{L1} + I_{L2} & DT \\ I_{L2} & \frac{(1-D)T}{2} \\ 0 & \frac{(1-D)T}{2} \end{cases} \quad (37)$$

$$I_{T1} = I_{T2} = \sqrt{(I_{L1} + I_{L2})^2 D + I_{L2}{}^2 \frac{(1-D)}{2}} \quad (38)$$

$$I_{T1av} = I_{T2av} = (I_{L1} + I_{L2})D + I_{L2}\frac{(1-D)}{2} \quad (39)$$

$$i_{D3} = i_{D4} = \begin{cases} 0 & DT \\ I_{L2} & \frac{(1-D)T}{2} \\ 0 & \frac{(1-D)T}{2} \end{cases} \quad (40)$$

$$I_{D3} = I_{D4} = I_{L2}\sqrt{(\frac{1-D}{2})} \quad (41)$$

$$I_{D3av} = I_{D4av} = I_{L2}(\frac{1-D}{2}) \quad (42)$$

Figure 6 shows the voltage gain versus the duty ratio of the proposed converter and cascaded boost converter. As it is shown in Fig. 6, the presented converter has higher voltage gain.

2.2. DC-AC converter

The T-type three-level inverter is formed by the nine main power devices. A capacitor voltage divider, formed by C_1 and C_2 provides a half supply voltage point, node n in Fig. 1. The auxiliary switches, formed by the controlled bidirectional switch S1, S2 and S3, connect the center point of the left hand three-level inverter to the node n.

The required three voltage output levels for one phase (V_{An}) are generated as follows:

1) Zero output: The auxiliary switches S1 is on, short-circuiting the V_{An} and V_{An}=0. All other controlled switches in phase A are off.

2) Maximum positive output, V_{dc}: S4 is on, connecting the node A to V_{dc} and $V_{An}=V_{dc}$. All other controlled switches are off.

3) Maximum negative output, -V_{dc}: S5 is on, connecting the node A to V_{dc} and $V_{An}=-V_{dc}$. All other controlled switches are off.

In the switching strategy, in one phase at any time only one switch is on. The T-type inverter is able to generate five voltage levels in the line-to-line output voltage. The line-to-line output voltages are given as:

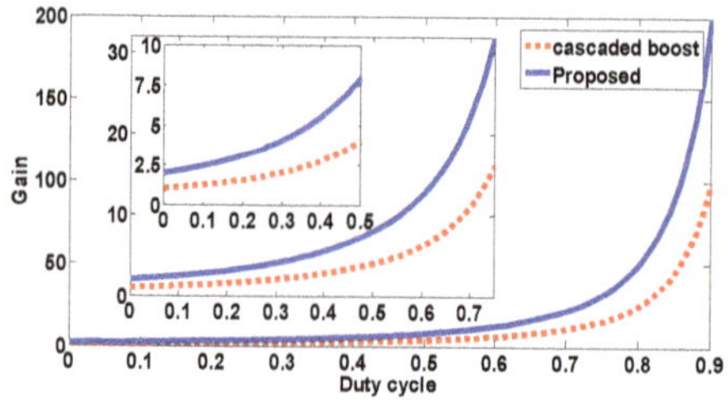

Fig. 6. The voltage gain versus the duty ratio of the proposed converter and cascaded boost converter.

$$\begin{cases} V_{AB} = V_{An} - V_{Bn} \\ V_{BC} = V_{Bn} - V_{Cn} \\ V_{CA} = V_{Cn} - V_{An} \end{cases} \tag{43}$$

Table 1 shows the on switches lookup table of five-level multilevel inverter. It is clear that switches in one phase can't be on simultaneously, because a short circuit across the voltage V_{dc}.

Since only three power electronic switches are used in the proposed inverter in one phase, the power circuit is significantly simplified. The states of the power electronic switches of the five-level inverter, as detailed previously, are summarized in Table 1. It can be seen that only one power electronic switch is switched for each switching operation.

The PIV on switches are given as:

$$V_{S1} = V_{S2} = V_{S3} = \frac{V_O}{2} \tag{44}$$

$$V_{S4} = V_{S5} = V_{S6} = V_{S7} = V_{S8} = V_O \tag{45}$$

Table 1. Lookup table of a three-phase inverter.

On Switches	V_{AB}	V_{BC}	V_{CA}
S4 S7 S8	$2V_{dc}$	$-2V_{dc}$	0
S4 S7 S3	$2V_{dc}$	$-V_{dc}$	$-V_{dc}$
S4 S7 S9	$2V_{dc}$	0	$-2V_{dc}$
S4 S2 S9	V_{dc}	V_{dc}	$-2V_{dc}$
S4 S6 S9	0	$2V_{dc}$	$-2V_{dc}$
S1 S6 S9	$-V_{dc}$	$2V_{dc}$	$-V_{dc}$
S5 S6 S9	$-2V_{dc}$	$2V_{dc}$	0
S5 S6 S3	$-2V_{dc}$	V_{dc}	V_{dc}
S5 S6 S8	$-2V_{dc}$	0	$2V_{dc}$
S5 S2 S8	$-V_{dc}$	$-V_{dc}$	$2V_{dc}$
S5 S7 S8	0	$-2V_{dc}$	$2V_{dc}$
S1 S7 S8	V_{dc}	$-2V_{dc}$	V_{dc}

Although the S1, S2 and S3 are bidirectional switches but the PIV of these switches are half of other switches. The voltage changing of switches is lower than conventional three-level inverter. The dv/dt is given as:

$$\frac{dv}{dt} = \frac{V_O}{2} \tag{46}$$

3. CASE STUDY

The DG units can be operated in stand-alone mode (independent of the grid) or at grid tied mode (connected to the grid). In both modes, DG unit feeds load or power system through an inverter. In this paper, a system shown in Fig. 7 is a case study. In the case study operation of converter in grid tied mode is discussed. In the case study, generated power by FC injects to the grid. The basic control schemes used in the DC-DC and DC-AC converters are shown in Fig. 7. The DC-DC converter control scheme consists of one loop control. The I_{in} is the measured input current and I^*_{in} is the reference current provided by

$$I^*_{in} = \frac{P^*_{FC}}{V_{in}} = \frac{P^*_{in}}{V_{in}} \tag{47}$$

where, P^*_{FC} is reference power of FC. The control loop consists of a conventional PI controller.

Regarding the DC-AC converter, it can operate in a grid tied mode. In the case of grid-connected mode, a decoupled control strategy is used.

The voltage equations in the stationary frame are:

$$e_a = E\cos(\omega t) = V_{Ao} - L_f \frac{di_a}{dt}$$
$$e_b = E\cos(\omega t - \frac{2\pi}{3}) = V_{Bo} - L_f \frac{di_b}{dt} \quad (48)$$
$$e_b = E\cos(\omega t - \frac{4\pi}{3}) = V_{Co} - L_f \frac{di_c}{dt}$$

where, E and ω are the maximum phase voltage and angular frequency of the grid, respectively. The voltage equations in the synchronous frame (*dq*) are given by

$$\begin{bmatrix} e_d \\ e_q \end{bmatrix} = -L_f \frac{d}{dt}\begin{bmatrix} i_d \\ i_q \end{bmatrix} - \omega L_f \begin{bmatrix} -i_q \\ i_d \end{bmatrix} + \begin{bmatrix} V_d \\ V_q \end{bmatrix} \quad (49)$$

where, e_d and i_d are the *d*-axis output voltage and current and e_q and i_q are the *q*-axis output voltage and current. The grid voltages of *dq*-axis are *E* and zero.

For a unity power factor, it is desirable that the *q*-axis current is zero. Then the *q*-axis current is controlled with the zero reference current. The active power supplied to the grid is:

$$P = \frac{3}{2}(e_d i_d + e_q i_q) = \frac{3}{2} E i_d \quad (50)$$

Since the active power is directly proportional to the *d*-axis current, the d-axis reference current is generated from the PI voltage controller for the DC-link voltage regulation. The conventional PI voltage controller is:

$$i_d^* = k_p(V_{dc}^* - V_{dc}) + k_i \int (V_{dc}^* - V_{dc})dt \quad (51)$$

where, V_{dc}^* and V_{dc} are the reference DC-link voltage and the DC-link voltage k_p and k_i are the proportional and integral control gains of the PI voltage controller. The following decoupling control is:

$$V_d = E - \omega L_f i_q + \Delta V_d \quad (52)$$
$$V_q = \omega L_f i_d + \Delta V_q \quad (53)$$

The output signals ΔV_d and ΔV_q of the current controllers generate transient additional voltages required to maintain the sinusoidal input currents.

$$\Delta V_d = k_{pd}(i_d^* - i_d) + k_{id}\int (i_d^* - i_d)dt \quad (54)$$
$$\Delta V_q = k_{pq}(i_q^* - i_q) + k_{iq}\int (i_q^* - i_q)dt \quad (55)$$

k_{pd} and k_{id} are proportional control gains and k_{pq} and k_{iq} are integral control gains.

3. SIMULATION RESULTS

To confirm the feasibility of the proposed converter and the analyses done above, simulation results are carried out by using MATLAB/SIMULINK software. Two parts of simulations are carried out.

In the first simulation the operation of DC-DC boost converter is studied and in the second simulation operation of the proposed converter in the case study is shown. There are several modulation strategies for inverters [21-24].

3.1. Operation of boost DC-DC converter

Table 2 shows parameters of DC-DC converter. This converter is used to increase the 96 V input voltage to 768 V at the output. According to Eq. (15), duty cycle is 0.5. The frequency of switching is 25 kHz. $\Delta i_{in}/I_{in}$ is lower than 5% and is $\Delta v_o/V_O$ lower than 1%. Fig. 8 depicts the output voltage and current. The output voltage is 768 V.

Figure 9 shows voltage of semiconductors. Fig. 9 (a) shows voltage of switches (T_1 and T_2). The voltages of diodes are shown in Fig. 9(b).

The voltages of semiconductors are lower than output voltage. It should be noted that, the voltage stress value of the diode D_1 and D_2 is lower than other semiconductors. Fig. 10 shows the voltages and currents of inductors. The input current (I_{L1}) is shown in Fig. 10(a). The ripple of input current is 1.925 A. Fig. 11 depicts the currents of semiconductors.

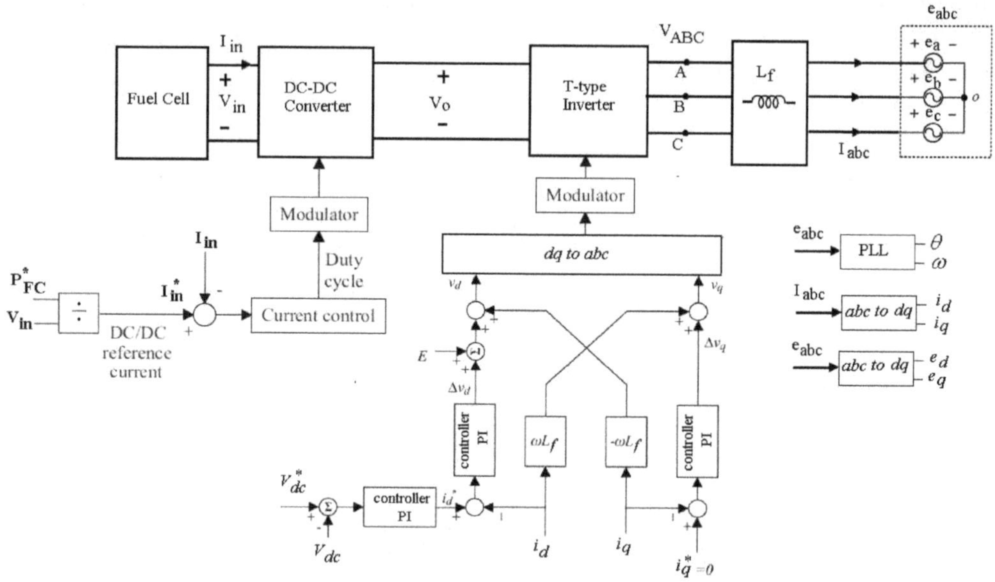

Fig. 7. Case study.

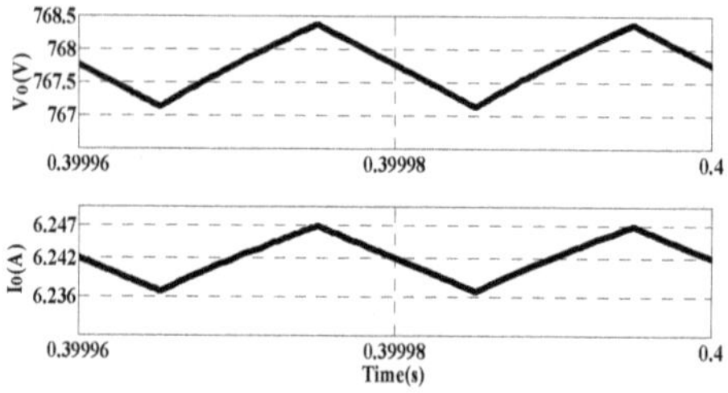

Fig. 8. Output voltage and current.

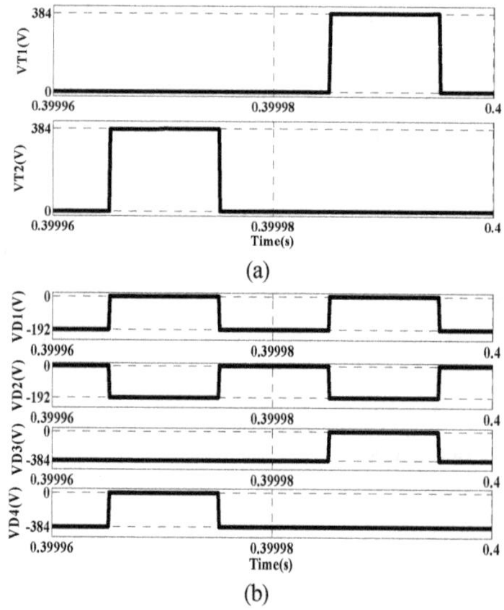

Fig. 9. Voltage of semiconductors (a) voltage of switches (T_1 and T_2) and (b) voltages of diodes.

Table 2. Parameters of DC-DC converter.

DC source	96 V
L_1, L_2	500 μH
C_1, C_2 and C_3	100 μF
Switching frequency	25000 Hz
D	0.5
Load	102 Ω

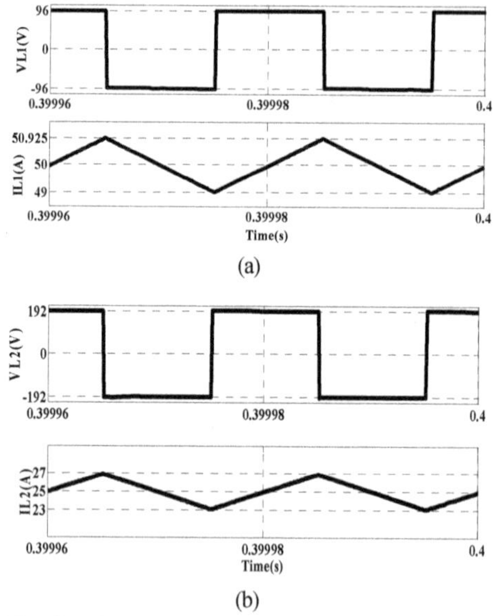

Fig. 10. Voltages and currents of inductors (a) Voltage and current of L_1 and (b) Voltage and current of L_2.

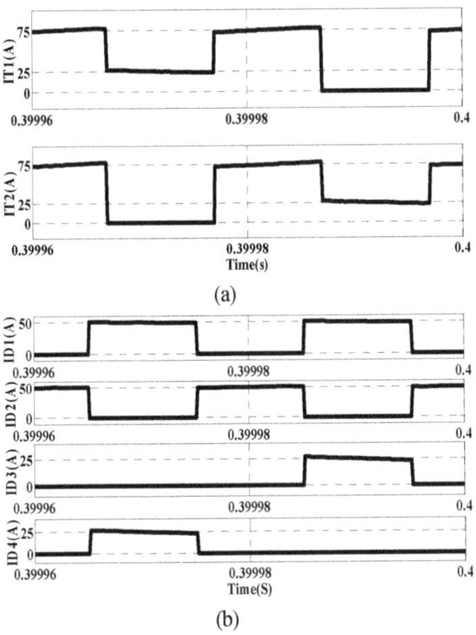

Fig. 11. Currents of semiconductors (a) currents of switches (T_1 and T_2) and (b) currents of diodes.

A low current ripple is an important factor in fuel cell applications.

3.2. Case study simulation

Table 3 shows the parameters of case study simulation. Injection of generating power from FC into the grid is the main goal of the case study. Two cells of FC are connected as series to create DC source. In the case study simulation, the generated power change in 0.4 s.

Figure 12 shows input voltage and current of DC source. The generated power by FC (P_{in}) and injected power to the grid (P) is shown in Fig. 13. The exchanged power between FC and grid is shown in Fig.13. The voltage of DC-link is shown in Fig. 14. Fig. 15 shows the line to line voltage. This voltage has five levels.

Table 3. Parameters of Case study.

Parameters	values
Fuel cell	AFC, 2.4 KW, 48 V
Voltage of DC bus(V_O)	700 V
Nominal frequency	50Hz
Grid voltage	380 V
L_f	750 μH
Inverter frequency	5000 Hz

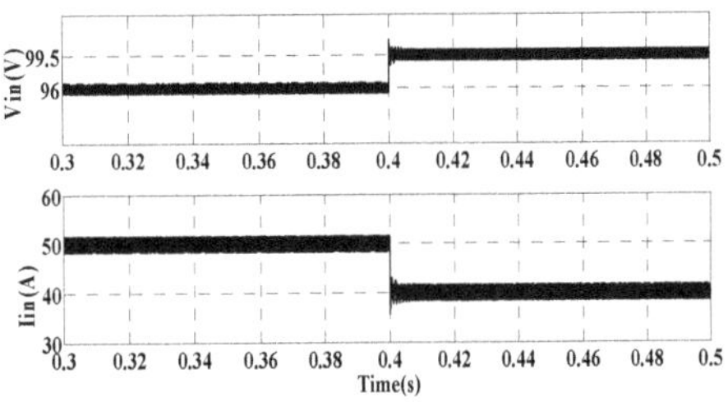

Fig. 12. Input voltage and current of source.

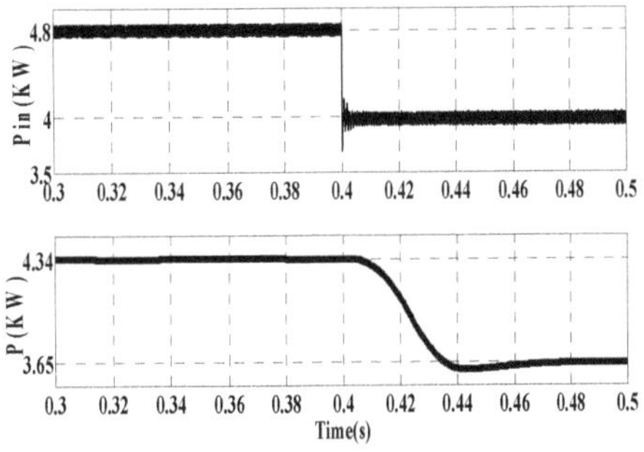

Fig. 13. The exchanged power between FC and grid.

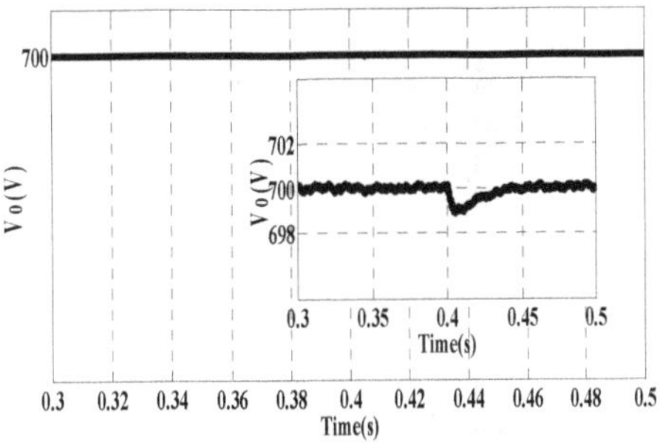

Fig. 14. Voltage of DC-link.

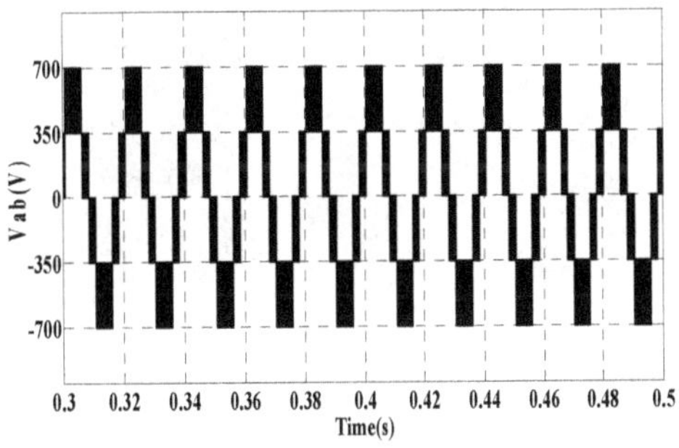

Fig. 15. line to line voltage.

4. EXPRIMENTAL RESULTS

To verify the performance of the converter, a prototype based on the DSP28335 is developed and tested. The parameters of the prototype are listed in Table 4. The DSPTMS320F28335 is used to implement controller and the opto coupler TLP250 is used to drive switches. The fundamental switching frequency is used for inverter part. Fig. 16 shows photographs of the different parts of prototype. The DC-DC converter elements are shown in Fig. 16(a). Fig. 17 shows gate pulse of DC-DC converter. Fig. 18 shows waveforms of DC-DC converter. The input current is shown in Fig. 18(a). Fig. 18(b) shows voltage of capacitors. The dc-link voltage (output voltage of DC-DC converter) is sum of voltage of C_1 and C_2. Fig. 18(c) shows dc-link voltage. Fig. 19 shows the measured output voltage waveforms of inverter part. This is a 50Hz staircase waveform. As it can be seen, the results verify the ability of the proposed inverter in generation of desired output voltage waveform. Fig. 19(a) shows the output voltage of the phases to node n. Each phase generates a quasi-square waveform. Fig. 19(b) shows the line to line output voltage.

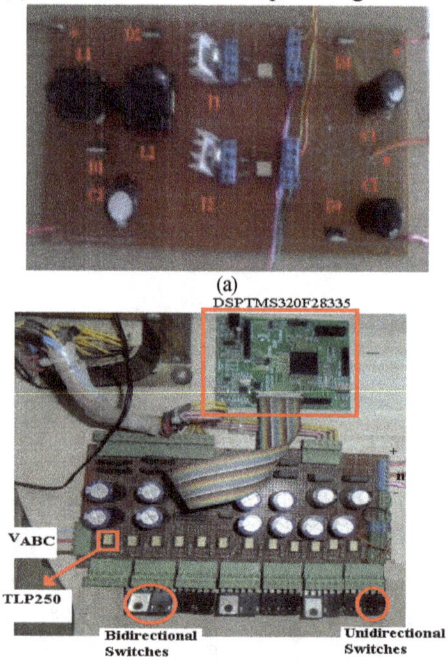

Fig. 16. Laboratory prototype of proposed converter (a) DC-DC converter and (b) DC-AC inverter.

Table 4. Parameters of experimental results.

DC source	12 V
L_1, L_2	500, 500 μH
C_1, C_2 and C_3	1000 μF
DC-DC Switching frequency	15800 Hz
D	0.5
Load	75 Ω

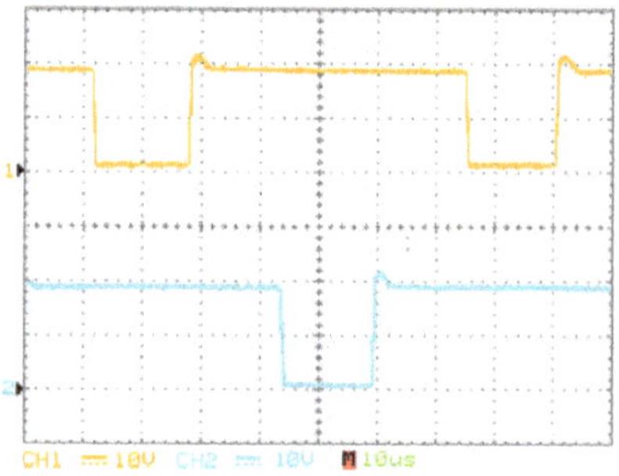

Fig. 17. Laboratory Gate pulses of DC-DC converter.

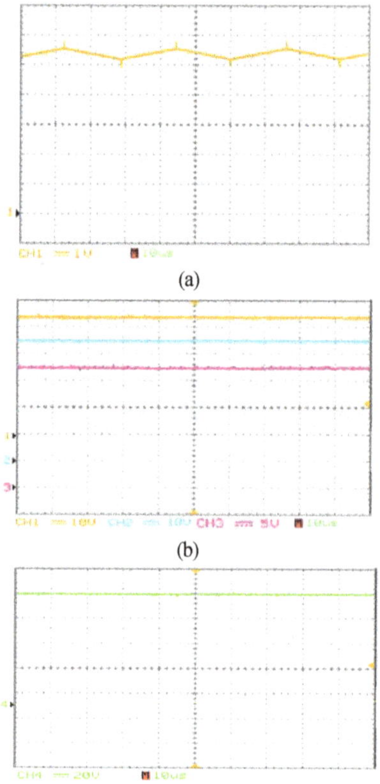

(c)

Fig. 18. Waveforms of DC-DC converter (a) input current (b) voltage of capacitors and (c) output voltage.

5. CONCLUSIONS

This paper proposes a multi-stage inverter for FC applications. The high step-up DC-DC converter is employed to provide a high voltage gain. The first stage is a boost converter that has a higher gain than the conventional boost converter to enable reducing the number of series connected FC modules. In order to investigate the performance of the converter in fuel cell systems, parameters such as voltage transfer gain and input current ripple are calculated.

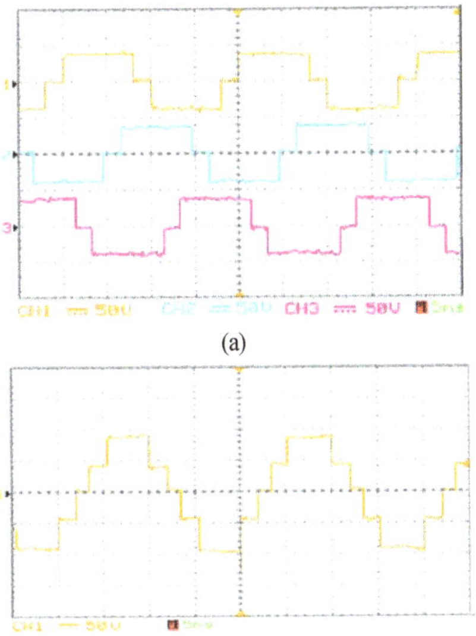

(a)

(b)

Fig. 19. Measured output voltages (a) V_{An}, V_{Bn}, V_{Cn} and (b) V_{AB}.

The simulation results demonstrate considerable increase in voltage transfer gain and reduction in input current ripple. These parameters are the most important features in fuel cell DC-DC converters. The second stage is a T-type inverter. This inverter is based on the two level inverter topology where it consists of a main inverter switches and an auxiliary three bidirectional switches. Equations of converters have been presented for DC-DC and DC-AC part. Simulation and experimental results are given to show the overall system performance.

REFERENCES

[1] M. Allahnoori, Sh. Kazemi, H. Abdi and R. Keyhani, "Reliability assessment of distribution systems in presence of microgrids considering uncertainty in generation and load demand," *Journal of Operation and Automation in Power Engineering*, vol. 2, no. 2, pp. 113- 120, 2014.

[2] K.N. Reddy and V. Agarwal, "Utility interactive hybrid distributed generation scheme with compensation feature," *IEEE Transactions on Energy Conversion*, vol. 22, no. 3, pp. 666-673, 2007.

[3] D. Sera, R. Teodorescu, J. Hantschel and M. Knoll, "Optimized maximum power point tracker for fast-changing environmental conditions," *IEEE Transactions on Industrial Electronics*, vol. 55, no. 7, pp. 2629-2637, 2008.

[4] U.S. Selamogullari, D. A. Torrey and S. Salon, "A systems approach for a stand-alone residential fuel cell power inverter design," *IEEE Transactions on Energy Conversion*, vol. 25, no. 3, pp. 741-749, 2010.

[5] Z. Zhao, M. Xu, Q. Chen, J.S Jason Lai and Y. H. Cho, "Derivation, analysis, and implementation of a boost-buck converter-based high-efficiency pv inverter," *IEEE Transactions on Power Electronics*, vol. 27, no. 3, pp.1304-1313, 2012.

[6] J.M. Shen, H.L. Jou and J.C. Wu, "Novel transformer-less grid-connected power converter with negative grounding for photovoltaic generation system," *IEEE Transactions on Power Electronics*, vol. 27, no. 4, pp.1818-1829, 2012.

[7] D.C. Lu, K.W. Cheng and Y.S. Lee, "A single-switch continuous-conduction-mode boost converter with reduced reverse-recovery and switching losses," *IEEE Transactions on Industrial Electronics*, vol. 50, no. 4, pp. 767-776, Aug. 2003.

[8] J.E. Baggio, H.L. Hey, H. A. Grundling, H. Pinheiro and J. R. Pinheiro, "Discreate control for three-level boost pfc converter," *in Proceedings of the 24th International Telecommunications Energy Conference*, pp.627-633, 2002.

[9] J.M. Kwon, B. H. Kwon and K.H. Nam, "Three-phase photovoltaic system with three-level boosting mppt control," *IEEE Transactions on Power Electronics*, vol. 23, no. 5, pp.2319-2327, 2008.

[10] L.S. Yang, T.J. Liang and J.F. Chen, "Transformerless DC-DC converters with high step-up voltage gain," *IEEE Transactions on Industrial Electronics*, vol. 56, no.8, pp. 3144-3152, 2009.

[11] X. Ruan, B. Li, Q. Chen, S. Tan and C.K. Tse, "Fundamental considerations of three-level DC–DC converters: topologies, analyses, and control," *IEEE Transactions on Circuits and Systems*, vol. 55, no. 11, pp. 3733-3743, 2008.

[12] W. Li and X. He, "Review of non-isolated high-step-up DC/DC converters in photovoltaic grid-connected applications," *IEEE Transactions on Industrial Electronics*, vol. 58, no. 4, pp. 1239-1250, 2011.

[13] Y. Cheng, C. Qian, M.L. Crow, S. Pekarek and S. Atcitty, "A comparison of diode-clamped and cascaded multilevel converters for a STATCOM with energy storage", *IEEE Transactions on Industrial Electronics*, vol. 53, no. 5, 1512-1521, 2006.

[14] S. Laali, E. Babaei and M.B.B. Sharifian, "Reduction the number of power electronic devices of a cascaded multilevel inverter based on new general topology," *Journal of Operation and Automation in Power Engineering* ,vol. 2, no. 2, pp.

81-90, 2014.

[15] M.R. Banaei and E. Salary, "New multilevel inverter with reduction of switches and gate driver", *Energy Conversion and Management*, vol. 52, pp. 1129-1136, 2011.

[16] N.A. Rahim and J. Selvaraj, "Multistring five-level inverter with novel PWM control scheme for PV application," *IEEE Transactions on Power Electronics*, vol. 57, no. 6, pp. 2111-2123, 2010.

[17] B. Axelrod, Y. Berkovich and A. Ioinovici, "Switched-capacitor/switched-inductor struct-ures for getting transformer less hybrid DC-DC pwm converters," *IEEE Transactions on Circuits and Systems,* vol. 55, no. 2, pp.687-696, 2008.

[18] J.M. Shen, H.L. Jou, J. C. Wu and K. D. Wu, "Five-level inverter for renewable power generation system," *IEEE Transactions on Energy Conversion,* vol. 28, no. 2, pp. 257-266, 2013.

[19] S.R. Pulikanti, G. Konstantinou and V.G. Agelidis, "Hybrid seven-level cascaded active neutral-point-clamped-based multilevel converter under SHE-PWM," *IEEE Transactions on Industrial Electronics*, vol. 60, no. 11, pp. 4794-4804, 2013.

[20] Y. Ounejjar, K. Al-Hadded and L.A. Dessaint, "A novel six-band hysteresis control for the packed u cells seven-level converter: experimental validation," *IEEE Transactions on Industrial Electronics*, vol. 59, no. 10, pp. 3808-3816, 2012.

[21] S. Khomfoi and L.M. Tolbert, Multilevel power converters. Power electronics handbook. Elsevier; 2007, pp. 451-82 [chapter 17].

[22] K.A. Corzine, M.W. Wielebski, F.Z. Peng and J. Wang, "Control of cascaded multi-level inverters," *IEEE Transactions on Power Electronics*, vol. 19, no. 3, pp. 732-738, 2004.

[23] E.A. Mahrous, N.A. Rahim, W.P. Hew and K.M. Nor, "Proposed nine switches five level inverter with low switching frequencies for linear generator applications", *in Proceedings of the Internati-onal Conference on Power Electronics and Drives Systems*, pp. 648-653, 2005.

[24] E. A. Mahrous, N.A. Rahim and W. P. Hew, "Three-phase three-level voltage source inverter with low switching frequency based on the two-level inverter topology", *IET Proceedings on Electric Power Applications*, vol. 1, Issue 4, pp. 637-641, 2007.

Permissions

All chapters in this book were first published in JOAPE, by University of Mohaghegh Ardabili; hereby published with permission under the Creative Commons Attribution License or equivalent. Every chapter published in this book has been scrutinized by our experts. Their significance has been extensively debated. The topics covered herein carry significant findings which will fuel the growth of the discipline. They may even be implemented as practical applications or may be referred to as a beginning point for another development.

The contributors of this book come from diverse backgrounds, making this book a truly international effort. This book will bring forth new frontiers with its revolutionizing research information and detailed analysis of the nascent developments around the world.

We would like to thank all the contributing authors for lending their expertise to make the book truly unique. They have played a crucial role in the development of this book. Without their invaluable contributions this book wouldn't have been possible. They have made vital efforts to compile up to date information on the varied aspects of this subject to make this book a valuable addition to the collection of many professionals and students.

This book was conceptualized with the vision of imparting up-to-date information and advanced data in this field. To ensure the same, a matchless editorial board was set up. Every individual on the board went through rigorous rounds of assessment to prove their worth. After which they invested a large part of their time researching and compiling the most relevant data for our readers.

The editorial board has been involved in producing this book since its inception. They have spent rigorous hours researching and exploring the diverse topics which have resulted in the successful publishing of this book. They have passed on their knowledge of decades through this book. To expedite this challenging task, the publisher supported the team at every step. A small team of assistant editors was also appointed to further simplify the editing procedure and attain best results for the readers.

Apart from the editorial board, the designing team has also invested a significant amount of their time in understanding the subject and creating the most relevant covers. They scrutinized every image to scout for the most suitable representation of the subject and create an appropriate cover for the book.

The publishing team has been an ardent support to the editorial, designing and production team. Their endless efforts to recruit the best for this project, has resulted in the accomplishment of this book. They are a veteran in the field of academics and their pool of knowledge is as vast as their experience in printing. Their expertise and guidance has proved useful at every step. Their uncompromising quality standards have made this book an exceptional effort. Their encouragement from time to time has been an inspiration for everyone.

The publisher and the editorial board hope that this book will prove to be a valuable piece of knowledge for researchers, students, practitioners and scholars across the globe.

List of Contributors

R. Baghipour and S.M. Hosseini
Department of Electrical Engineering, Babol Noshirvani University of Technology, Babol, Iran

S. Laali, E. Babaei and M.B.B. Sharifian
Faculty of Electrical and Computer Engineering,University of Tabriz,Tabriz, Iran

M. Moazen and M. Sabahi
Faculty of Electrical and Computer Engineering, University of Tabriz, Tabriz, Iran

M. Darabian, A. Jalilvand and R. Noroozian
Department of Electrical Engineering, University of Zanjan, Zanjan, Iran

Z. Moravej and S. Bagheri
Faculty of Electrical & Computer Engineering, Semnan University, Semnan, Iran

M. Sedighizadeh
Faculty of Electrical and Computer Engineering, Shahid Beheshti University, Tehran, Iran

M.M. Mahmoodi
Department of Electrical Engineering, College of Engineering, Saveh Branch, Islamic Azad University, Saveh, Iran

N. Ghorbani
Eastern Azarbayjan Electric Power Distribution Company, Tabriz, Iran

E. Babaei
Faculty of Electrical and Computer Engineering, University of Tabriz, Tabriz, Iran

N. Zendehdel and R. Asgarian Gannad Yazdi
Department of Electrical Engineering, Faculty of Engineering, Ferdowsi University of Mashhad, Mashhad, Iran

S.M. Mohseni-Bonab, A. Rabiee and S. Jalilzadeh
Departemant of Electrical Engineering, University of Zanjan, Zanjan, Iran

B. Mohammadi-Ivatloo and S. Nojavan
Faculty of Electrical and Computer Engineering, University of Tabriz, Tabriz, Iran

F. Namdari and R. Sedaghati
Department of Electrical Engineering, Faulty of Engineering, Lorestan University, Khorram abad, Iran

F. Namdari, L. Hatamvand, N. Shojaei and H. Beiranvand
Faculty of Engineering, Lorestan University, Khorram abad, Iran

M. Bigdeli
Department of Electrical Engineering, Zanjan Branch, Islamic Azad University, Zanjan, Iran

D. Azizian
Department of Electrical Engineering, Abhar Branch, Islamic Azad University, Abhar, Iran

E. Rahimpour
ABB AG, Power Products Division, Transformers, Bad Honnef, Germany

E. Dehnavi, H. Abdi and F. Mohammadi
Department of Electrical Engineering, Engineering Faculty, Razi University, Kermanshah, Iran

A. Badri and K. Hoseinpour Lonbar
Faculty of Electrical Engineering, Shahid RajaeeTeacher Training University, Tehran, Iran

A. R. Nafar Sefiddashti and A. Elahi
Department of Electrical Engineering, Shahrekord University, Shahrekord, Iran

G. R. Arab Markadeh
Member of Center of Excellence for Mathematics, Department of Electrical Engineering, Shahrekord University, Shahrekord, Iran

R. Pouraghababa
Esfahan Regional Electrical Company

H. Khorramdel, B. Khorramdel, M. Tayebi Khorrami and H. Rastegar
Department of Electrical Engineering, Safashahr Branch, Islamic Azad University, Safashahr, Iran

H. Bagheri Tolabi
Faculty of Engineering, Islamic Azad University, Khorramabad Branch, Khorramabad, Iran

M. H. Ali
Department of Electrical and Computer Engineering, University of Memphis, Tennessee, United States

M. Rizwan
Department of Electrical Engineering, Delhi Technological University, Delhi-110042, India

R. Kazemzadeh, E. Najafi Aghdam and M. Fallah
Renewable Energy Research Center, Faculty of Electrical Engineering, Sahand University of Technology, Tabriz, Iran

Y. Hashemi
Department of Electrical Engineering, University of Mohaghegh Ardabili, Ardabil, Iran

E. Salary, M. R. Banaei and A. Ajami
Department of Electrical Engineering, Azarbaijan Shahid Madani University, Tabriz, Iran

Index

www.ingramcontent.com/pod-product-compliance
Lightning Source LLC
Chambersburg PA
CBHW061945190326
41458CB00009B/2788
9781682853634